Lecture Notes in Artificial Intelligence   754

Subseries of Lecture Notes in Computer Science
Edited by J. Siekmann

Lecture Notes in Computer Science
Edited by G. Goos and J. Hartmanis

Heather D. Pfeiffer   Timothy E. Nagle (Eds.)

Conceptual Structures: Theory and Implementation

7th Annual Workshop
Las Cruces, NM, USA, July 8-10, 1992
Proceedings

Springer-Verlag
Berlin Heidelberg New York
London Paris Tokyo
Hong Kong Barcelona
Budapest

Series Editor

Jörg Siekmann
University of Saarland
German Research Center for Artificial Intelligence (DFKI)
Stuhlsatzenhausweg 3
D-66123 Saarbrücken, Germany

Volume Editors

Heather D. Pfeiffer
Computing Research Laboratory, New Mexico State University
Box 30001/3 CRL, Las Cruces, NM 88003-0001, USA

Timothy E. Nagle
Center for Research in Learning, Perception and Cognition
at the University of Minnesoty, and Unisys Corporation
1641 E Old Shakopee Road, Bloomington, MN 55425, USA

CR Subject Classification (1991): I.2, G.2.2, H.2.1

ISBN 3-540-57454-9 Springer-Verlag Berlin Heidelberg New York
ISBN 0-387-57454-9 Springer-Verlag New York Berlin Heidelberg

© Springer-Verlag Berlin Heidelberg 1993
Printed in Germany

Typesetting: Camera ready by author
Printing and binding: Druckhaus Beltz, Hemsbach/Bergstr.
45/3140-543210 - Printed on acid-free paper

Preface

This book was drawn from the papers presented at the Seventh Annual Workshop on Conceptual Graphs. The workshop was held at New Mexico State University in Las Cruces, New Mexico, USA and was sponsored by the American Association for Artificial Intelligence and the NMSU Computer Science Department. The authors were invited to submit expanded versions of their papers for this book. The contents of this volume roughly generalize into the areas of: representation issues, reasoning, data modeling and databases, algorithms and tools, and applications and natural language.

One of the highlights of this workshop was the landmark meeting of the first PEIRCE Project Workshop. The PEIRCE Project was announced earlier this year and its participants gathered this summer for the first time to discuss their work and the project. The PEIRCE Project aims to "... build a state-of-the-art, industrial strength conceptual graphs workbench. PEIRCE is integrating the conceptual graphs development efforts that are taking place around the world." (Ellis & Levinson, this volume).

The workshop as a whole was successful in viewing what is currently being explored in several areas of research in Conceptual Structures. The areas of active research include: sharing and structuring knowledge, graph matching and projection algorithms, constraint propagation, type hierarchy processing, uncertain and plausible reasoning, and natural language issues. As demonstrated by the PEIRCE project, there is also significant interest in the practical application of Conceptual Structures. Included are reports on the PEIRCE project, bridging accounting and business planning, open systems interconnection notation and the analysis of radiology reports. In all, this volume is fairly well rounded, covering several dimensions with both theoretical and practical papers.

Heather Pfeiffer
Timothy E. Nagle
September, 1993

Table of Contents

Databases and Modeling

Tools and Algorithms

Natural Language and Applications 273

I. Knowledge Representation Issues - General

The Scope of Coreference in Conceptual Graphs*

John W. Esch[1]

[1]Paramax Systems Inc., A Unisys Company, PO Box 64525   MS U1N28, St. Paul, MN
55164, (612) 456-3947, esch@email.sp.unisys.com

Abstract. To modularize a large semantic net you need to show how the
pieces relate to each other. Coreference allows concepts in two different graphs
to refer to the same thing. This paper reviews the subject of coreference in
the theory of conceptual graphs.[2] It explains basic concepts, gives current
definitions, gives examples of different uses, and discusses some advanced applications, extending the application "scope" of coreference in the process.

The work reported on in this paper is based on two efforts. First, several
discussion sessions on coreference were lead by the author at the University
of Minnesota's Conceptual Graph Discussion Group organized by Professor
Slagle. These were followed by extensive e-mail discussion among the Conceptual Graph community on most of the subjects described below.[3] Many
people contributed to those discussions and this paper is significantly better
due to their participation. In short, this paper organizes and presents the
results of those two efforts.

1 Introduction

All notational schemes suffer from complexity as the subject matter grows. Semantic
nets are no exception and rapidly become too large to comprehend when applied
to even modest problems. Different kinds of modularization and abstraction are
commonly used to help solve this problem. To break off a manageable piece of a large
conceptual graph, one makes a cut through one or more concept nodes. However, a
way is needed to show that the two parts of each split concept are really part of the
same concept. Coreference is used for this purpose. It indicates that two concepts
refer to the same thing. Since graphs can not cross context boundaries, coreference
is also used, like static scope rules, to show that one concept is the same thing as
another concept in a nested or dominate context.

An example is state governments. A semantic net covering both the revenue and
motor vehicle departments would be impractically large. It could be modularized
by noting that people in them are the same, but having different links in the two
contexts. The basic idea would be to make the PERSON concepts in the two contexts coreferent to the same government wide PERSON concept as in the following
conceptual graph.

---

* For the Seventh Annual Workshop on CONCEPTUAL GRAPHS, July 8-10, 1992, New
  Mexico State University, Las Crues, NM 88003-0001.
[2] Conceptual Graphs are a modernized and extended version of Charles Peirce's existential
  graphs that were defined by John Sowa.[SOWA84] The reader is assumed to be familiar
  with basic conceptual graph theory.
[3] All e-mail referred to in this paper was sent to address cg@cs.umn.edu

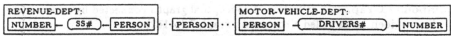

In the theory of conceptual graphs a context can be thought of as a conjunction of the assertions made by a set of graphs. These contexts may be properly nested to any depth. Formal definition are given in Section 2. These are followed by examples in Section 3 to clarify the basic ideas and definitions. To better understand coreference, Sections 4 reviews the algorithms for converting compound graphs with coreference links to simple graphs without them.

Section 5 reviews the mapping of conceptual graphs to predicate logic expressions with equality. One of the uses of coreference is to represent universal quantification. The algorithm for converting a concept with a universal quantifier as its referent to a graph without it, but with a coreference link between nested negative contexts, is reviewed and explained with an example in Section 6.

A very interesting application, discussed in Section 7, is the use of coreference in the differentia graphs of type definitions. If the genius node of a differentia graph is coreferent with another node, that node is also a genius node. For example, one could define a SQUARE as a coreferent ROMBUS and RECTANGLE. This kind of definition is compared with other ways of expressing multiple inheritance in conceptual graphs and other system like Common Lisp. And that old favorite, the Nixon Diamond problem is also revisited in Section 7.

An advanced application of coreference is in the definitions of individuals and aggregates which is discussed in Section 8. It raises a number of issues as to whether coreferent concepts can have different individual markers. These issues are discussed and partially resolved. Lastly, Section 9 summarizes the paper.

2 Definitions

This section briefly lists the pertinent ideas and definitions from [SOWA84]. They have been updated to reflect changes discussed via e-mail, in particular, Assumption 4.2.5. A few others that are needed are also included.

Coreference Links Intuitively, "Coreference links show which concepts refer to the same entities. In a sentence these links are expressed as pronouns and other anaphoric references." (p 10)

Line of identity It is a connected, undirected graph g whose nodes are concepts and whose arcs are *coreference links* connecting pairs of concepts.[4] In any coreference link, either both concepts occur in the same context or the context of one concept dominates the context of the other.

Display Form "The dotted line joining the two concepts is called a line of identity, which shows that two nodes represent exactly the same individual." (p 141)

Simple Graph "A conceptual graph without lines of identity or nested contexts is called a simple graph." (p 142)

---

[4] As Professor Slagle has pointed out, it is not really a line and probably should be called something like "graph of identity" or "identity graph".

Compound Graph "A collection of conceptual graphs connected by one or more lines of identity is called a compound graph." (p 142)

Context Dominance "If a context y occurs in a context x, then x is said to dominate y." (p 141) If x dominates y, then y is said to be *nested* in x.[5] In effect, a context dominates all nodes of the graph(s) in it, any nested contexts, and any nodes of graphs in those nested contexts.

Context of a Concept The context of a concept is the most closely containing context; that is, it contains the concept and no other context exists which dominates the concept and is nested in it.

Scope of a Concept "A concept b is within the scope of a concept a only if the context of a dominates the context of b." (p 141) Or they are in the same context.

Uniqueness "No concept may belong to more than one line of identity." (p 142)

Coreference Path It is a path in a line of identity $< a_1, ..., a_n >$ in g where for each i, either a_i and a_{i+1} both occur in the same context or the context of a_i dominates the context of a_{i+1}. Note that where a line of identity may zig-zag through contexts, coreference paths can not. Also note that a line of identity may be composed of several coreference paths.

Concept Dominance A concept a is said to dominate another concept b if there is a coreferent path from a to b.

Coreference "Two concepts a and b are coreferent if either a dominates b or b dominates a." (p 142) More simply, two concepts are coreferent if they are on the same coreferent path.

Dominant Concept "A concept a is dominant if a dominates every concept that dominates a." It is worded this way because two coreferent concepts in the same context dominate each other. So, if one is dominant, the other is also dominant.

Understanding dominance and how to find a concept's dominant concept or concepts is important because it is essential to converting display forms to linear forms and understanding the scope of variables in both forms. The key is that coreference paths are chains of coreference links that strictly follow context nesting or are in the same context. Consequently, two concepts can only be coreferent if one dominates the other. The next section gives some examples to help explain.

A good analogy is static scope in block structured programming languages where visibility is determined by the nesting of blocks.

3 Basic Examples

This section gives some basic examples to help explain the ideas and definitions in the previous section. In the first example, CAT ⊢··CAT , the two CAT concepts are coreferent because their is a line of identity between them and they are in the same context. In the second example, CAT ···⎡···CAT⎤ , the two CAT concepts are also coreferent because their is a line of identity between them and it is a coreferent path. It is a coreferent path because the first CAT concept dominates the CAT concept inside the context.

---

[5] Note that dominance is independent of whether the contexts are negated or not.

In the third example, `CAT` ··· │ ··· `CAT`, the two `CAT` concepts are not coreferent. Even though they are on the same line of identity, their is no coreferent path between them because neither dominates the other. As John Sowa suggested, if someone tries to create a graph like this, the editor should signal an error and, perhaps, suggest the following graph `CAT` ··· │ ··· `CAT` ··· │ ··· `CAT`. In this fourth example, there is one line of identity but two coreference paths. The middle `CAT` concept dominates both coreference paths. So each of the two `CAT` concepts in the contexts are coreferent with the middle `CAT` concept. Note, however, that they are not coreferent with each other. It is similar to the situation in programming languages where two nested blocks share a declaration in a containing block and each has independent visibility of that declaration.

A practical application of this example is disjunctions. For example,

`CAT:*x` `:*x`→ (ATTR)→ `BLACK` → (OR)→ `:*x`→ (ATTR)→ `WHITE`

The two `:*x` concepts inside the disjunctive contexts are each coreferent with the `CAT:*x` concept outside, but they are not coreferent with each other.

In the fifth and last example, `CAT` ··· │ ··· `CAT` ··· │ ··· `CAT`, there is again one line of identity but two coreferent paths. In this example each end `CAT` concept dominates one of the paths and the middle `CAT` concept is coreferent with both of them. A practical application occurs when it is combined with negative contexts. For example, the graph for the sentence, "There exists at least two people." is `PERSON` ¬ │ ··· `PERSON` ··· │ ··· `PERSON`. The only way that the `PERSON` concept in the negative context can be coreferent with both of the `PERSON` concepts outside the context is if it doesn't exist at all, which is what the graph says; i.e. that there does not exist a person which is the same as both the persons in the outer context.

4 Converting Compound to Simple Graphs

Aliases are different referents for concepts of the same individual. The referent field notation is extended to allow multiple referents separated by equal signs, with the concept's own identifier, if any, first. There can be both name aliases, as in `PERSON: Rosann=Rosalie`, and variable aliases, as in `PERSON: *x=*y`.

The procedure defined by [SOWA84, p143] says to assign a unique variable name to every generic concept in the graph (i.e. gives it its own identifier); copy the variables of the dominate concepts to the concepts they dominate as aliases; and erase the coreference links. (Remember, coreferent concepts in the same context dominate each other.) For example, `PERSON:Rosalie` ¬ │ ··· `PERSON:Rosann`

converts to `PERSON:Rosalie` ¬ `PERSON:Rosann=Rosalie`.

5 Mapping to Predicate Logic

The basic mapping of conceptual graphs with coreference links to predicate logic is done in two steps. First lines of identity are replaced by aliases and then phi, extended to handle these, is applied.[6]

Extending phi to cover aliases is straight forward. If b is a concept of the form `t:x1=x2=...=xn`, then phi(`t:x1=x2=...=xn`) has the form $t(x1) \land x1=x2=...=xn$. For example, phi(`cat:felix=morris`) = cat(felix) $\land$ felix=morris. Variable aliases require an existential quantifier. For example, , phi(`cat:*x=*y`) = $\exists x(cat(x) \land x=y)$.

The result of replacing lines of identity with aliases and applying an extended phi is that conceptual graphs with coreference links are mapped to predicate logic with equality. Conversely, "any formula in first order logic can be expressed by nested negative contexts and lines of identity." [SOWA84, p147]

The following are a couple of interesting examples. For each I give: English, display form, converted liner form, logic, and comment.

1. "Rosalie is Rosann."   `PERSON:Rosalie`‥‥`PERSON:Rosann`

   [PERSON:Rosalie=Rosann] [PERSON:Rosann=Rosalie]

   PERSON(Rosalie) $\land$ PERSON(Rosann) $\land$ Rosalie=Rosann $\land$ Rosann=Rosalie.
   This simplifies to PERSON(Rosalie) $\land$ Rosalie=Rosann which says, in effect, that Rosalie has an alias or nickname Rosann and that the referent of the concept can be referred to by either name.

2. "Rosalie is not Rosann."   `PERSON:Rosalie`‥¬‥`PERSON:Rosann=Rosann`

   [PERSON:Rosalie] ¬[[PERSON:Rosann=Rosalie]]

   PERSON(Rosalie) $\land$ ¬(PERSON(Rosann) $\land$ Rosann=Rosalie)
   This graph is saying that the referent of the concept can't be referred to as Rosann.

6 Existential and Universal Quantification

This section addresses the interpretation of coreference when existential and universal quantifiers are involved. The theory of conceptual graphs allows the referent to be a ∀ quantifier. For example, the sentence, "Every man loves a woman.", has the following graph as one interpretation.

`MAN:∀`←`( EXPR )`←`LOVE`→`( OBJ )`→`WOMAN`

The algorithm for converting this linear form to display form has three steps. [SOWA84, p146]

First, draw a double negation around the graph or graphs. For the example this leads to the following graph. ¬ ¬ `MAN:∀`← `( EXPR )`—`LOVE`— `( OBJ )`—`WOMAN`

Second, replace the ∀ with the concept's identifier and insert a copy of the concept between the outer and inner contexts. This can be done by the rule [SOWA84, p150] which allows any graph to be copied to odd nested negative contexts without changing the truth of the graph as follows

¬ `MAN:*x` ¬ `MAN:*x`← `( EXPR )`—`LOVE`— `( OBJ )`—`WOMAN`

---

[6] phi is the function defined in [SOWA84, p143] that maps simple conceptual graphs to predicate logic.

Third, convert to display form by replacing the variable coreference with a coreference link to obtain the following

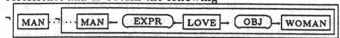

The intuition to take from this is that a coreference link from between nested negative contexts denotes a universal quantifier. That is, the reverse of the above algorithm is to convert existential conceptual graphs in nested negative contexts with coreference links to universally quantified concepts.

An alternative shorthand or extended linear form is to replace the nested negative contexts with IF...THEN constructs. For example, ¬| graph1(*x) ¬| graph2(*x) | | converts to IF graph(*x) THEN graph2(*x). The above example could also be expressed as IF [MAN:*x] THEN [MAN:*x]←(EXPR)←[LOVE]→(OBJ)→[WOMAN].

7 Type Definitions and Contractions

This section covers the use of coreference in type definitions and contractions. An example is used to bring out some interesting possibilities. The example involves using the definitions of a rombus and a rectangle to simplify a graph, corresponding to a square, to that square. Suppose we are given an example of a quadrilateral:

[SIDES:Equal]←(CHAR)←[QUAD]→(CHAR)→[ANGLES:@90] .

First we want to split [QUAD] into two coreferent concepts to get the equivalent graph:

[SIDES:Equal]←(CHAR)←[QUAD]··[QUAD]→(CHAR)→[ANGLES:@90]

Assume that the cannon containing this graph also contains the following type definition:

TYPE ROMBUS(g) IS [QUAD:*g]→(CHAR)→[SIDES:Equal]

We can now use the type definition to contract to obtain:

[ROMBUS]··[QUAD]→(CHAR)→[ANGLES:@90]

In a similar fashion, assume a definition for a rectangle:

TYPE RECTANGLE(g) IS [QUAD:*g]→(CHAR)→[ANGLES:@90]

We can now use this definition to contract again to obtain: [ROMBUS]··[RECTANGLE]. Now for the really interesting step. This corresponds to specifying multiple inheritance in a "mix-in" fashion. Assume we allow lines of identity (coreference links) in definitions as in

TYPE SQUARE(g) IS [ROMBUS:*g]··[RECTANGLE:*g]

The last step in the example is a final contraction to obtain: [SQUARE] .

Aren't conceptual graphs fun and exciting! Allowing lines of identity or coreference links in type definitions, as used above, is equivalent to having the meta language statements SQUARE < ROMBUS < QUAD and SQUARE < RECTANGLE < QUAD.

7.1 Nixon Diamond Revisited

Here is another interesting example of coreference in type definitions. It is a new way to treat the classic AI problem called the Nixon Diamond Problem.[7]

The Nixon Diamond problem has to do with slot/attribute clash in multiple inheritance where the same slot is inherited from different superclasses. In such situations, the system normally has some sort of rule for either selecting one of the attributes or combining them. A strength of CGs is that they do not force such a rule. Here's an example illustrating this point. We start with [NIXON:Dick] and expand it using the type definition TYPE NIXON(g) IS [QUAKER:*g]···[REPUBLICAN] to obtain [QUAKER:Dick]···[REPUBLICAN:Dick]. Similarly, we expand that result using the type definition

TYPE REPUBLICAN(g) IS [PERSON:*g]→(CHAR)→[WARISHNESS:Hawk] to obtain

[QUAKER:Dick]···[PERSON:Dick]→(CHAR)→[WARISHNESS:Hawk]

For the last expansion we use the type definition

TYPE QUAKER(g) IS [PERSON:*g]→(CHAR)→[WARISHNESS:Pacifist] to obtain

[PERSON:Dick]→(CHAR)→[WARISHNESS:Pacifist]

⋮

[PERSON:Dick]→(CHAR)→[WARISHNESS:Hawk]

Now the two [PERSON] concepts are joined to obtain

[PERSON:Dick] -
→(CHAR)→[WARISHNESS:Pacifist]
→(CHAR)→[WARISHNESS:Hawk]

Lastly, the two [WARISHNESS] concepts are joined to obtain

[PERSON:Dick]→(CHAR)→[WARISHNESS:{Pacifist Hawk}]

This is a very interesting sequence of operations and results. If there is an assertion in the canon that something can not be both a Pacifist and Hawk, then we have the result that [PERSON:Dick] is self contradictory, which is fine. More significantly, the conceptual graph notation and operations have not forced the adoption of some arbitrary rule for handling slot clashes in multiple inheritance situations.

8 Aggregations and Individuals

Individuals of some type are defined as an aggregate of other individuals or concepts according to the pattern of the type's definition. This section is very similar to the previous section on Definitions and Contraction, except their are some additional interesting issues. These issues have to do with what we mean in CG theory by "individual markers". John Sowa writes:[8]

---

[7] I agree with others that the Nixon Diamond Problem is not a good example for discussing multiple inheritance. But many people are familiar with it and it makes a good example of applying coreference.

[8] cg@cs.umn.edu e-mail Feb 20, 1992.

The purpose of the individual markers is to provide an internal "surrogate" for external entities. Some database systems make a distinction between "lexical object types" (LOTs) and "nonlexical object types (NOLOTs). A LOT is something that can be represented uniquely by a string of symbols, such as a number or a character string. A NOLOT is anything that cannot be represented by symbols, such as physical objects, actions, states, and abstractions.... To represent distinct instances of such things, you need a surrogate that can be stored on a disk. The individual markers #1, #2, #3, ..., are surrogates within a conceptual system....

Consider the context of some state's government. Most states have a revenue department. These departments will probable have a table for people indexed by social security number (SS#). If we use conceptual graphs to represent this information, we would have a type like PERSON. Its extension consists of the individuals in the revenue department's table.

Because these individuals are indexed by SS#, those numbers can be used as the individual markers for those persons inside the CG system. Pretend your SS# is 15, then $\boxed{\text{PERSON:\#15}}$ would represent a concept of you in the CG system.

State's also have motor vehicle departments and they will probably have a table of drivers indexed by driver's license number. Inside a CG system we would have a type like DRIVER. Its extension consists of the individuals in the driver's license table. And, if your driver's license number is 32, then $\boxed{\text{DRIVER:\#32}}$ is also a concept of you in the CG system. The question is, does $\boxed{\text{PERSON:\#15}} \cdots \boxed{\text{DRIVER:\#32}}$ make sense?

Before addressing this question, consider another example leading to the same question. Say we have SQUARE #203 which is defined as an aggregation of an individual ROMBUS and an individual RECTANGLE.

TYPE SQUARE(g) IS $\boxed{\text{ROMBUS:*g}} \cdots \boxed{\text{RECTANGLE}}$

INDIVIDUAL SQUARE(#203) IS $\boxed{\text{ROMBUS:\#1520}} \cdots \boxed{\text{RECTANGLE:\#12304}}$

Given $\boxed{\text{SQUARE:\#203}}$, we can make a coreferent copy to get

$\boxed{\text{SQUARE:\#203}} \cdots \boxed{\text{SQUARE:\#203}}$. Then we can expand the second of these, using the individual definition to get

$\boxed{\text{SQUARE:\#203}} \cdots \boxed{\text{ROMBUS:\#1520}} \cdots \boxed{\text{RECTANGLE:\#12304}}$

Very interesting! In effect, this graph says that individual SQUARE #203 is the same as ROMBUS #1520 and RECTANGLE #12304, just as PERSON #15 was the same as DRIVER #32.

Now expand $\boxed{\text{SQUARE:\#203}} \cdots \boxed{\text{SQUARE:\#203}}$ using the type definition instead of the individual definition to get

$\boxed{\text{SQUARE:\#203}} \cdots \boxed{\text{ROMBUS:\#203}} \cdots \boxed{\text{RECTANGLE:\#203}}$

The interpretation here is that individual #203 can be conceptualized as a SQUARE, ROMBUS, or RECTANGLE. This second' interpretation is consistent with each individual having a base type to which is conforms. It also conforms to all supertypes of that base type. Thus, if the base type of individual #203 is SQUARE and if SQUARE is a subtype of both ROMBUS and RECTANGLE, then individual #203 also conforms to both of those types.

So the issue comes down to, do we allow aggregations like $\boxed{\text{PERSON:\#15}} \cdots \boxed{\text{DRIVER:\#32}}$? They, in effect, allow referents to have different individual labels for different supertypes of the base type.

Other examples are giving Dr. Jekell and Mr. Hyde different individual markers before realizing that they are the same, and giving the morning and evening stars different individual markers before discovering that they are both really Venus.

The issue comes down to, how unique do individual markers have to be? There are at least three levels of uniqueness possible.

Total They have to be totally unique (within the canon). In this case, when two individuals markers are different, they refer to different individuals. So none of the above examples would be allowed.

Type They have to be unique within each type. In this case, two different markers, that conform to the same type, refer to different individuals. So `PERSON:#15`· · ·`DRIVER:#32` would be allowed but `STAR:#morning`· · ·`STAR:#evening` would not.

Equivalence Class They have to be unique across equivalence classes over each type. In this case, if #evening and #morning were put in the same equivalence class for STARs, then `STAR:#evening`· · ·`STAR:#evening` is allowed.

Role types, like DRIVER, and specializations, like SQUARE < RECTANGLE, permeate knowledge engineering applications. That there would be different indexing schemes in different contexts, like the revenue and motor vehicle departments, seems quite likely.

My personal opinion is that, in real systems, it will be normal for extensions of types to be identified as individuals and later discover that they are really the same. This is common in natural language, analytic, diagnostic and assessment systems that are putting facts together, trying to make sense out of them, and gradually putting referents together. Most of us have many individual labels that are each unique in some particular context. Consequently, I think we need at least the second level of "uniqueness" described above.

9 Summary and Conclusions

In summary, this paper reviewed and refined the concept of coreference and explained many applications through examples. Coreference links show that two concepts refer to the same thing. This is indicated in sentences via pronouns, anaphoric references and being verbs. In linear form graphs it is indicated by variables and in display forms by dotted lines called lines of identity. When two concepts share the same variable or line of identity and one dominates the other, they are coreferent.

Section 3 gave a number of basic examples. Section 4 explained how to convert compound graphs with lines of identity to equivalent simple graphs without them. Section 5 showed how to map simple graphs to predicate logic with equality. Section 6 showed the relationship between Existential and Universal quantifiers and IF THEN rules. Section 7 explored some of the interesting applications that are possible when coreference links re allowed in type definition. And Section 8 extended that idea to coreferent individuals.

There are many more aspects and applications of coreference that are just beginning to be explored. These include copying graphs that participated in lines of identity, extracting subgraphs, and relating Conceptual Graph coreference to other theories which define coreference, such as Guha's Microtheories.

10   Acknowledgment

I am especially thankful to John Sowa for his theory of conceptual graphs, the clarity of his many writings and explanations, and the many e-mail message to the CG group. I would also like to thank a number of regular e-mail contributors for their incites: Mark Willems, Fritz Lehmann, Harry Delugach, Gerard Ellis, and Jan Schmidt. My apologies if I missed anyone.

References

[SOWA84] John F. Sowa, *Conceptual Structures: Information Processing in Mind and Machine*, Addison Wesley, 9184.

Using World Structures for Factoring Knowledge

Bernard Moulin, Guy W. Mineau

Université Laval, Département d'Informatique
Ste-Foy, Quebec, Canada, G1K 7P4

Abstract. When we understand a story, we not only recognize the various concepts and conceptual relations involved in the sentences, but we also build a mental model in which we visualize the scene with the agents involved in the story. We link the speech acts with the agent who performs them, we memorize the feelings and personal beliefs of each agent as we perceive them to be, etc. Hence, we build several models. In this paper we introduce the notion of agent-world, a support structure which contains the agent's knowledge which is known to be true under some hypotheses. Since different hypotheses may generate different hypothetical worlds in which the agent is involved or which emanates from her, an agent can be associated with several worlds, and different kinds of inferences may be carried on the knowledge included in these worlds. Agent-worlds extend Sowa's notion of "world". We propose mechanisms for reasoning with agent-worlds (inheritance, inference rules, inter-world reasoning). We also suggest how, for reasoning purposes, illocutionary acts can be expressed as a special form of modality that extends Sowa's notation. We also propose a way of generating agent worlds from discourse conceptual structures.

1. Introduction

Let us consider the following short story.

In a car dealer's showroom, John meets Peter, the salesman.
John asks Peter: "How much costs the blue car?"
Peter says: "The price of this car is 30 000$".
John thinks that it is very expensive.
John says: "If I were rich, I would buy it".
Peter thinks that John is charming.
Peter proposes a good loan to John.

When we understand such a story, we not only recognize the various concepts and conceptual relations involved in the sentences, but we also build a mental model in which we visualize the scene with the two agents, John and Peter, we link the speech acts with the agent who performs them, we memorize the feelings and personal beliefs of each agent as we perceive them to be, etc. Hence, we build several models: the narrator's model which records information about the scene, two models, one for Peter, the other for John, which contain respectively Peter's and John's feelings, ideas and beliefs. We then use these models in order to reason about the scene, the characters and their actions. For instance, we could have the following reasoning: "if John knew that Peter found him charming, then John could ask for a better interest rate.

We consider that a story results from the description of several situations in which various agents (human or artificial) are involved. Depending on how relevant are these situations to the sequence of events, the story line is said to be more or less logical. Usually, consistent texts describe stories based on logical story lines through out the narration. The analysis of such texts can then proceed by first determining which situations are part of the story, and how they relate to the story line. As a first step in that direction, Moulin [7] [8] proposed an extension of Sowa's Conceptual Graph approach (1984) in order to represent, in an integrated framework, the situations described by a text, as well as the agents' perspectives, the temporal localizations and the temporal relations relating these structures together. This framework can be used to describe the semantic content of a text, as it pertains temporally to a story line.

Sowa's conceptual graph theory provides a powerful approach for representing knowledge, and integrates several modelling techniques from artificial intelligence, linguistics and logics. As an extension of Peirce's existential graphs, conceptual graphs are embedded in contexts and can be mapped to predicate calculus formulas [10, 11,12,13]. Reasoning can be carried on these formulas (predicates associated with truth values), using appropriate inference rules which are usually derived from first order predicate logic. The mapping of simple CGs is quite straightforward, but the mapping of embedded CGs is more subtle and can raise some difficulties. For instance, if we consider the sentence: "In a car dealer's showroom, John meets Peter, the salesman", we can associate the truth value 'true' to the proposition represented by the conceptual graph: [PERSON: John] <- (AGNT) <- [MEET] -
          (PAT)-> [SALESMAN: Peter]
          (LOC) -> [SHOW-ROOM].

The representation issue is more subtle in the case of the sentence "Peter thinks that John is charming". The truth value 'true' is associated to the proposition represented by the graph: [PERSON: Peter] <- (AGNT) <- [THINK] -> (OBJ)-> [PROPOSITION: p1].

where p1 represents the proposition "John is charming".

However, we cannot associate any truth value to proposition p1, since we do not know if Peter is right or wrong in his thinking about John. Supposing that one believes his/her thoughts to be true facts, proposition p1 has a truth value 'true' in

"Peter's world". But the text narrator and the reader cannot take for granted that "John is charming" (hence making the proposition p1 'true' in their own worlds), if they don't make the assumption that "Peter's thoughts are right".

Intuitively, there is a need for specialized structures that will enable us to partition knowledge according to different points of view (i.e. John's, Peter's, the narrator's) and to reason on this partitionned knowledge. In this paper we introduce the notion of *agent-world*, a support structure which contains the agent's knowledge which is known to be true under some hypotheses. Since different hypotheses may generate different hypothetical worlds in which the agent is involved or which emanates from her, an agent can be associated with several worlds, and different kinds of inferences may be carried on the knowledge included in these worlds.

Section 2 summarizes the characteristics of "worlds" as they were introduced in [10] and refers to Filman's and Guha's notions of world and context. Section 3 presents the notion of agent-worlds that extends Sowa's proposal. Section 4 presents mechanisms for reasoning with agent-worlds (inheritance, inference rules, inter-world reasoning). Section 5 proposes that for reasoning purposes, illocutionary acts can be expressed as a special form of modality that extends Sowa's notation. Section 6 discusses a way of generating agent worlds from discourse conceptual structures [9].

2. Related works
2.1 Worlds in the Conceptual Graph Theory
In section 4.4 of his book, Sowa [10] discusses the model theory that provides the formal logical foundations of the Conceptual Graph theory. Discourses contain a great variety of utterances referring to wishes, possibilities, hypothetical events, desires. People create imaginary universes, dream or reason on possible worlds etc. Each of those possible worlds can be represented by a collection of conceptual graphs."Hintikka [4] criticized the infinite, closed-world of standard logic"... "Instead of infinite models, Hintikka proposed open-ended, finite, *surface models*. Understanding a story would consist of building a surface model containing only those entities that were explicitly mentioned. The model would then be extended in a 'step-by-step investigation' of all implicit entities that must exist to support or interact with the ones that were mentioned". Sowa defines the notion of open-world.
"An *open world* W is a triple <T, F, I> where T and F are sets of simple graphs and I is a set of individual markers:
. T is called the *true set*, F is called the *false set*, and the elements of I are called the *individuals* of world W.
. The graphs in T and F may contain either generic concepts or individual concepts with referents in I.
. No individual marker in I occurs in more than one concept in T, but there is no such

restriction on the referents of F"...

"Let W = <T, F, I> be an open world. The *denotation operator* δ for W maps sets of conceptual graphs to one of the three *truth values* true, false or unknown, and maps n-adic abstractions to n-tuples of individual markers in I"...

Following Dunn's proposal [1], Sowa defines a world basis.

"A *world basis* consists of an open world W together with a set L of conceptual graphs called *laws* and a set S of schemata or prototypes. Let u be an arbitrary conceptual graph.

. u is *necessary* if it is provable from L.

. u is possible if it is consistent with L (i.e. the empty clause is not provable from L and u).

. u is plausible if u is possible and du becomes true when T is replaced with some set of graphs that are canonically derivable from S".

"Understanding a story involves the selection of laws and defaults L and S, and the construction of a possible world W that is consistent with L and guided by S"... Sowa adds (p186):" A coherent story, either fictional or true, develops a miniature world. Unless the story is treated as myth or fantasy, the laws and defaults for the normal world are assumed. Understanding the story requires the reader or listener to construct a consistent world basis"... However, Sowa described the worlds' content only in terms of simple CGs, and the models he proposed did not include nested contexts. In this paper, we propose an extension of Sowa's notion of world in order to reason with graphs including nested contexts.

2.2 Other related works

Filman [2] describes an extension of the KEE system which includes a system for manipulating contexts (called 'worlds') and an assumption-based truth maintenance system (ATMS) [5]. The ATMS incorporates three basic concepts: facts (or propositions), assumptions (which correspond to primitive decisions or choices), justifications (sets of assumptions and/or propositions). Using the ATMS as a foundation, a context mechanism has been built by Filman's team. A world (or context) is characterized by a set of assumptions (both the assumptions of the existence of that world and its ancestor worlds, and the assumptions of facts explicitly asserted and deleted in that world). In addition, "the KEE system has two kinds of rules: deduction rules and action rules. Deduction rules express the theories of a particular domain representation, truths believed in every world. Action rules create contexts and change the assumptions of particular contexts".

Guha [3] proposes to extend first-order predicate logic using contexts: "formulas are not just true or false; they are true or false in some context". The author also proposes a formal definition of the operation of 'lifting' which enables the system to lift a formula from one context to another, using 'partial relative decontextualization of formulas'... A context can be used as a means for referring to a group of related assertions about which something can be said, called a theory (for example, a theory of mechanics, a theory of the weather in winter, etc.). Contexts used in this sense are called 'microtheories'. Different microtheories make different assumptions and simplifications about the world using contexts which, among other things, provide a mechanism for recording and reasoning with these assumptions.

In this paper we propose a notion of world which is compatible with Filman's approach. In addition, we suggest a mechanism that helps to derive the structure of worlds from the analysis of a text structure. Our approach presents several similarities with Guha's model of contexts. We suggest an extension which aims also at dealing with modal and temporal knowledge found in texts.

3. Agent worlds

We refine Sowa's definition of a world basis (section 2.1) by the introduction of the notion of agent-world.

An *agent world A-W* is a triple <H, W, L> where H is a set of conceptual graphs called *hypothesis of the agent world*, W is an *open world*, L is a set of conceptual graphs called *laws of the agent world*.

The **set** H of hypothesis of the agent world A-W is a set of conceptual graphs which state the conditions (axioms) under which the agent world A-W can be activated: all facts (the conceptual graphs contained in W) and laws (the conceptual graphs contained in L) are dependent on H, the set of hypotheses.

Since the set of hypotheses of an agent-world is the basis from which any fact or law related to A-W can be inferred, the following notations will be used:

$$A\text{-}W \vdash f \qquad \text{and} \qquad A\text{-}W \vdash l$$

to state the conditional existence of fact f and law l to the existence of agent-world A-W.

Fact f and law l exist as long as agent-world A-W exists, which is dependent upon H. At times, this alternative notation may be used in order to explicitly link particular hypotheses to certain facts or laws:

$$A\text{-}W: h_1, h_2, ..., h_i \vdash f \qquad \text{and} \qquad A\text{-}W: h_1, h_2, ..., h_i \vdash l$$

Given that PROPOSITION is the type of a concept whose referent is a conceptual graph identified by prop.id, we will indicate, when needed, the whole conceptual graph instead of prop.id. The notation $A\text{-}W \vdash$ [PROPOSITION: prop.id]; is an abbreviation for the rule

$$h_1 \wedge h_2 \wedge \ldots \wedge h_n \Rightarrow [\text{PROPOSITION: prop.id}]$$

where $\wedge$ denotes a logical 'and', and $\Rightarrow$ denotes a logical implication.

Extending Sowa's definition (see section 2.1), an *open world* W is a tuple <T,F,U,I> where T, F and U are sets of conceptual graphs and I is a set of individual makers. The conceptual graphs of T, F and U are called *facts*:
. The elements of I are called the *individuals* of world W.
. T is called the *true set*. The conceptual graphs included in T (i.e. [PROPOSITION: prop.id]) are associated with the truth value 'true' in an agent-world A-W. Their existence in the agent-world A-W is expressed by

$$A\text{-W} \vdash [\text{PROPOSITION: prop.id}].$$

. U is called the *unknown set*. U represents the set of conceptual graphs encoding knowledge available to the agent but to which no truth value can yet be associated. The conceptual graphs included in U are associated with the truth value 'unknown'. Their existence with regard to A-W will be denoted by

$$A\text{-W} \vdash^? [\text{PROPOSITION: prop.id}].$$

. F is called the *false set* : the conceptual graphs included in F are associated with the truth value 'false'. Their existence in the agent-world A-W is expressed by

$$A\text{-W} \vdash \neg [\text{PROPOSITION: prop.id}].$$

The conceptual graphs included in sets T, U and F may be associated with modalities. In section 5 we precise further the characteristics of such graphs in relation with the representation of illocutionary acts.

The *set L of laws of the agent world A-W* is a set of conceptual graphs representing the rules that characterize the domain microtheory of A-W. The *rules* are denoted by

A-W $\vdash$ (Premise $\Rightarrow$ Conclusion);
where Premise and Conclusion are any combination of propositions [PROPOSITION: prop.id], using the operators $\wedge$ ('and'), $\vee$ ('or'), $\neg$('not'). For clarification purposes, we use the notation using the implication symbol instead of Sowa's notation using nested contexts and combinations of 'not' operators.

The notation A-W $\vdash$ (Premise $\Rightarrow$ Conclusion); is an abbreviation for the rule

$$h_1 \wedge h_2 \wedge \ldots \wedge h_n \wedge \text{Premise} \Rightarrow \text{Conclusion}.$$

4. Reasoning with agent worlds
4.1 Agent worlds manipulation
Quite often, several agent worlds share some common knowledge and only differ by some specific hypothesis, facts or rules. Filman's system [2] allows the specification

of a network of worlds ("a directed, acyclic graph over the 'parent-child' relation"). Guha does not provide such mechanism for manipulating his contexts: the only way of transferring knowledge from one context to another is to use lifting rules.

Both approaches have advantages and limitations. Inheritance enables the designer to concentrate on the similarities and differences existing between worlds, but it is expensive in terms of computing resources when the system has to navigate in the inheritance structure during processing. Using rules to transfer knowledge from one world into another is quite effective, but it can be painful for the designer to have to state all transfer rules when defining a new world which is slightly different from another one: Guha [3 p 34] indicates that "we need some general default lifting axioms" for this purpose. Default reasoning may also be considered at that point.

We will combine both approaches. We will use *transfer rules* for transferring knowledge from one world into another (see section 4.3), and we will provide some inheritance mechanisms that are discussed in the present section.

The agent-worlds are organized in an inheritance structure similar to Filman's world structure [2]. The *inheritance relation*, called *SPECLZ* (for "specialize"), is defined between two agent-worlds.

Given two agent-worlds A-W1= <H1, W1, L1> and A-W2= <H2, W2, L2>, we can say that A-W1 specializes A-W2, denoted [A-W1] -> (SPECLZ) -> [A-W2], iff H1 ... H2.

. **Properties of the inheritance relation**
1. Given two agent-worlds A-W1= <H1, W1, L1> and A-W2= <H2, W2, L2> such as [A-W1] -> (SPECLZ) -> [A-W2], facts and rules from A-W2 are valid in A-W1.

The proof is simple. For all facts of A-W2 we have : $(\forall i \ f_i \ \varepsilon \ W2) \ (H2 \vdash f_i)$.

Since H1 $\supset$ H2, $(\forall i \ f_i \ \varepsilon \ W2) \ (H1 \vdash f_i)$ and $(f_i \ \varepsilon \ W1)$.

We have a similar proof with rules of set L2.
2. If an agent-world A-W1= <H1, W1, L1> specializes several other agent-worlds A-Wi= <Hi, Wi, Li>, ..., A-Wn= <Hn, Wn, Ln>. Then $H_1 \supset (H_i \cup H_{i+1} \cup ... \cup H_n)$.

4.2 **Manipulating knowledge within a world**
Whenever an agent needs to reason with the knowledge contained in an agent-world A-Wi, she may use the assert and retract commands, plus the conventional inference operators on three value logic systems[1]. For example

. *asserting* fact f becomes equivalent to adding the logical expression A-$W_i \vdash f$.
. *asserting* the law (Premise => Conclusion) becomes equivalent to adding the

logical expression A-W$_i$ ⊢ (Premise => Conclusion).

retracting fact f becomes equivalent to deleting the logical expression A-W$_i$ ⊢ f.
. *retracting* the law (Premise => Conclusion) becomes equivalent to deleting the
logical expression A-W$_i$ ⊢ (Premise => Conclusion).

Note that the retract command is not equivalent to asserting the negation of the fact or
of the law; *retract* fact f is not equivalent to *assert* ¬fact f because an agent-
world is an open world.

. **Inference rules**

Any agent-world A-W$_i$ inherits inference rules from first order predicate calculus such
as the modus ponens inference rule which is equivalent to:

> A-W$_i$ ⊢ (Premise => Conclusion)
>
> A-W$_i$ ⊢ f
>
> ----------------------
>
> *assert* A-W$_i$ ⊢ Conclusion

if and only if f subsumes premise.

Some inference rules are related to the truth values that are attached to facts (true, false
or unknown). For example

1. We cannot have at the same time a fact and its negation in the same world

> A-Wi ⊢ f
>
> A-Wi ⊢ ¬ f
>
> ------------------
>
> The world is inconsistent

2. If in a given world a fact (or its negation) appears together with the unknown
   version of the fact, the latter should be discarded.

A-Wi ⊢ f$_j$	A-Wi ⊢ ¬f$_j$
A-Wi ⊢ ?f$_j$	A-Wi ⊢ ?f$_j$
------------------	------------------
retract A-Wi ⊢ ?f$_j$	*retract* A-Wi ⊢ ?f$_j$

4.3 Inter-world reasoning

Depending on the application, we may need rules to transfer some knowledge from
one agent-world into another, as long as one does not specialize the other. This will

be particularly useful for reasoning on illocutionary acts. So, we define inter-world transfer rules that provide a way for transferring facts or rules from one agent-world into another.

Given a source agent-world $A\text{-}W_u$ and a destination agent-world $A\text{-}W_v$, given a fact f_i to be transferred from $A\text{-}W_u$ to $A\text{-}W_v$ and a triggering fact f_j from $A\text{-}W_w$ (where w being equal either to u or v), an *interworld transfer rule itr* is defined as follows:

itr: $A\text{-}Wu \vdash g(f_i)$

  $A\text{-}Ww \vdash f_j$

  $\Rightarrow$ $A\text{-}Wv \vdash h(f_i)$

where the symbol $\Rightarrow$ separates the premises and the conclusion of the inter-world transfer rule itr; g and h are functions that possibly express modalities over facts in different worlds (see section 5.1).

This generalizes to laws:

itr: $A\text{-}Wu \vdash$ (Premise $\Rightarrow$ Conclusion)

  $A\text{-}Ww \vdash f_j$

  $\Rightarrow$ $A\text{-}Wv \vdash$ (Premise $\Rightarrow$ Conclusion)

Examples of these rules are given in section 6.

5. Illocutionary acts and modal reasoning

5.1 Representation of illocutionary acts

As we mentioned in section 2.2, Sowa [10] distinguished *extensional verbs* (like 'see', 'carry', 'take' etc.) from *intensional verbs* (like 'believe', 'hope', 'want', 'seek' etc.). Usually, intensional verbs express a mental act or an illocutionary act that is performed by an agent. An illocutionary act is also called a speech act. These acts are performed by agents when they utter sentences. Requests, orders, wishes are examples of illocutionary acts. For more details see [8].

"John believes that Mary travels during the summer holidays" is an example of a mental act performed by John. "John suggests to Peter that Mary should visit Quebec" is an example of a speech act performed by John.

We refer in this research to Searle's and Vanderveken's illocutionary logic [14] which provides the foundations for a formal theory of speech acts. According to these authors, most illocutionary acts represent the meaning of sentences in the context of utterance together with an illocutionary force and a propositional content. For example the two utterances "send me a letter !" and "you may send me a letter", expressed in the same context, can serve to perform two illocutionary acts with the same propositional content, but with different forces, since the first utterance has the illocutionary force of a directive and the second utterance has the illocutionary force of a prediction.

The *propositional content* of a speech act is the sense in context[2] of the clause corresponding to the sentence. It can be represented by the use of formalisms such as modal logic or other intensional extensions of predicate calculus.

The *context of an utterance*[2] in which illocutionary acts are performed consists of five elements : the speaker(s), the hearer(s), the time and the place of the utterance and the "world of utterance". The *world of utterance* contains basic facts which are relevant to understand the speaker's utterance in that context : especially important are the speaker's and hearer's beliefs, desires and intentions, since illocutionary acts are rational kinds of language use which serve to achieve certain conversational purposes.

In the present research, we use an extension of conceptual graph theory for representing mental and speech acts. Let us take some examples and their representation using Sowa's conceptual graphs. "John believes that Mary travels during the summer holidays" is represented by:

[PERSON: John]<-(AGNT)<-[BELIEVE]->(OBJ)->
   [PROPOSITION:
   [PERSON:Mary]<-(AGNT)<-[TRAVEL]->(DUR)->[SUMMER-HOLIDAYS]].

"John suggests to Peter that Mary should visit Quebec" is represented by:

[PERSON: John]<-(AGNT)<-[SUGGEST]-
      (PAT)->[PERSON: Peter]
      (OBJ)->[PROPOSITION:
         [PERSON: Mary]<-(AGNT)<-[VISIT]->(OBJ)->[QUEBEC]].

In each case the main clause of the sentence is represented by a conceptual graph that indicates which act is performed by agent John: it provides some information about the "context of utterance" of the corresponding act. In each case, the relative clause of the sentence is represented by a nested conceptual graph introduced by the concept PROPOSITION. Sowa remarks: "intensional verbs always have a nested context for their object, and that context may refer to a different possible world from the world that includes the agent of the verb". From a logical point of view, we cannot associate the same truth values with the propositions corresponding to the main and relative clauses in the sentence. In our example we can say that "it is true that John believes *proposition 1*", but we don't know if *proposition 1: Mary travels during the summer holidays*" is true or false.

Modal logics provides the formal apparatus for reasoning on the propositions involved in mental or illocutionary acts. Sowa [10] adopted Filmore's approach for representing modalities (like 'possibility' and 'necessity') attached to the propositions: Sowa represents these modalities as unary conceptual relations.

We will consider modalities from a larger point of view, as a simplified representation of mental and speech acts. We propose to represent mental and speech acts as extended forms of modalities.

Let us consider a particular *mental or illocutionary act* mp_i (referent) of type MT_i (concept type of that act) performed by an agent X, directed towards an agent Y at time $TINT_i$. The propositional content of mp_i is PROPOSITION: p_j associated with a time interval $TINT_j$. This mental or illocutionary act is considered as an extended form of modality applied on PROPOSITION: p_j, and is represented in the following way using the special conceptual relation MOD, indicating the modality:

$$[MT_i: mp_i <<X, Y, TINT_i>>] \rightarrow (MOD) \rightarrow [PROPOSITION: p_j << TINT_j>>].$$

This representation extends the way that Sowa denotes modalities using monadic conceptual relations. It makes explicit the main parameters of the mental and illocutionary acts and of their propositional contents: the act type, the agents involved in the act, the time intervals respectively associated with the act and its propositional content. Note that we indicate between the symbols $<< ... >>$ the relevant parameters for reasoning on the acts and their propositional content.

For instance the sentence corresponding to a mental act, "John believes that Mary travels during the summer holidays" is represented by:

[BELIEVE: bel1 <<John, -,TINT1>>]->(MOD)->[PROPOSITION: pr2 <<TINT2>>]
[PROPOSITION: pr2 :=

    [PERSON:Mary]<-(AGNT)<-[TRAVEL]->(DUR)->[SUMMER-HOLIDAYS]]

Note that for clarification purposes we separate the representation of the mental act from the description of its propositional content (the symbol := introduces the conceptual graph corresponding to proposition pr2).

As another example, the sentence corresponding to the illocutionary act, "John suggests to Peter that Mary visit Quebec" is represented by:

[SUGGEST: sug3<<John, Peter, TINT3>>] ->(MOD)->

               [PROPOSITION: pr4 <<TINT4>>]

[PROPOSITION: pr4 :=

    [PERSON: Mary]<-(AGNT)<-[VISIT]->(OBJ)->[PROVINCE: Quebec]].

Note that the temporal information is implicit in these sentences. However, in other examples verb tenses or temporal adverbs may appear. They would be resolved using temporal relations according to the model proposed in [7].

5.2 Agent-worlds and modalities

We can now introduce modalities in agent-worlds descriptions as seen in section 4.

Let us give an example illustrating how worlds can be used for reasoning with illocutionary acts. Consider two agents John and Mary. Suppose that John says to Mary:"I will visit you on April 22 1992". In Mary's agent-world A-W.Mary we have the fact:

A-W.Mary ⊢ [PROMISE: pr1 <<John, Mary, TINT1>>]->(MOD)->
                [PROPOSITION: pr4 <<April 22 1992>>]

with the description of proposition pr4 as a conceptual graph:

[PROPOSITION: pr4 :=
           [PERSON: John]<-(AGNT)<-[VISIT]->(PAT)->[PERSON: Mary]].

We can state a rule related to promises that says that if "proposition p1 is promised at time TINT1, its truth value is unknown":

A-W_y ⊢ [PROMISE: pr_i <<X, Y, TINT1>>]->(MOD)->
                [PROPOSITION: prj <<TINTj>>]

⇒   A-W_y ⊢ $^?$[PROPOSITION: prj <<TINTj>>].

Hence the reasoning system can add in Mary's agent-world the fact:

A-W.Mary ⊢ $^?$[PROPOSITION: pr4 <<April 22 1992>>]

Now suppose that John did not visit Mary on April 22 1992. We add in Mary's agent-world the following fact

A-W.Mary ⊢ ¬[PROPOSITION: pr4 <<April 22 1992>>]

The second inference rule of section 4.2 makes the system retract the unknown fact:

retract A-W.Mary ⊢ $^?$[PROPOSITION: pr4 <<April 22 1992>>]

We can also have an inference rule about "promise" modalities stating that if "agent X promises at time TINT1 to do proposition p2 at time TINT2, and if proposition p2 does not hold at TINT2, then the promise is unkept":

A-W_y ⊢ [PROMISE: pr_i <<X, Y, $TINT_j$>>]->(MOD)->
                [PROPOSITION: prj <<TINTj>>]

A-W_y ⊢ ¬[PROPOSITION: prj <<TINTj>>]

⇒ A-W_y ⊢ [PROMISE: pr_i] -> (CHRC) -> [UNKEPT].

Hence we can add in Mary's agent-world the fact:

A-W.Mary ⊢ [PROMISE: pr_4] -> (CHRC) -> [UNKEPT].

Suppose that we have another inference rule stating that: if "agent X promises at time TINT1 to do proposition p2 at time TINT2, and if the promise is unkept, then agent X is unreliable":

$$A\text{-}W_y \vdash [\text{PROMISE: } pr_i \ll X, Y, TINT_i \gg] \text{->(MOD)->}$$
$$[\text{PROPOSITION: } pr_j \ll TINT_j \gg]$$
$$A\text{-}W_y \vdash [\text{PROMISE: } pr_i] \text{-> (CHRC) -> [UNKEPT]}$$
$$=> \quad A\text{-}W_y \vdash [\text{AGENT: X] -> (CHRC) -> [UNRELIABLE]}.$$

Then the reasoning system can add in Mary's agent-world the fact:

$$A\text{-}W.Mary \vdash [\text{AGENT: John] -> (CHRC) -> [UNRELIABLE]}.$$

The representation of illocutionary acts that we propose on the basis of conceptual graphs, augmented with our form of modality along with the notion of agent-world, enables us to reason on speech acts.

6. Generating worlds from discourse conceptual structures

We proposed in [7] [8] and refined in [9] an approach for representing the conceptual structure of discourses. We consider that a discourse describes temporal situations (events, states, processes etc.) and eventually words that are uttered by agents which are involved in the discourse. This approach is an extension of the conceptual graph model which makes explicit the temporal structure of discourses, thanks to the introduction of the notion of time coordinate system. We distinguish two main types of time coordinate systems: the agent's perspectives and the temporal localizations. The narrator of the discourse is always present in our model, since she is the agent who creates the discourse. The main temporal coordinate system is the *narrator's perspective*. If the narrator reports the words of other agents, as it is the case in our short story (see section 1), a new *agent perspective* is created when each agent says something within the discourse. What is said by an agent includes temporal situations and eventually new agent perspectives.

A temporal localization (corresponding to utterances like "last year", "today", "next week") indicates a temporal frame within which temporal situations and agent perspectives can be situated. Temporal situations, temporal localizations and agent perspectives are linked together by temporal relations (see [9] for details about the representation conventions).

Figure 1 gives a representation of the content of our short story that explicits the perspective of each agent. The embedding perspective is the narrator's perspective, since each sentence of the text is uttered according to the narrator's point of view,

Figure 1 : Conceptual representation of our sample text

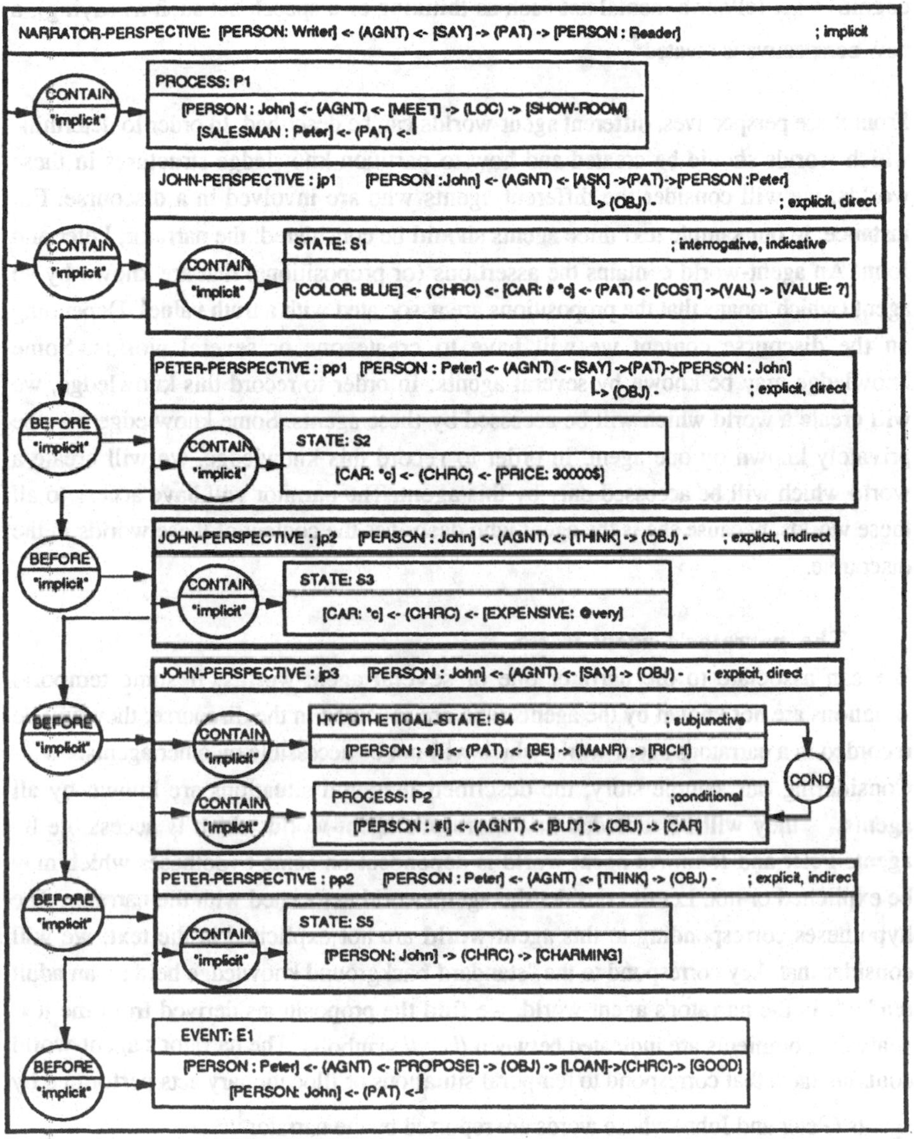

NARRATOR-PERSPECTIVE: [PERSON: Writer] <- (AGNT) <- [SAY] -> (PAT) -> [PERSON : Reader]   ; implicit

CONTAIN "implicit"

PROCESS: P1

[PERSON : John] <- (AGNT) <- [MEET] -> (LOC) -> [SHOW-ROOM]
[SALESMAN : Peter] <- (PAT) <-

CONTAIN "implicit"

JOHN-PERSPECTIVE : jp1   [PERSON :John] <- (AGNT) <- [ASK] ->(PAT)->[PERSON :Peter]
-> (OBJ) -   ; explicit, direct

CONTAIN "implicit"

STATE: S1   ; interrogative, indicative

[COLOR: BLUE] <- (CHRC) <- [CAR: # °c] <- (PAT) <- [COST] ->(VAL) -> [VALUE: ?]

BEFORE "implicit"

PETER-PERSPECTIVE : pp1   [PERSON : Peter] <- (AGNT) <- [SAY] ->(PAT)->[PERSON : John]
-> (OBJ) -   ; explicit, direct

CONTAIN "implicit"

STATE: S2

[CAR: °c] <- (CHRC) <- [PRICE: 30000$]

BEFORE "implicit"

JOHN-PERSPECTIVE : jp2   [PERSON : John] <- (AGNT) <- [THINK] -> (OBJ) -   ; explicit, indirect

CONTAIN "implicit"

STATE: S3

[CAR: °c] <- (CHRC) <- [EXPENSIVE: @very]

BEFORE "implicit"

JOHN-PERSPECTIVE : jp3   [PERSON : John] <- (AGNT) <- [SAY] -> (OBJ) -   ; explicit, direct

CONTAIN "implicit"

HYPOTHETICAL-STATE: S4   ; subjunctive

[PERSON : #I] <- (PAT) <- [BE] -> (MANR) -> [RICH]

COND

CONTAIN "implicit"

PROCESS: P2   ;conditional

[PERSON : #I] <- (AGNT) <- [BUY] -> (OBJ) -> [CAR]

BEFORE "implicit"

PETER-PERSPECTIVE : pp2   [PERSON : Peter] <- (AGNT) <- [THINK] -> (OBJ) -   ; explicit, indirect

CONTAIN "implicit"

STATE: S5

[PERSON: John] -> (CHRC) -> [CHARMING]

BEFORE "implicit"

EVENT: E1

[PERSON : Peter] <- (AGNT) <- [PROPOSE] -> (OBJ) -> [LOAN]->(CHRC)-> [GOOD]
[PERSON: John] <- (PAT) <-

which believes them to be true. Each time an agent (i.e. John or Peter) performs a cognitive act (either a mental act such as thinking or a speech act such as saying), a new perspective is created[3].

From these perspectives, different agent-worlds can be described. In order to determine which worlds should be created and how to partition knowledge structures in these worlds, we will consider the different agents who are involved in a discourse. For instance, in our sample text three agents should be considered: the narrator, Peter and John. An agent-world contains the assertions (or propositions) that are known by an agent (which means that the propositions are associated with a truth value). Depending on the discourse content we will have to create one or several worlds. Some knowledge may be known by several agents: in order to record this knowledge, we will create a world which will be accessed by these agents. Some knowledge may be privately known by one agent: in order to record this knowledge, we will create a world which will be accessed only by this agent. The narrator will have access to all these worlds, because she is the agent who describes the content of these worlds in the discourse.

. The narrator's agent-world

We can associate to the narrator one or several agent-worlds. If some temporal situations are not known by the agents who are involved in the discourse, they will be recorded in a narrator's agent-world which will not be accessible by other agents.
Considering our sample story, the described temporal situations are known by all agents: they will be recorded in a narrator's agent-world which is accessible by agents Peter and John. An agent-world is dependent on some hypotheses which may be explicited or not. Let us consider the agent-world associated with the narrator. The hypotheses corresponding to this agent-world are not explicited in the text: we will consider that they correspond to the "standard background knowledge held by an adult reader". In the narrator's agent-world, we find the propositions derived from the text analysis (comments are indicated between // ... // symbols). The narrator's agent-world contains facts that correspond to temporal situations or illocutionary acts performed by agents (Peter and John) whose words are reported by the narrator[4]:

A-W-NARRATOR

          // indicates the name of the agent-world of the NARRATOR //
  ACCESS: John, Peter
  // note that the agent narrator is not specified since she has access to every world//
  HYPOTHESIS: implicit
  FACTS:

// the following fact represents a situation known by the three agents//
-     [PERSON: John] <- (AGNT) <- [MEET] -
           (PAT)-> [SALESMAN: Peter]
           (LOC) -> [SHOW-ROOM].

//the two following facts correspond to illocutionary acts known by the three agents//

-     [ASK: jp1 <<John, Peter, ->>] ->(MOD)->
           [PROPOSITION: S1 :=
               [COLOR: blue]<-(CHRC)<-[CAR:#c36]->(PAT)
                 ->[COST]->(VAL)->[VALUE:?]].

-- [SAY: pp1 <<Peter, John, ->>] ->(MOD)->
         [PROPOSITION: S2 :=
            [COLOR: blue]<-(CHRC)<-[CAR:#c36]->(CHRC)->[PRICE: 30000$]].

// the following fact represents a situation known by the three agents//
-     [PERSON: Peter] <- (AGNT) <- [PROPOSE] -
           (PAT)-> [PERSON: John]
           (OBJ) -> [LOAN]-> (CHRC) -> [GOOD].

The perspective (jp3) associated with agent John in figure 1 is more subtle. John considers explicitly an hypothetical world by saying "if I were rich". Hence, a new agent-world "A-W-Narrator-John1 specializes the narrator's agent-world under this hypothesis. Since John evoked that hypothesis when speaking to Peter, both agents have access to this world.

A-W-NARRATOR-JOHN1   // hypothetical world evoked by John //
    SPECIALIZE: A-W-NARRATOR
    ACCESS: John, Peter
    HYPOTHESIS:
         HYP: hp1 :=   // states the hypothesis //
            [PERSON:John]->(CHRC)->[RICH]
    FACTS:
    --[PERSON:John]<-(PAT)<-[BUY]-
                 (OBJ)->[CAR:#c36]->(CHRC)->[COLOR:blue]

Note that the anaphoric references found in the text, "I" corresponding to John and "it" corresponding to "the blue car", have been resolved before the creation of the fact in A-W-Narrator-John1.

. Other agent-worlds

When we read a story we build independent models of beliefs and hopes which characterize the different characters in the story. In the representation of our sample text (figure 1) two perspectives provide some information about John's and Peter's mental states: perspectives jp2 and pp2. Each perspective corresponds to an agent's belief which is not known by the other agent. In each case we will create a world which is accessed only by the agent that possesses the corresponding belief. These worlds are A-W-Peter1 and A-W-John1

A-W-PETER1   // Peter's world 1 accessed only by Peter and the narrator//
    ACCESS: Peter
    HYPOTHESIS: implicit
    FACTS:
        –   [THINK: pp2 <<Peter, -, ->>] ->(MOD)->
                [PROPOSITION: S5 :=
                    [PERSON: John]->(CHRC)->[CHARMING]].

A-W-JOHN1   // John's world 1 accessed only by John and the narrator//
    ACCESS: John
    HYPOTHESIS: implicit
    FACTS:
        –   [THINK: jp2 <<John, -, ->>] ->(MOD)->
                [PROPOSITION: S3 :=
                    [COLOR: blue]<-(CHRC)<-[CAR:#c36]
                        ->(CHRC)->[EXPENSIVE: @very]].

. Reasoning with agent-worlds

In addition to hypotheses, agent-worlds may contain **inference rules** which are used to manipulate the agent-world content. Suppose that we have a rule R1 saying that

> IF a person x says proposition prop1 to a person y
> AND person x is honest
> > THEN person x believes prop1.

Now if we consider an agent-world $A\text{-}W_x$ of agent X, this rule can be expressed by:

$A\text{-}W_x$ ⊢ [SAY: sy_i <<X, Y, $TINT_i$>>]->(MOD)->
                [PROPOSITION: prj <<TINTj>>]

$A\text{-}W_x$ ⊢ [PERSON: X]->(CHRC)->[HONEST]

⇒ $A\text{-}W_x$ ⊢ [PROPOSITION: prj <<TINTj>>].

The narrator may not know if Peter, as a salesman, is honest. But she could make the hypothesis that "Peter is honest". Hence, we could create another hypothetical world which specializes the world A-W-Peter1:

    A-W-PETER1.1   // An hypothetical world accessed only by the narrator//

    SPECIALIZE: A-W-Peter1

    ACCESS: Narrator

    HYPOTHESIS:

        [PERSON: Peter]->(CHRC)->[HONEST]

    FACTS:

    -- [BELIEVE: pp3 <<Peter, -, ->>] ->(MOD)->

        [PROPOSITION: S2 :=

        [COLOR:blue]<-(CHRC)<-[CAR:#c36]->(CHRC)->[PRICE:30000$]].

Inter-world transfer rules can be used to transfer knowledge from one agent-world into another. Let us consider an example with the agent-world W-NARRATOR. Suppose that we have a rule R2 saying that

        IF a person x says proposition prop1 to a person y

        AND person y is credulous

                THEN person y believes prop1.

Now if we consider an agent-world $A\text{-}W_z$ of agent Z, and the agent-world $A\text{-}W_y$ corresponding to the model of agent Y viewed from agent Z's perspective, this rule can be expressed by:

    $A\text{-}W_z \vdash$ [SAY: sy_i <<X, Y, $TINT_i$>>]->(MOD)->

                [PROPOSITION: pr_j <<TINTj>>]

    $A\text{-}W_z \vdash$ [PERSON: Y]->(CHRC)->[CREDULOUS]

    $\Rightarrow$ $A\text{-}W_y \vdash$ [PROPOSITION: pr_j <<TINTj>>].

Suppose that we add in the narrator's agent-world that "John is credulous", namely the belief represented by the conceptual graph:

        [PERSON: John]->(CHRC)->[CREDULOUS],

which is now part of the narrator's background knowledge about John. Rule R2 would be triggered and the system could add in the agent-world A-W-JOHN1 the belief:

    [BELIEVE: jp3 <<John, -, ->>] ->(MOD)->

        [PROPOSITION: S2 :=

        [COLOR:blue]<-(CHRC)<-[CAR:#c36]->(CHRC)->[PRICE:30000$]].

Agent-worlds can then evolve non monotonically as the result of the triggering of rules that are associated with them.

7. Conclusion

In this paper we have presented an approach for factoring knowledge into structures called agent-worlds. Agent-worlds can be viewed as knowledge bases representing "microtheories". We have described several mechanisms for reasoning with the knowledge contained in the agent-worlds, as well as mechanisms for transferring knowledge from one world into another. The propositions that are contained in agent-worlds are associated with a special form of modality. This modality reflects the kind of illocutionary act that was performed by the agent who utterred the corresponding sentence. This approach will enable us to reason on illocutionary acts.

Further research is needed in that area. We will have to refine our approach for generating the content of agent-worlds on the basis of the temporal structure of discourses. A special study will be devoted to the analysis of temporal relations and their impact on the generation of propositions in agent-worlds. A detailed study of different types of illocutionary acts will be undertaken in order to characterize the reasoning mechanisms (inference rules, inter-world transfer rules) needed to manipulate them in a multiagent system.

ACKNOWLEDGMENTS

This research is supported by the Natural Sciences and Engineering Research Council of Canada (grants OGP 05518 and OGP 0105365) and by FCAR.

NOTES

1. An agent-world A-W1 may specialize another agent-world A-W2 so that most of its facts and rules are inherited from A-W2, but some of them may be redefined within A-W1. This question will not be detailed in the present paper.

2. Here the term "context" is used in the sense defined by Searle and Vanderveken. It should not be confused with Sowa's and Guha's definitions of "context". Similarly, "the world of utterance" is a term that was defined by Searle and Vanderveken. Its sense is not as precise as our notion of agent-world.

3. Notice that the parameters following the semi-colon in fig.1, give some linguistic information related to the text: 'explicit' indicates that the text contains a sentence describing explicitly the perspective ('implicit' corresponds to the contrary), 'direct' indicates that the speech acts that correspond to the situations that are included in the perspective are reported in a direct way in the text ('indirect' corresponds to the contrary).

4. In order to simplify the representation, agents' beliefs will be represented without specifying the "belief" illocutionary act (beliefs are considered as default modalities).

Hence if an agent's belief holds on a time interval that is not explicited, the representation of the mental act of believing proposition pr_j will be simply

represented by   A-W$_y$ ⊢ [PROPOSITION: prj <<TINTj>>]

if we don't need to give any detail about the act of believing;

or by   A-W$_y$ ⊢ BEL:bl$_i$:=

[PROPOSITION: prj <<TINTj>>]

if we need to identify the act of believing ('BEL' which is an abbreviation of 'BELIEVE' indicates the type of the mental act whose identifier is bl$_i$).

BIBLIOGRAPHY

1. DUNN M.G. (1973), A truth value semantics for modal logic, in H. Leblanc ed., *Truth, Syntax and Modality*, North Holland.

2. FILMAN R. E. (1988), Reasoning with worlds and truth maintenance in a knowledge-based programming environment, in *Communications of the ACM*, vol 31 n4, pp 382-401.

3. GUHA R. V. (1991), Contexts: a Formalization and some applications, MCC tech.report ACT-CYC-423-91.

4. HINTIKKA J. (1973), Surface semantics: definition and its motivation, in H. Leblanc ed., *Truth, Syntax and Modality*, North Holland.

5. de KLEER J. (1986), An assumption-based truth maintenance system, in *Artificial Intelligence*, vol 28, n2, pp 127-162.

6. MOULIN B. (1990), Côté D. , Extending the conceptual graph model for differentiating temporal and non-temporal knowledge, in proceedings of the Fifth Annual Workshop on Conceptual Structures, AAAI conference, Boston July 1990.

7. MOULIN B. (1991), A conceptual graph approach for representing temporal information in discourse, in proceedings of the Sixth Annual Workshop on Conceptual Structure, Binghamton, New York, July 1991, to appear in *Knowledge-Based Systems* journal.

8. MOULIN B. (1991), D. Rousseau, D. Vanderveken, Speech acts in a connected discourse, a computational representation based on conceptual graph theory, in proceedings of the Sixth Annual Workshop on Conceptual Structures, Binghamton, New York, July 1991, to appear in the *Journal of Experimental and Theoritical Artificial Intelligence*.

9. MOULIN B. (1992), Modelling temporal knowledge in discourse: a refined approach, in proceedings of the Seventh Annual Workshop on Conceptual Structures, July 1992.

10. SOWA J.F. (1984), *Conceptual Structures : Information Processing in Mind and Machine*, Addison Wesley.

11. SOWA J. F. (1987), Semantic networks, in S. C. Shapiro ed., *Encyclopedia of Artificial Intelligence*, Wiley and Sons , New York, pp 1011-1024.

12. SOWA J. F. (1990), Towards the expressive power of natural language, in proceedings of the Fifth Annual Workshop on Conceptual Structures, AAAI conference, Boston July 1990.

13. SOWA J. F. (1991), Matching logical structure to linguistic structure, in N.Houser, J. Van Evra eds., *Studies in the Logic of Charles Sanders Peirce*, Indiana University Press, Bloomington.

14. SEARLE J. R. (1985), and D. Vanderveken, *Foundations of Illocutionary Logic*, New York, Cambridge University Press.

Sharing Knowledge: Starting with the Integration of Vocabularies

Guy William Mineau

Computer Science Department, Université Laval
Quebec City, Quebec, Canada, G1K 7P4

Abstract[1]. This paper addresses the knowledge sharing problem from an operational perspective: how and when can one use somebody else's knowledge. It proposes the implementation of some control mechanism over the knowledge acquisition process. This limits the expressiveness of the notation but will improve the shareability of knowledge structures among different but related knowledge domains. This control mechanism is based on the construction and use of a part-of hierarchy built with the descriptions of the terms composing the vocabulary.

1 Introduction

Improving the accessibility and availability of various computerized knowledge sources is the ultimate goal behind the knowledge sharing efforts being deployed in the artificial intelligence community nowadays [1, 2]. As knowledge acquisition is a current bottleneck in the development of knowledge based systems, reuseability of already acquired knowledge becomes of the outmost importance. The development of a knowledge base could then benefit from other people's efforts [3]. Also there may be applications whose inference engine would need the expertise of other systems' knowledge in order to carry out their own reasoning. Co-operating agents display such a behaviour [4]. For these reasons, knowledge reuseability is thus a prime objective within the knowledge sharing research community [2].

Unfortunately, mainly due to their informal definitional frameworks, knowledge-based systems have been developed using good sense and good will. Knowledge engineers have had to make systems work in order to solve particular problems. To that effect, they each developed a certain *know-how* which contributed to the development of incompatible systems. Consequently, knowledge sharing is now facing major challenges. Among them let us cite: the standardization of knowledge representation languages (as with the KIF project [5]), the development of knowledge exchange protocols, the sharing of ontologies (as with the Ontolingua project [6]), and the useability of the exchanged knowledge [1]. This paper discusses this last issue: the useability of exchanged knowledge, also called *imported knowledge*.

In effect, imported knowledge will often become useful to an application which requires it to take part in some inference process. Also, as the output of knowledge-

---

1. This work is sponsored by a NSERC grant (Natural Sciences and Engineering Research Council of Canada) #OPG0105365.

based systems requires to be interpreted in order to be of any help to the end user, the imported knowledge must be understood. This understanding guarantees a correct interpretation of its meaning. Without it, the newly accessible knowledge sources would remain useless, and the whole point of sharing knowledge becomes irrelevant. The problem of understanding imported knowledge structures is thus central to all knowledge sharing efforts. Of course, the understanding of knowledge structures depends greatly on the content of the ontology used to describe them. Consequently, it is vital to allow ontologies to be portable from one system to the other. Knowledge-based systems willing to share their knowledge will have to share their ontologies as well.

Gruber proposes to allow portability of ontologies based on the use of a common set of ontology description primitives, called Ontolingua [6]. This set is said to be the common subset of what can be found in diverse knowledge representation languages today. However, it is our personal belief that this set of primitives is less complete than the actual tools offered by the conceptual graph formalism of Sowa [7], which we are going to discuss in a forthcoming paper. Consequently, the whole ideas of sharing knowledge and ontologies as being developed in this paper are based on the conceptual graph formalism.

Furthermore, though Ontolingua allows the expression of standardized descriptions of ontologies, and thus, their sharing among systems being able to interpret such descriptions, nothing has been said yet on the *integration* of ontologies. In effect, it may often be the case that an application already developed using a particular ontology, requires knowledge from some other knowledge-based system. In order to interpret the knowledge structures it will receive, the ontology upon which they were built must also be imported. However, it may conflict with the ontology of the receiving system. In that case, both ontologies must be integrated.

Of course, there could be some universal ontology upon which every system's own ontology could be based. Due to multiple and diverse needs of knowledge-based applications, due to their evolving nature, and considering the tremendous effort one already has to put on knowledge acquisition, it is quite unlikely that such an ontology could be built. Rather, it is more likely that independently built knowledge systems will need to communicate and use each other's knowledge. This more realistic situation introduces the problem of *ontology integration*.

An ontology is described using many knowledge description and structuring mechanisms, such as class description, instance creation, type hierarchies formation, etc. This article will concentrate on the class description mechanism, otherwise named *vocabulary description mechanism*, and its link to the shareability of knowledge structures built with this vocabulary. Thus, it presents *vocabulary integration* as a first step to ontology integration, and to ultimately knowledge sharing.

2 The Semantics Behind

When explaining a term, one can only use what is commonly known. Otherwise the use of some other unknown terms in an explanation will necessitate other explanations. In fact, this is the major problem of any dictionary: everything is defined in terms of everything else. Despite these recursive descriptions, people do learn the meanings of words. For instance, natural language is acquired through a long training period which spans over a lifetime, and which enriches one's knowledge with experience, examples, counter-examples, links to concrete concepts through sensory devices, etc. This forms a basis upon which some terms, though defined by other terms, are familiar enough to be called *primitive*: their meaning is commonly known and accepted.

Knowledge-based systems handle symbols which represent concepts and their semantical relationships to one another. These symbols are conceptually linked to some interpretation. The knowledge engineer establishes such conceptual links, and decides upon mechanisms to validate and enforce them. Generally, this interpretation can not be guessed nor inferred through experimentation, except for a few systems which explore extremely simple domains [8, 10]. Consequently, understanding a new term is directly related to the discovery of the conceptual link relating it to the semantics it represents.

Also, as stated earlier, the understanding of a term depends upon its definition which must use commonly known terms. In other words, the integration of a new term to a vocabulary is feasible through the establishment of a description made out of terms already in the integrated vocabulary. Basically, the integration of a new term depends upon the integration of the terms composing its definition. We can view the integration process as a reinforcement mechanism: the integration of a term will have an impact on the integration of other terms. Furthermore, the integration of *key terms*, i.e., of particularly important terms in the ontology, will have more impact than the integration of arbitrarily chosen terms.

The following section summarizes the vocabulary description mechanism associated with the conceptual graph formalism, and defines the problem of vocabulary integration more formally. Section 4 proposes control and definitional mechanisms needed to improve the portability and the integration of terms from a source to a target vocabulary. Section 5 presents related problems and areas of research were active work is needed.

3 The Vocabulary Description Mechanism

In the conceptual graph notation, terms belonging to the vocabulary are of two kinds: concept or relation. Since concepts and relations are used in different ways, they each have their own definitional mechanism which is based on lambda expressions. For concepts it is called *type definition*, for relations, *relational definition*. They appear below.

> "A *type definition* declares that a type label t is defined by a monadic abstraction $\lambda a\ u$. It is written, **type** $t(a)$ **is** u."

> "A *relational definition*, written **relation** $t(a_1,...,a_n)$ **is** u, declares that the type label t for a conceptual relation is defined by the n-adic abstraction, $\lambda a_1,...,a_n\ u$."

The body u of each description of type t is a conceptual graph where the formal parameters $a_1,...,a_n$ ($\forall n \in N^*$) appear as non-instantiated concepts in u. From its description, t becomes a term in the vocabulary, available for knowledge encoding conceptual graphs to be acquired at a later stage.

Furthermore, all types are declared in relation to some other type. A hierarchy of types can be built according to Aristotle's model of object classes, since each class is a subclass of a more general one, and since the concept which is at the heart of the class description is the concept identifying the superclass. This concept is called the *genus*, and the specialization of the description in regard to the description of the superclass is called the *differentia*. This hierarchy, called *generalization hierarchy*, is defined over all abstractions. Because the vocabulary is composed of two kinds of terms, concepts and relations, two generalization hierarchies can be built.

According to this Aristotelian model, let type t_1 be defined by the monadic abstraction $\lambda a\ u$, and let type t_2 be defined by the monadic abstraction $\lambda b\ v$. In order to have a generalization hierarchy over concept types, we must have one of the following conditions: $t_1 < t_2$, $t_2 < t_1$ or $t_1 < t_3$ and $t_2 < t_3$, where t_3 is a type defined in the same manner as before, or is the *universal concept type* $\top$ (defined in the notation along with the *absurd concept type* $\bot$, which is a specialization of everything), and where the $<$ operator signifies *is a subclass of*. The $<$ operator can be evaluated from the abstractions defining types t_1 and t_2. From a set of two monadic abstractions t_1 and t_2, t_1 is said to be more specific than t_2 if and only if these following two conditions hold:

1) u is a specialization of v;
2) there exists a projection π of v into u, which maps the parameter b of v into the parameter a of u.

A similar generalization hierarchy can be built with relations, as one relation can be defined based on others. As in the concept type hierarchy, there is a general unifying element called LINK (also defined in the notation), which states semantical linkage between a certain number of concepts depending on the arity of the relation.

With these generalization hierarchies, one can see that there already exists a structure, a knowledge structuring classification scheme, which indicates what type is defined in terms of what other type. This classification will prove to be useful but insufficient for the integration of terms belonging to a vocabulary, as the rest of this section shows.

Based on the preceding definitions, the integration of a term t_1 from a source vocabulary V_1 to a destination vocabulary V_2, can be described by the following algorithm.

1) find a term t_1 from V_1 whose definition u is such that, for all concepts c and for all relations r in u, c and r both belong to V_2;
2) if such t_1 exists then:

4 Searching for Semantical Equivalence

When semantical equivalence must be established between terms encoding similar but different semantics in different systems, or for terms from the source vocabulary having no counter-parts in the target vocabulary, the knowledge engineers must intervene. They must define precisely how a term from one system can be defined using the other system's vocabulary. The term to be transferred then needs to be closely scrutinized in order to capture its correct and complete meaning. To help the knowledge engineers in performing this task, all sorts of tools should be available: on-line documentation of terms in textual format, examples of correct use of the term in different conceptual graphs (p.e., using a schematic cluster [7]), a prototype definition for the term (when appropriate), the sets of terms which represent respectively its superclasses and its subclasses, its definition in terms of a lambda expression, etc. Of course, not all those tools may be available. A type hierarchy structure is available from the vocabulary description mechanism; the lambda expression defining a type is also available; but a prototype may or may not exist for the definition of a particular concept, depending on the nature of this concept. Also, schematas may or may not have been acquired while defining a term.

In order to benefit from these knowledge structuring data (prototypes and schematas), the acquisition process should be reinforced with additional information extracting mechanisms such as conceptual clustering techniques [9]. These techniques could generate empirical prototypes and schematas, as found within the data, as they are oriented toward learning from observation [10]. At last year's workshop on Conceptual Structures, we presented a tool based on conceptual clustering of semantically related conceptual graphs [11]. This tool was based on some earlier research on conceptual clustering of conceptual graphs [12, 13]. Conceptual clustering offers a wide range of techniques which can detect regularity within the data. Consequently, prototypes or schematas can be detected if they keep on reoccurring within the data, eventhough the knowledge engineer had not expected their occurrences. Thus the knowledge acquisition process should attempt to apply such techniques to terms definitions and to knowledge encoding conceptual graphs. That way, the proposed tool could prompt the two knowledge engineers to correlate prototypes and schematas from different systems, when they are detected in graphs encoding similar semantical situations.

But even in cases where much information about a term is available, the knowledge engineers are faced with the following interrogations. What terms can be translated? What terms should be translated next? What terms will be easy to translate? What terms are important to translate? The knowledge engineers must determine answers for these questions in order to continue the integration of the vocabularies of the two systems, since subsequent translation of knowledge structures depends greatly upon this integration.

Based on the recursive algorithm given above, a term is translatable from a source vocabulary to a target vocabulary if all the components of its description (or at least the most important ones) are already translated. Consequently, we need to build a part-of hierarchy over type definitions. That way, terms already translated into the target vocabulary could be identified, and their possible contribution to the translation of other terms, whose definitions they take part in, could also be pinpointed. The impact of the translation of a term could then be evaluated in terms of the number of translations made possible by this initial translation. *Key terms* (*elements*) could also be identified. They are those that have greater impact on the translation of other terms. They can either be identified when acquiring knowledge, whenever the knowledge engineer knows beforehand what terms are important, or they could be detected as *key elements* in this part-of hierarchy. The following example shows this point. Figures 1 thru 3 shows three type definitions, and figure 4 shows the corresponding part-of hierarchy. From figure 4, it is obvious that the concept type perform and the relation types loc and agnt are important in this vocabulary, and consequently, in this knowledge domain.

type circus-elephant(x) is

[elephant:*x]<-(agnt)<-[perform]->(loc)->[circus]

Figure 1: Type definition for circus-elephant.

type elephant-circus(x) is

[elephant]<-(agnt)<-[perform]->(loc)->[circus:*x]

Figure 2: Type definition for elephant-circus.

type circus(x) is

[location:*x]<-(loc)<-[people]<-(agnt)<-[perform]->(obj)->[act]
———(loc)<-[people]<-(agnt)<-[look]->(obj)———

Figure 3: Type definition for circus.

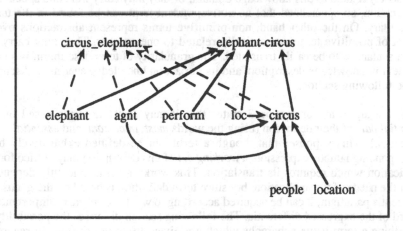

Figure 4: Subset of the part-of hierarchy built from the type definitions of figures 1 thru 3.

More formally, this part-of hierarchy can be described as follows. Let t be a type described by **type** t **is** u. Naturally, all concepts c and relations r of u must already belong to the vocabulary V. Then, for all elements e of u, either c or r, $t \supseteq e$, where $\supseteq$ is the inclusion operator defined over the components of type definitions. $\supseteq$ is reflexive $(e \supseteq e)$, anti-symetrical $(t \supseteq e$ and $e \supseteq t <=> t = e)$, transitive $(t \supseteq e$ and $e \supseteq f => t \supseteq f)$. By definition, the part-of hierarchy is an oriented graph structure. Cycles may exist. Additional control over the knowledge acquisition process is required in order to avoid these cycles, as explained in the next paragraph.

Of course, the algorithm of section 3 being recursive, one has to ensure that the recursion termination conditions will be reached. In terms of the part-of hierarchy which is at the heart of this algorithm, it restricts its topology to an acyclic graph. This limits the power of expression of the notation, since type definitions can not be recursive, but is not incoherent with the philosophy underlying knowledge description with conceptual graphs. In effect, one of the great advantages of the conceptual graph formalism is that the epistemological level [14] is in no way incorporated within the notation itself. Concretely, the knowledge engineer decides upon a set of primitive concepts and relations, as needed according to the analysis of the knowledge domain to describe. The vocabulary is built with these terms and with abstractions of these terms. Primitive terms are those which are left without a definition; non-primitive terms are those described in terms of other terms. Consequently, fewer representational biases are due to the notation than with other kowledge reprensentation languages [14]. This allows great flexibility when encoding knowledge. Then, translation of knowledge structures from one domain of semantics to another depends on semantical equivalences between corresponding *primitive sets*, the sets of primitive terms of each system [11].

Unfortunately, establishing semantical equivalences between different primitive sets, though the conceptually easiest solution, is not so simple. Primitive terms carry semantics quite related to the knowledge domain they describe. They may either be too closely related to this knowledge domain, or they may carry little and specialized information, so specialized that no correspondent term can be found in the target vocabulary. On the other hand, non-primitive terms represent abstractions over a subset of primitive terms semantically related to one another. They may carry too much semantics to be easily translated. The granularity of term description is a basic problem in knowledge description, and its impact on knowledge sharing is discussed in the following section.

Also, the algorithm of section 3 states that it may translate terms based on the *essential part* of their definition (using the words *most*, *important*, and *essential* in the algorithm). This supposes that though a term can be defined exhaustively by a corresponding lambda expression, a *working set* of this description may suffice for the application which requires its translation. This working set is evidently dependent upon the needs of the application, but since term definition is based on the genus and differentia paradigm, it can be acquired according to what is of primary importance in regard of the expressed differentia. The following section discusses the possibility of describing a term using subgraphs which are given different priorities in regard of their relative importance to the differentia.

5 Related Problems and Further Developments

As seen in the previous section, the vocabulary description methodology influences the knowledge sharing capability of the system. Terms carrying too little semantics in a source vocabulary may have no correspondence in the target vocabulary; terms carrying too much semantics may not be translatable as a whole. Granularity of term definitions is important in this regard.

Of course, there will be more important terms in a vocabulary. These terms represent the main concepts needed to describe the core of a knowledge domain. They can be defined and identified as *key terms*. In order to allow their translation to the target vocabulary, it would be wise to ensure that their understanding could be established through their definition in terms of simpler and less specialized terms, and in a non-recursive manner as seen in the previous section. Doing so introduces other terms, primitive and non-primitive. A *description subgraph* for these key terms can be extracted from the part-of hierarchy. Some of the terms in the description subgraph will find counterparts in the target vocabulary. The propagation of their translation will ease the translation process of key terms. It is expect that semantical compatibility between the two knowledge domains will be established at different levels within the part-of hierarchy, leaving lower-level terms untranslated, helping the translation of upper-level terms. Consequently, key terms, once identified by the knowledge engineer, must be defined in such a way to introduce other terms, which in turn will also be defined, etc. The resulting part-of hierarchy will display a certain average height between the key terms and their primitive component terms. We will correlate this average height with the granularity of the definitions of the key terms, hence giving a metric for granularity measurement. Its impact on the translation of key terms will be harder to measure since it depends greatly on the semantics carried by the terms composing each definition in description subgraphs. We hope that empirical analysis of different description subgraphs for the same key terms may give hints about the impact of granularity on term translation.

Secondly, when describing a term t_1 using a lambda expression $\lambda a\ u$, some description methodology incorporating the importance factor of a subgraph of u in regard of the whole differentia represented by u, should be introduced in the description mechanism. Consequently, we propose to build u in layers, each layer having a different importance factor. Thus we define a type in terms of a conceptual graph u, which is defined as a maximal join over sets of subgraphs, each set defining a layer. More formally, let us define a type t_1 as **type** t_1 **is** u, where $u = \mathrm{join}(u_0, u_{1,1},...,$ $u_{1,n1}, u_{2,1},...,u_{2,n2},...,u_{l,1},...,u_{l,nl})$, where u_0 represents the genus of type t_1, where $u_{i,j}$ is a conceptual graph describing the jth subgraph of layer i ($\forall i \in [1,l]$, $\forall j \in [1,ni]$), where join is the maximal join operator between conceptual graphs [7], and where ni is the number of subgraphs composing layer i. Furthermore, some overlay mechanism could be implemented to associate numerical *relevance factors* of the different layers, to the genus of the definition. For example, some vector $<f_1,f_2,...,f_l>$ associating a relevance factor f_i with each layer i ($\forall i \in [1,l]$) could either be inputted directly by the knowledge engineer, computed according to f_{i-1}, or deducted using some heuristics and f_j ($\forall j \in [1,i-1]$), etc.

Naturally, this new type definition mechanism is identical to **type** t_1 **is** u, when only one layer composed of only one conceptual graph is used (when $1 = 1$ and $n1 = 1$). That is, **type** t_1 **is** u, **where** $u = \text{join}(u_0, u)$ is identical to **type** t_1 **is** u because $\text{join}(u_0, u) = u$ since u_0 must belong to u according to the Aristotelian model of object classes based on genus and differentia. Nevertheless it allows more power to describe types in terms of layers, each of which being a specialization of the preceding one, i.e., an addition to the differentia part of the definition of the term. With relevance factors set according to predetermined rules by the knowledge engineer, working sets of term definitions could be determined as well. For example, it could arbitrarily be decided by the knowledge engineer that all term definitions found in layers 1 to 5 form the working set of a term. Less work would then be needed to translate terms whose working set could already be translated. Though not as exhaustive and complete as it could be, this translation could permit the use of the translated term in order to import knowledge, as long as the working set used to define t_1 is the same working set needed by the requesting target application (from the target domain). That is, what is important to the target application should be perceived as such by the source domain in order to adjust its translation process in terms of what element of the vocabulary should be translated next. This should orient all the translation efforts.

As seen previously, it is possible to associate relevance factors to each term. The target application, upon translation of a term from the source vocabulary, could indicate to the source domain what is the relevance factor of this newly translated term according to its actual needs. The challenge facing the source domain would then be to identify the *key terms* according to the needs of the target application. Terms of interest to the target application could be identified if relevance factors would somehow be combined by some propagation mechanism through the part-of hierarchy built to associate different terms to one another. This combination mechanism necessary to determine translation priorities among terms of the source vocabulary is currently under study. Upward propagation seems to fit a reinforcement function: key terms compose the definition of other important terms; downward propagation is still speculation: key terms are not strictly composed of important terms. Other heuristic information is expected to be needed in order to make wise decision about translation priorities. For example, an 80% frequency of reoccurrence of a term in other term definitions may signify that this term may also be a key term (or *key subterm*). In summary, different information is needed to complete downward propagation of relevance factors, and general questions like: how many levels in the part-of hierarchy should the propagation reach? Should it lose weight with each level? Should it be combined with other relevance factors? Should it be exhaustive before another term is tentatively translated? What effect has granularity of term definition on the propagation? And so on. Again, we hope that empirical analysis of some real experiences will shed light on these interrogations, giving us at least some guidelines for the description of a vocabulary whose elements are to be translated to some target vocabulary in order to allow knowledge sharing between these two systems.

6 Conclusion

Knowledge sharing starts with the integration of the vocabularies the knowledge structures to share are built with. It is more realistic to talk about vocabulary integration than to hope for some universal vocabulary to fit most of the needs of diverse applications. The integration of terms from a source vocabulary into a target vocabulary may require human assistance to establish semantical equivalence among terms, or among term definitions.

Since a term is defined by other terms, we presented a data structure which encodes hierarchical inclusion of term definitions. This part-of hierarchy, built with the content of each term definition, shows the inclusion of definitions of terms into other definitions. Consequently, it is possible to forsee the impact of the translation of a term on the translation of other terms. We presented an algorithm which uses this hierarchical structure to identify which terms should and could be translated next. This algorithm is recursive; consequently the part-of hierarchy has been defined as an acyclic graph structure. This limits the expressivity of the vocabulary description mechanism, but allows knowledge sharing algorithms to be implemented. Together with other tools, this part-of hierarchy should help the knowledge engineers in their work of integrating the two vocabularies together. For example, it could be used to support a browsing mechanism through the various type definitions, allowing the exploration and the discovery of definition inclusions.

In order to facilitate knowledge sharing among systems described by conceptual graphs, the type definition mechanism needs to be modified. The new type definition mechanism introduces layers of subdefinitions. Each layer can be associated with a relevance factor which gives it a relative importance in regard of the whole definition. Whether statistically computed, heuristically deducted, user inputted, or application related, these factors can be seen as an overlay over the part-of hierarchy. They can be used in order to determine what *should* be translated next (determination of *key terms* through relevance factors propagation), and what *could* be translated next (if the essential part of a definition, called working set, is ready to be translated).

These notions of *definition layers, relevance factors, working set*, and *key terms* modify the vocabulary description mechanism associated with the conceptual graph formalism [7]. It is expected that other description mechanisms may also be added, such as example and counter-example description, and characteristic and functional definition.

This more complete ontology description mechanism not only improves the capability of knowlege-based systems to exchange knowledge structures, but it also gives a more formal methodology in order to describe a knowledge domain, which is evidently lacking in today's know-how on knowledge-based systems. Consequently, this paper also exposed the close relationship between knowledge modelization and knowledge sharing.

References

1. Wileden, J.C., Wolf, A.L., Rosenblatt, W.R. & Tarr, P.L., (1991). "Specification-Level Interoperability". In: CACM, vol.34, no.5. 73-87.
2. Neches, R., Fikes, R., Finin, T., Gruber, T., Patil, R., Senator, T. & Swartout, W.R., (1991). "Enabling Technology for Knowledge Sharing". In: AI Magazine, vol.12, no.3. 36-56.

3. Prieto-Diaz, R., (1991). "Implementing Faceted Classification for Software Reuse". In: CACM, vol.34, no.5. 89-97.

4. Rosenschein, J.S., (1988). "Synchronisation of Multi-Agent Plans". In: *Readings in Distributed Artificial Intelligence*. A.H. Bond & Les Gasser (eds). 187-191.

5. Genesereth, M.R. & Fikes, R., (1991). "Knowledge Interchange Format, Version 2.2, Reference Manual". Technical Report #Logic-90-4, Logic Group, Stanford University, Stanford, CA, USA. March.

6. Gruber, T.R., (1991). "Ontolingua: A Mechanism to Support Portable Ontologies". Technical report #KSL 91-66, Knowledge Systems Laboratory, Stanford University, Stanford, CA, USA. November.

7. Sowa, J. F., (1984). *Conceptual Structures: Information Processing in Mind and Machine*. Addison-Wesley Publishing Co.

8. Winston, P.H., (1975). "Learning Structural Descriptions from Examples". In: *The Psychology of Computer Vision*. P.Winston (ed). Computer Science Series. McGraw-Hill. 157-209

9. Stepp, R. E. & Michalski, R. S., (1986). "Conceptual Clustering: Inventing Goal-Oriented Classifications of Structured Objects". In: *Machine Learning: An Artificial Intelligence Approach, Volume II*. Michalski, R.S., Carbonell, J.G. & Mitchell, T.M. (eds). Morgan Kaufmann Publishers. 471-498.

10. Michalski, R. S., Carbonell, J.G. & Mitchell, T.M., (1983). *Machine Learning: An Artificial Intelligence Approach*. Morgan Kaufmann Publishers.

11. Mineau, G., (1991). "Toward Compatible Primitive Sets". In: Proceedings of the 6th Annual Workshop on Conceptual Structures. Way E. (ed.). July. State University of New York at Binghamton (SUNY), Binghamton, USA. 53-66.

12. Mineau, G., (1989). "Induction on Conceptual Graphs: Finding Common Generalization and Compatible Projections". In: Proceedings of the 4th Annual Workshop on Conceptual Structures. Nagle, T. & Nagle J. (eds). July. IJCAI-89. Detroit, USA. 3.07.1-3.07.19.

13. Mineau, G., Gecsei, J. & Godin, R., (1990). "Structuring Knowledge Bases Using Automatic Learning". In: Proceedings of the 6th International Conference on Data Engineering. February. Los Angeles, USA. IEEE Computer Society Press. 274-280.

14. Brachman, R.J., (1985). "On the Epistemological Status of Semantic Network". In: *Readings in Knowledge Representation*. 191-216.

A Conceptual Graphs Approach to Information Systems Design

Linda Campbell and Peter Creasy

Key Centre for Information Technology
Department of Computer Science
University of Queensland, Queensland, Australia 4072

Abstract. There is a growing tendency and need in information systems to model applications using a number of different languages, with a subsequent integration of these models. One possibility would be to provide a mapping between each pair of these modelling languages. Alternatively, a core conceptual language could be used as a standard translation intermediary, thereby reducing the total mappings required from n^2 to $2n$. This language could be the centre of a CAiSE (computer aided information systems engineering) tool with a mapping provided between each modelling language and the core language. This idea corresponds to the ANSI proposal for the IRDS (Information Resource Dictionary System) conceptual schema. ANSI have proposed conceptual graphs as a normative language which corresponds to such a core language. This paper discusses some of the requirements of this normative language by examining the semantics provided by the NIAM language. The paper also proposes conceptual graph abstractions to provide direct equivalents to the NIAM constructs.

1 Introduction

In order to facilitate the effective and economic production of information systems, formal modelling tools (schema or modelling languages) and associated methodologies are used. This modelling process, which aims to capture as much as possible of the Universe of Discourse (UoD), is central to the implementation of database systems.[7]

Several of these tools (e.g. ER [14] and NIAM [11]) have been used commercially for a number of years. The languages capture the static aspects of the UoD, providing the ability to represent a few (but important) constraints, and permit the simple translation into a relational database schema. There have been attempts to expand these languages to capture more of the UoD, in particular to capture the dynamic aspects. However no tool has been as successful as the static languages.

A number of these languages have been examined as candidates for the IRDS normative language. The IRDS describes and manages an organisation's information system resources. The normative language is part of the IRDS layered framework. In this framework, each layer provides a higher level of abstraction in terms of the layer directly below it. Within the normative schema layer the normative language can define all constructs that would be used in any modelling schema language. Thus the normative language "will be rich enough to include a superset of the semantics of all existing schema languages" [1]. ANSI has proposed conceptual graphs for the normative language.

This paper focusses on the conceptual schema aspects of the normative language. It examines several languages and briefly comments on their suitability as a normative language. As NIAM is a language which encompasses the semantics of most static conceptual schema languages we examine this in moderate detail in section 2 so we can establish the requirements of a normative language in conceptual schema terms.

Conceptual graphs are much more general than NIAM. In particular it is much easier to express the dynamic aspects of a UoD with conceptual graphs [6]. However if conceptual graphs are to be used as a normative language they must be able to express NIAM's static constraints relatively easily. In section 4, we present some of the difficulties involved in expressing NIAM constraints using conceptual graphs.

A few previous papers have shown very simple mappings between ER and conceptual graphs, but have avoided any examples that could not be simply represented in conceptual graphs. We [3] have previously pointed out some of the problems of using the basic form of conceptual graphs as a conceptual modelling language due to the relatively complex nature of the specialised constraints required. Section 5 proposes a number of abstractions to enable the constraints provided by NIAM to be expressed in a readable form using conceptual graphs.

2 NIAM

NIAM is a strongly typed, fact-based, static modelling language similar to the better known Entity-Relationship (ER) model. However it is semantically richer (in the sense that more of the UoD can be captured) than, and is essentially a superset of, ER. NIAM has an associated methodology in which atomic sentences from the UoD are used to identify entity types and general syntactic rules (fact types or predicates) between entity types. An example of a NIAM schema is shown in figure 1. The circles of the diagram represent entity types. The joined boxes (e.g. has-firstname) represent fact types. The individual boxes represent the role that entity types play in a fact type. The fact type and entity type of NIAM roughly correspond to the conceptual relation and set-referenced concept (respectively) of conceptual graphs.

Every NIAM (non-lexical) entity type should be identifiable in terms of special lexical entity types, called label types. The fact type between a (non lexical) entity type and a label type is called a reference type. Whenever an entity type can be uniquely identified by a single label type, the reference type can be abbreviated by showing the label type label within the entity type's circle. An example of the abbreviation is shown in figure 1 where ADDRESS has a label type of "address code" in parentheses (i.e. each address is represented by a unique code).

In addition, NIAM has a number of (syntactic) constraints, some of which are synonymous with the relational model semantics. The intrafact uniqueness constraint (e.g. C1) corresponds to the key of a relation. Although it should be noted that, since the NIAM fact types are derived from atomic sentences, they are irreducible. Hence a uniqueness constraint represents a functional dependency. C1 (arrow) indicates that each person has only one surname. C2 (dot), the mandatory role constraint, specifies that each person must have a surname. Constraints C1 and C2 are also expressible in ER and SQL.

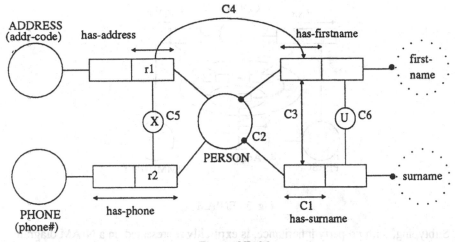

Fig. 1. NIAM

Of the additional constraints, C3, C4 and C5 are set constraints between roles. The roles in NIAM are considered as sets (in terms of populated fact types). C4 is a subset constraint (relational referential integrity). In this case every person with an address has a first name. C3, the equality constraint, indicates that everyone with a first name also has a surname and vice versa. An astute reader will have realised, that in this particular example, constraints C3 and C4 are superfluous. C5, the exclusion constraint, indicates that no one can have both an address and a phone.

The interfact uniqueness constraint, C6, specifies that for any combination of first name and surname there is at most one person. This construct is useful in identifying the equivalent of ER's weak entity type.

A simple algorithm permits a relational schema to be derived from the NIAM diagram. The relational schema corresponding to figure 1 is

PERSON(first-name , surname , address)
PHONE(first-name , surname , phone-num).

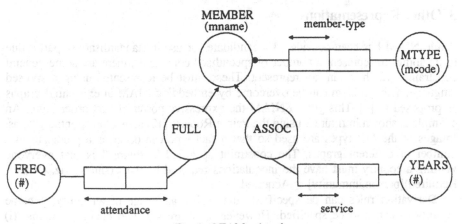

Fig. 2. NIAM subtyping

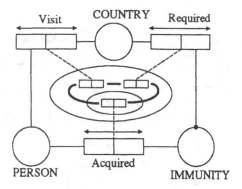

Fig. 3. ENIAM

Subtyping, with property inheritance, is explicitly represented on a NIAM diagram. Any fact types used in the subtype defining rules must be at a supertype level. The rules can only be represented using stylised English. The example in figure 2 shows two subtypes of member, viz. full and associate. The fact type "member-type" would be used in the subtype rules (which have not been shown). In some versions of NIAM it is possible to specify overlapping and disjoint subtype constraints.

We have only shown simple examples here. The structures can be made arbitrarily large. It is only practical limitations which dictate the size of each diagram. They can be joined via the entity and label types, assuming there are no duplicate fact types in the structures to be joined. It should be noted that it is not necessary to write down the fact type names, role names or constraint names unless they are needed for clarification.

As mentioned, ER is similar to NIAM. We can approximate the NIAM and ER concepts as follows: the NIAM reference type corresponds to the attribute of ER, the fact type corresponds to the relationship type and the label type to the value set. Since the diagrammatic form of ER does not specifically represent a role (a box in NIAM), it is not possible to neatly represent a number of NIAM's constraints.

3 Other Representations

While NIAM has been considered a candidate for use in standardisation, partly due to its ability to represent a number of specialised constraints, there are some general constraints which it can not represent. These must be represented using a stylised language. This problem can be overcome by embedding NIAM in existential graphs as proposed in [4]. This gives ENIAM the expressive power of first order logic. An example is shown in figure 3 using the Peirce/Roberts [12] style of existential graphs. Images of the fact types are used so that a fact type can occur as a predicate in a number of different graphs. The constraint specified in figure 3 is that everyone visiting a country must have the inoculations required by that country; viz. (Visit * Required)[person,immunity] $\subseteq$ Acquired.

Derivation rules can be specified, and, using an image of an entity, subtype definitions can also be specified. However there are still two major problems: (i) there is no means for abstraction/generalisation or decomposition/specialisation and (ii) the model is essentially a static model.

"SQL [8] has been proposed by ISO as a normative language. It has been extended to a specification language and uniqueness, referential and check constraints added. However, although SQL is powerful, it is verbose, and awkward to read schema specifications. In addition there is no abstraction mechanism.

4 Conceptual Graphs as a Conceptual Schema Language

We have previously [5] discussed the use of conceptual graphs as a conceptual schema language and concluded that conceptual graphs were not suitable in their basic form. The problems arose when expressing constraints. While it is possible to express constraints in first order predicate logic, it is highly desirable to express them in a more readable graphical form. In addition, ANSI's IRDS proposal requires schemas be expressed in (readable) graphical forms. The most important constraints we need to represent are the mandatory and functional dependency (NIAM's intra fact uniqueness constraint) constraints.

Sowa [14] proposed using a data flow graph to represent functional dependencies. In linear form, these dependencies can be expressed using E-R type mappings (n to 1, n to m or 1 to 1). For example, given the conceptual graph

[PERSON: ∀] ← (AGNT) ← [WORK] → (BENF) → [DEPT]

the functional dependency could be expressed by the data flow graph

[PERSON] - n - <PER-DEPT> - 1 - [DEPT],

i.e. each person works in exactly one department.

However if an entity type plays the role of source and sink in separate dataflow graphs, different referent field quantifiers are required for each case.

In [9] we pointed out that the quantifiers, which can readily express a range of natural language ideas, are not suited to the specific role of expressing the mandatory constraint and functional dependencies. We proposed attaching the mapping information to the links of the conceptual relation. However there are disadvantages with this method and it does not have the versatility of the method proposed in the next section. Certain combinations of mandatory and functional dependency constraint can be easily expressed in conceptual graphs. For example figure 4(a) expresses in NIAM that "every box contains exactly one cookie." This can be expressed in conceptual graphs as:

[BOX: ∀] → (CONT) → [COOKIE: $\exists^1$].

In order to additionally express that "all cookies are in boxes" we can use one diagram in NIAM (figure 4(b)), but require the additional statement in conceptual graphs:

[BOX] → (CONT) → [COOKIE: ∀].

Using universal quantifiers on both referents in the same expression does not express these constraints.

To represent "boxes which contain cookies contain exactly one cookie" we could attempt to use:

[BOX: {*}] → (CONT) → [COOKIE: $\exists^1$]

However, this only expresses that some boxes have exactly one cookie; others may in fact contain more, as shown by the mapping:

$(\exists x)(set(x) \land ((\forall u)(u \in x \Rightarrow (box(u) \land (\exists^1 y)(cookie(y) \land cont(u,y))))))$.

51

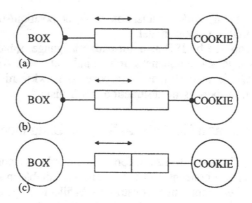

Fig. 4. Typical NIAM constraints

The NIAM diagram of figure 4(c) expresses more than this, it indicates that no box can contain more than one cookie. Alternatively, we could state that every box contains less than 2 cookies:

[BOX: ∀] → (CONT) → [COOKIE: {*}<2] i.e.

(∀x)(box(x) ⇒ (∃s)(set(s) ∧ count(s) < 2 ∧

(∀y)(y ∈ s ⇒ (cookie(y) ∧ cont(x,y))))).

But this indicates that there exists a set with no more than 2 elements - there may, however, be other sets with more than two elements.

The first column of figure 5 expresses the NIAM constraints which it is imperative that we can express in conceptual graphs. A number of them, in fact, require two conceptual graphs. For example figure 5(e) expresses that some A's may have one or more B's, while some B's may have one or more A's. The following two conceptual graphs describe this situation:

[A: Col{*}] → (R) → [B: Dist{*}]

[A: Dist{*}] → (R) → [A: Col{*}]

Another issue which affects the complexity of the conceptual graph representation is that one cannot, in the same graph, express that "every box contains one cookie" and "some boxes are heavy." This is essential in a unified conceptual schema. In the next section we show a number of abstractions, which have been proposed in [2], that overcome the problems covered above.

5 Extending the Conceptual Graph Notation

One possible way to represent the required constraints is to create other specialised set operators with mappings which provide the correct semantics. However, this would require "meaningful" operators (like Dist) and a relatively large number of them. The proposed course of action results in the introduction of fewer such abstractions.

As we saw in the example in the previous section we couldn't express "boxes which contain cookies, contain exactly one cookie". To express this, in addition to stating that some boxes contain one cookie we need to state that no box contains more than one cookie. Thus we need the two expressions:

[BOX: {*}] → (CONT) → [COOKIE: $\exists^1$]

$\neg$[[BOX: {*}] $\rightarrow$ (CONT) $\rightarrow$ [COOKIE: Col{*}>1]]

In fact, as the main part of the constraint is in the second expression, we do not need to make the first expression as explicit, e.g.

[BOX: {*}] $\rightarrow$ (CONT) $\rightarrow$ [COOKIE: {*}]

$\neg$[[BOX: {*}] $\rightarrow$ (CONT) $\rightarrow$ [COOKIE: Col{*}>1]].

As a number of the NIAM expressions of figure 5 require this type of constraint (viz. an A is associated with no more than one B), it is proposed to abbreviate these two expressions to

[BOX: {*}] $\rightarrow$ (CONT) $+\!\!\!\!\nmid$ [COOKIE:{*}].

The (perpendicular) dash (which we call the *dependency dash*) implies a functional dependency, i.e. box $\rightarrow$ cookie. This dependency dash construct is defined as follows:

[A: ref1] $\rightarrow$ (R) $+\!\!\!\!\nmid$ [B:ref2]

expands to

[A:ref1] $\rightarrow$ (R) $\rightarrow$ [B:ref2]

$\neg$[[A] $\rightarrow$ (R) $\rightarrow$ [B:Col{*}>1]],

while   [A:ref1] $+\!\!\!\!\nmid$ (R) $\rightarrow$[B:ref2]

expands to

[A:ref1] $\rightarrow$ (R) $\rightarrow$ [B:ref2]

$\neg$ [[A: Col{*}>1] $\rightarrow$ (R) $\rightarrow$ [B]]

and   [A:ref1] $+\!\!\!\!\nmid$ (R) $+\!\!\!\!\nmid$ [B:ref2]

expands to

[A:ref1 *x] $\rightarrow$ (R) $+\!\!\!\!\nmid$ [B:ref2 *x]

[A:ref1 *x] $+\!\!\!\!\nmid$ (R) $\rightarrow$[B:ref2 *x].

We can thus express that "boxes which contains cookies contain exactly one cookie" as

[BOX: {*}] $\rightarrow$ (CONT) $+\!\!\!\!\nmid$ [COOKIE: {*}]

This dependency dash can be likewise used to specify dependency constraints on n-ary relationships.

   We also noted in the last section that to express that "every box contains exactly one cookie and every cookie is in a box" we need two conceptual graphs. Using the dependency dash we can express these as:

[BOX: $\forall$] $\rightarrow$ (CONT) $+\!\!\!\!\nmid$ [COOKIE]

[BOX] $\rightarrow$ (CONT) $\rightarrow$ [COOKIE: $\forall$]

which is much less readable than the equivalent NIAM diagram (figure 4(b)).

   We can also use the collective set referent in place of the generic referent. It produces the same mapping to logic, since it has a narrower scoping. We can take advantage of this to define the *dual relationship arrow* which will permit us to express these dual relationships. We define the dual relationship arrow as follows:

[A: ref1] $\Rightarrow$ (R) $\Rightarrow$ [B:ref2]

expands to

[A:ref1] $\rightarrow$ (R) $\rightarrow$ [B: Col{*}>0]

[A: Col{*}>0] $\rightarrow$ (R) $\rightarrow$ [B: ref2].

   The dual relationship arrow can be combined with the dependency dash. We can now express the above constraint (equivalent to figure 4(b)) as

[BOX: $\forall$] $\Rightarrow$ (CONT) $\Leftrightarrow\!\!\!\!\nmid$ [COOKIE: $\forall$]

The dual relationship arrow can also be generalised to n-ary relations.

   Figure 5 shows the ten possible combinations of mandatory role and intra-fact uniqueness constraints, together with the conceptual graphs equivalents expressed

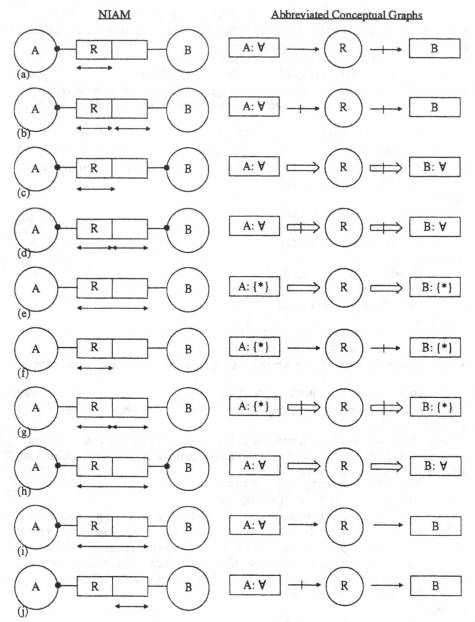

Fig. 5. NIAM and equivalent abbreviated conceptual graphs

using the dependency dash and dual relationship arrow. There is a simple methodology for expressing the abstractions.

In [9] we proposed a two level approach to information modelling. This approach involves a first level design using NIAM followed by a more detailed design using conceptual graphs. This would define each fact type in terms of concepts and conceptual relations. For example, the NIAM diagram of figure 6(a) can be expanded to that of

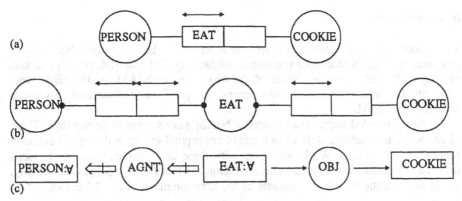

(a)

(b)

(c)

Fig. 6. Expanded forms

figure 6(b) or its extended conceptual graph's equivalent of figure 6(c). Similar expansions can be made for other forms of the NIAM binaries of figure 5. An interfact uniqueness constraint is also required where an intrafact uniqueness constraint covers both roles. These interfact uniqueness constraints can be represented on a unified conceptual graph schema by abstracting the constraint definition into a conceptual relation of type "constraint".

All conceptual relations used in the schema should be inserted into a relational hierarchy, which defines the level of generalisation (category) of this relation. Similarly, concepts should be inserted into a type hierarchy. This allows the constructs of modelling languages, such as NIAM, to be reconstructed from the conceptual graph representation, permitting a two-way mapping.

The ternary fact types of NIAM can also be represented using the abbreviated conceptual graphs. Figure 7(a) shows a NIAM ternary with the equivalent extended conceptual graphs version in figure 7(b).

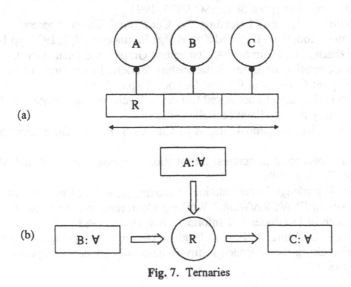

(a)

(b)

Fig. 7. Ternaries

6 Conclusion

This paper examines conceptual graphs as a normative language for ANSI's IRDS proposal. It suggests that other possible candidates, NIAM and SQL are not suitable, although ENIAM overcomes some of the objections to NIAM. At the same time, conceptual graphs appear to be too general to specify the specialised constraints available in NIAM.

We have provided abstractions to enable NIAM's constraints to be matched. These abstractions are readable, defined in terms of conceptual graphs and keep the essential features of the graphs. We believe that, with these abstractions, conceptual graphs become much more usable as the universal language that ANSI appears to be seeking, while at the same time being suitable as the core language of a CAiSE tool.

References

1. ANSI Task Group X3H4.6 Working Paper on the Conceptual Schema, 1991.
2. L. Campbell: Conceptual Graphs in the IRDS Framework. Internal Report, University of Queensland, 1992.
3. P. Creasy: Conceptual Graphs from a Knowledge Systems Viewpoint.Proceedings Australian Joint Artificial Intelligence Conference. Adelaide, 1988.
4. P. Creasy: ENIAM: A More Complete Conceptual Schema Language. Proceedings of the Fifteenth International Conference on Very Large Databases. Amsterdam, August 1989.
5. P. Creasy, B. Moulin: Adding Semantics to Semantic Data Models. Proceedings Fourth Annual Workshop on Conceptual Graphs. Detroit, August 1989.
6. P. Creasy, B. Moulin: Approaches to Data Conceptual Modelling Using Conceptual Graphs. Sixth Annual Workshop on Conceptual Graphs. Binghamton, July 1991.
7. ISO: Concepts and Terminology for the Conceptual Schema and the Information Base. ISO TC97/SC5/WG3 Report. van Griethuysen, J.J. (Ed), 1983.
8. ISO: Database Language SQL. ISO 9075. 1991.
9. B. Moulin, P. Creasy: Extending the Conceptual Graph Approach for Data Conceptual Modelling. Data and Knowledge Engineering 8, 3. 1992. pp 165-194.
10. R. Muhlbacher, G. Neumann: A Conceptual Graph Base Dictionary as a Source for the Generation of Entity Relationship Models. Proceedings of the Fourth Workshop on Conceptual Graphs. Detroit, August 1989.
11. G. Nijssen, T. Halpin: Fact-Based Conceptual Schema and Relational Database Design Using NIAM. Prentice Hall, 1989.
12. D. Roberts: The Existential Graphs of Charles S. Peirce. The Hague: Mouton, 1973.
13. J. Sowa: Conceptual Structures: Information Processing in Mind and Machine. Addison Wesley, 1984.
14. J. Sowa: Knowledge Representation in Databases, Expert Systems, and Natural Languages. IFIP WG2.6/WG8.2 Working Conference on the Role of Artificial Intelligence in Databases and Information Systems. 1988.
15. T. Teorey, D. Yang, J. Fry: A Logical Design Methodology for Relational Databases Using the Extended Entity-Relationship Model. Computing Surveys 18, 2 (1986).

II. Knowledge Representation Issues - Special

Representing Knowledge About Substances

William M. Tepfenhart

AT&T Bell Laboratories
480 Red Hill Rd.
Middletown N.J 07748
908-615-1961
billt@vios.att.com

Abstract. This paper presents a new view of a usable definition of substance. The approach is incremental and the derivation employs conceptual graphs as the representation scheme for representing intermediate steps. The result is that a substance is defined as a set of unassociated instances of physical objects and substance properties are the congruences of the relationships that exist among the instances.

1 Introduction

If we are to reason about the physical world, we must represent knowledge about the substances of which things are made. Contour fing efforts to represent and reason about substances is the fact that we don't have a good definition for what constitutes a substance. For example, a dictionary [1] defines a substance as:

1. That which has mass and occupies space; matter; body, material of a particular kind; a constitution

while the second definition for a material is:

1. The substance or substances out of which a thing is or can be made.

Another definition is of much use in establishing representational requirements since they come frequently a material is a substance an Encyclopedia. A material (matter is defined as a material or that which has weight and occupies space).

A necessary precursor to establishing a means for representing and reasoning about substances is to establish a clear and usable definition for substance. That is the main topic of this paper. The approach is incremental with a dependence on how to proceed when building on the concept of a substance as a cognitive instance. Once that is firmly established we show how to facilitate reasoning about instances.

It should be mentioned that the work presented in this paper represents a particular view of many of the concepts and conceptual relations suggested in the Situation Data Model [The Situation Data Model has described for level concepts and conceptual relations required to reason about physical systems in very abstract and generalized terms. This model has been presented in three papers at previous Conceptual Graph Workshops [2-4].

Author and presenter questions concerning to the paper work.

Representing Knowledge About Substances

William M. Tepfenhart

AT&T Bell Laboratories
480 Red Hill Rd.
Middletown, NJ 07748
908 - 615 - 5996
bill@violin.att.com

Abstract. This paper presents a derivation for a useable definition of
substance. The approach is incremental and the derivation employes
conceptual graphs as the representation scheme for expressing
intermediate steps. The result is that a substance is defined as a set of
unspecified instances of physical objects and substance properties are
the consequences of the relationships that exist among the instances.

1 Introduction

If we are to reason about the physical world, we must represent knowledge about the
substances of which things are made. Confounding efforts to represent and reason about
substances is the fact that we don't have a good definition for what constitutes a
substance. For example, a dictionary [1] defines a substance as:
> 1.a That which has mass and occupies space; matter; b. A material of
>   a particular kind or constitution.

while the relevant definition for a material is:
> 1. The substance or substances out of which a thing is or can be
>   made.

Neither definition is of much use in establishing representational requirements since
they form a tautology - a material is a substance and a substance is a material (matter is
defined as a material or a substance or that which has weight and occupies space).

A necessary precursor to establishing a means for representing and reasoning about
substances is to establish a clear and usable definition for substance. That is the main
topic of this paper. Our approach is incremental with decisions on how to proceed based
on the idea that the concept of a substance is a cognitive necessity. Once this definition
has been achieved, we show how it facilitates reasoning about substances.

It should be mentioned that the work presented in this paper represents a quantification[1]
of many of the concepts and conceptual relations suggested in the Situation Data
Model. The Situation Data Model has described top level concepts and conceptual
relations required to reason about physical systems in very abstract and generalized
terms. This model has been presented in three papers at previous Conceptual Graph
Workshops [2-4].

---

[1] Albeit an abbreviated quantification due to the page limit.

2 Substance Definition

We start the definition process by considering a bounded locale that is initially empty of objects. The locale extends over a region of space and exists at a time relative to a reference frame. Asserting that the locale extends over a region of space means that it has a set of boundaries, each of which has a shape, orientation, extent, location, velocity, and acceleration relative to the location of the locale. We consider this locale to be our region of interest - this is the volume within which we explore the nature of substances. A conceptual graph that expresses our understanding of the region of interest, [ROI(x)] is,

```
type [ROI(x)] is
    [Reference_Frame]->
        (establishes_units)->[Units]
        (establishes_time)->[Time: T0=0]
        (establishes_location)->[Location: L0=0]
        (establishes_velocity)->[Velocity: V0=0]
        (establishes_acceleration)->[Acceleration: A0=0]
    [Local: x] ->
        (has_location) <- [Location: L0]
                      -> [Location: l]
        (has_velocity) <- [Velocity: V0]
                      -> [Velocity: v]
        (has_acceleration) <- [Acceleration: A0]
                          -> [Acceleration: a]
        (has_boundary) -> [Boundary:{*}]].
```

This conceptual graph is actually a simplification of a more detailed one that includes the time rate of change of the contour, orientation, and size of each boundary.

Although introducing this graph may seem unnecessary and complex, this graph is important. It serves as the vehicle by which we can explore what happens to any conceptual graph describing a region of the physical world as physical objects are introduced into it. We can use it to establish what computations can be performed on related graphs, what is required to perform these computations, and to establish how much memory is required to store the concepts and conceptual graphs. This knowledge allows us to select among different means of representing our understanding of the physical world. Our selection can be made on the basis of what questions we need to answer, what computational resources we are willing to apply in order to achieve answers, and how to do this within memory resources available to us.

Given a region of interest, we now introduce a single object into it and examine the consequences. For convenience, we shall assume that the object is of type physical object. The conceptual graph that results from this action is:

```
[[ROI: x]->
    (establishes_location)->[Location: l]
    (establishes_velocity)->[Velocity: v]
    (establishes_acceleration)->[Acceleration: a]
    (contains) -> [physical_object: y1] ->
        (has_location) <- [Location: l]
```

```
                                -> [Location: l1]
               (has_velocity) <- [Velocity: v]
                                -> [Velocity: v1]
               (has_acceleration) <- [Acceleration: a]
                                -> [Acceleration: a1]].
```

Additional concepts can be added to more fully describe the physical object that has been added. For example, we can include the substance of which it is composed, the orientation, the size, and the shape of the physical object. Of particular significance, this graph can be employed to compute the behavior of the physical object through the use of canonical graphs that describe how an object with a given position, velocity, and acceleration moves over time.

Since few graphs of any interest only contain a single object, we now add another physical object into the region of interest and examine the conceptual consequences. The conceptual graph that results from this action is:

```
     [[ROI: x]->
          (establishes_location)->[Location: l]
          (establishes_velocity)->[Velocity: v]
          (establishes_acceleration)->[Acceleration: a]
          (contains) -> [physical_object: y1] ->
               (has_location) <- [Location: l]
                                  -> [Location: l1]
               (has_velocity) <- [Velocity: v]
                                  -> [Velocity: v1]
               (has_acceleration) <- [Acceleration: a]
                                  -> [Acceleration: a1]
          (contains) -> [physical_object: y2] ->
               (has_location) <- [Location: l]
                                  -> [Location: l2]
               (has_velocity) <- [Velocity: v]
                                  -> [Velocity: v2]
               (has_acceleration) <- [Acceleration: a]
                                  -> [Acceleration: a2]
          [physical_object: y1] -> (location_relation) -> [physical_object: y2]
          [physical_object: y1] -> (velocity_relation) -> [physical_object: y2]
          [physical_object: y1] -> (acceleration_relation) -> [physical_object: y2]]
```

It is important to realize that by having placed two physical objects in the same region of interest has automatically established a number of relationships between them. In addition to the location, relative velocity and relative acceleration relationships shown in this graph, we have additional relationships such as size, mass, and substance relations. By explicitly representing these relationships, we have minimized the level of cognitive effort required to answer questions concerning what is occurring within the region of interest. However, we do so at the expense of memory resources required to store the conceptual graph - the graph has more than doubled in size.

If we add a third object to the region of interest, the conceptual graph that is produced by this action is:

```
     [[ROI: x]->
          (establishes_location)->[Location: l]
```

```
(establishes_velocity)->[Velocity: v]
(establishes_acceleration)->[Acceleration: a]
(contains) -> [physical_object: y1] ->
     (has_location) <- [Location: l]
                    -> [Location: l1]
     (has_velocity) <- [Velocity: v]
                    -> [Velocity: v1]
     (has_acceleration) <- [Acceleration: a]
                        -> [Acceleration: a1]
(contains) -> [physical_object: y2] ->
     (has_location) <- [Location: l]
                    -> [Location: l2]
     (has_velocity) <- [Velocity: v]
                    -> [Velocity: v2]
     (has_acceleration) <- [Acceleration: a]
                        -> [Acceleration: a2]
(contains) -> [physical_object: y3] ->
     (has_location) <- [Location: l]
                    -> [Location: l3]
     (has_velocity) <- [Velocity: v]
                    -> [Velocity: v3]
     (has_acceleration) <- [Acceleration: a]
                        -> [Acceleration: a3]
[physical_object: y1] -> (location_relation) -> [physical_object: y2]
[physical_object: y1] -> (velocity_relation) -> [physical_object: y2]
[physical_object: y1] -> (acceleration_relation) -> [physical_object: y2]]
[physical_object: y1] -> (location_relation) -> [physical_object: y3]
[physical_object: y1] -> (velocity_relation) -> [physical_object: y3]
[physical_object: y1] -> (acceleration_relation) -> [physical_object: y3]]
[physical_object: y2] -> (location_relation) -> [physical_object: y3]
[physical_object: y2] -> (velocity_relation) -> [physical_object: y3]
[physical_object: y2] -> (acceleration_relation) -> [physical_object: y3]]
```

The graph will continue to grow in complexity as a result of explicitly representing relationships created by introducing objects into the region of interest. It is important to remember that we have neglected to include many of the relations that exist among these objects. If these neglected relationships were important in the current problem solving endeavor, we would have to carry them along as well.

Explicitly representing relationships minimizes the effort required to answer questions concerning what is occurring within the region of interest. However, we do so by consuming memory resources to store the conceptual graph. The amount of memory used can be decreased by suppressing redundant information about our system since it can be retrieved from the instances that are stored. The resultant graph would be:

```
[[ROI: x]->
     (contains) -> [physical_object: y1]
     (contains) -> [physical_object: y2]
     (contains) -> [physical_object: y3]
[physical_object: y1] -> (location_relation) -> [physical_object: y2]
[physical_object: y1] -> (velocity_relation) -> [physical_object: y2]
```

        [physical_object: y1] -> (acceleration_relation) -> [physical_object: y2]]
        [physical_object: y1] -> (location_relation) -> [physical_object: y3]
        [physical_object: y1] -> (velocity_relation) -> [physical_object: y3]
        [physical_object: y1] -> (acceleration_relation) -> [physical_object: y3]]
        [physical_object: y2] -> (location_relation) -> [physical_object: y3]
        [physical_object: y2] -> (velocity_relation) -> [physical_object: y3]
        [physical_object: y2] -> (acceleration_relation) -> [physical_object: y3]]

Dropping the instance specific information in this manner reduces the memory required to store the specific graph while preserving the information that is more expensive to compute, namely, the information encoded in the relationships that exist between the objects.

As the number of physical objects introduced into the region of interest increases, the conceptual graph will grow in complexity and require ever increasing amounts of memory to represent it. In time, the graph will grow to a size where it requires too many memory resources to store it in such detail. Suppose that the number of objects at which this happens is 10. We could try to simplify the conceptual graph in the following manner,

        [[ROI: x]->
                (contains) -> [physical_object: y1]
                (contains) -> [physical_object: y2]
                (contains) -> [physical_object: y3]
                (contains) -> [physical_object: y4]
                (contains) -> [physical_object: y5]
                (contains) -> [physical_object: y6]
                (contains) -> [physical_object: y7]
                (contains) -> [physical_object: y8]
                (contains) -> [physical_object: y9]
                (contains) -> [physical_object: y10]]

However, any question that we ask about the region of interest shall require tremendous computational resources in order to achieve an answer. For example, to ask the question as to what physical objects are above [physical_object: y5] requires us to compute the relation with all of the others.

An alternative simplification is introduce the concept of an ensemble. An ensemble is an abstraction over a number of instances and is defined as:

        type Ensemble(x) is
        [[Reference_Frame: y]->
                (establishes_location)->[Location: l]
                (establishes_velocity)->[Velocity: v]
                (establishes_acceleration)->[Acceleration: a]
        [object(x)] ->
                (has_location) <- [Location: l]
                                -> [Location]
                (has_velocity) <- [Velocity: v]
                                -> [Velocity]
                (has_acceleration) <- [Acceleration: a]
                                -> [Acceleration]
                (has_members) -> [physical_object:{#1, #2,...,#VLN}]

```
(has_distribution) -> [distribution]
(has_dist_exp_rate) -> [dis_exp_rate]
(has_dist_accel_rate) -> [dist_accel_rate]
(has_orientation) -> [orientation]
(has_rotation_rate) -> [rotation_rate]
(has_rotational_accel) -> [rotational_accel]
(has_size) -> [size: Es1]
(has_mass) -> [mass: Em1]
...]
```

The portion of the graph that we have not included above contains the concepts and conceptual relations that describe other features and characteristics derived from the other relationships that exist among the objects that we have chosen not to represent explicitly.

Use of the ensemble concept has several advantages. The concepts that have been added and associated with the ensemble (for example, location, velocity, acceleration, distribution, size, and mass) have values that are computed from instance information about the members so no information has been lost. The distribution tells us how the objects are arranged in the region of interest. If the ordering of the instances corresponds to the order in which the distribution is defined, then we can answer most questions about relative locations with very little computational effort. Finally, the amount of memory required to store this graph is much less than when all of that information was explicit.

Expressing our conceptual graph using the ensemble concept results in:

```
[[ROI: x]->
        (establishes_location)->[Location: l]
        (establishes_velocity)->[Velocity: v]
        (establishes_acceleration)->[Acceleration: a]
        (contains) -> [Ensemble: E1] ->
                (has_location) <- [Location: l]
                                -> [Location: El1]
                (has_velocity) <- [Velocity: v]
                                -> [Velocity: Ev1]
                (has_acceleration) <- [Acceleration: a]
                                    -> [Acceleration: Ea1]
                (has_members) -> [physical_object:{#1, #2,...,#LN}]]
```

By introducing the concept of an ensemble, we have been able to maintain considerable information about the behavior of our objects as a group while retaining access to information about individuals. At this point, we might feel that we can continue to introduce additional physical objects into our region of interest without limit, or at least until our region of interest is filled and no more objects can be added. However, this is not the case since we are storing instance information about each object that we introduce and this consumes memory resources.

We must establish a means to reduce the memory resources required without overloading the computational resources. One way to accomplish this is to change from an enumerated set of instances to a set of unspecified instances in which only the number of instances is maintained. However, losing instance information has one significant consequence - we have lost the ability to maintain the distribution of physical objects and their time behavior. To counter this we can replace the distribution with a shape that contains all of the instances and replace the time behavior of the distribution with the concepts associated with the time behaviors of shapes. Other relationships and concepts can be associated with the set, as well. In particular, the relationships associated with the composition of the set, the structure of the elements within the set, and interactions among the elements of the set.

If we make the appropriate transformation from an enumerated set to a set of unspecified instances, the definition for the concept is ,

```
        type ???(x) is
            [[Reference_Frame: y]->
                (establishes_location)->[Location: l]
                (establishes_velocity)->[Velocity: v]
                (establishes_acceleration)->[Acceleration: a]
            [object(x)] ->
                    (has_location) <- [Location: l]
                                    -> [Location]
                    (has_velocity) <- [Velocity: v]
                                    -> [Velocity]
                    (has_acceleration) <- [Acceleration: a]
                                    -> [Acceleration]
                    (has_members) -> [physical_object:{*}] ->
                        (has_number) -> [Count]
                        (has_components) -> [Type_Of(physical_object){*}]
                        (has_internal_structure) -> [Structure]
                        ...
                    (has_shape) -> [shape]
                    (has_exp_rate) -> [rate]
                    (has_exp_accel) -> [exp_accel]
                    (has_orientation) -> [orientation]
                    (has_rotation_rate) -> [rotation_rate]
                    (has_rotational_accel) -> [rotational_accel]
                    (has_size) -> [size]
                    (has_mass) -> [mass]
                    ...]
```

We have not concerned ourselves with the type label for the new concept for a reason that will become clear. In making this transformation we are admitting that there are some questions that can not easily be answered about the region of interest.

In order to establish an appropriate type label for the new concept, let us compare the above definition with the type definition for a physical object,

```
        type physical_object(x) is
            [[Reference_Frame: y]->
                (establishes_location)->[Location: l]
                (establishes_velocity)->[Velocity: v]
                (establishes_acceleration)->[Acceleration: a]
            [object(x)] ->
                    (has_location) <- [Location: l]
```

```
                                       -> [Location]
                (has_velocity) <- [Velocity: v]
                                       -> [Velocity]
                (has_acceleration) <- [Acceleration: a]
                                             -> [Acceleration]
                (Made_Of) -> [substance] -> (has_quantity) -> [quantity]
                (has_shape) -> [shape]
                (has_exp_rate) -> [rate]
                (has_exp_accel) -> [exp_accel]
                (has_orientation) -> [orientation]
                (has_rotation_rate) -> [rotation_rate]
                (has_rotational_accel) -> [rotational_accel]
                (has_size) -> [size]
                (has_mass) -> [mass]
                ...].
```

When we contrast the type definition given above with the type definition for a physical object we can make an important observation. The definitions are identical with two exceptions: (1) the type label for the first has not been assigned, and (2) a physical object is made of some quantity of substance while the new concept is essentially composed of a large number of physical objects.

It is the second exception that is interesting for our purposes, particularly as the number of objects in the set grows large that it will become inevitable that they can not be counted directly or that the size of the physical objects can be so small that they can not be directly observed. When this happens the number of instances must be determined indirectly, using measures of mass or volume.

The only other concept that we treat in this manner is substance, which is typically employed in conceptual graphs in the following manner,

```
                (Made_Of) -> [Substance] ->
                    (has_quantity) -> [Quantity]
                    (has_components) -> [Type_Of(physical_object){*}]
                    (has_internal_structure) -> [Structure]
                    ...
```

We make the conceptual leap and assert that the graph fragment -

```
                (has_members) -> [physical_object:{*}] -> (has_number) -> [Count]
```

is equivalent to and can be replaced by

```
                (Made_Of) -> [substance] -> (has_quantity) -> [quantity].
```

If we perform the replacement in the new concept the result is that the definitions for the new concept and a physical object become indistinguishable and as such must be defining the same concept. In making the substitution we are defining a substance as,

```
        type substance(x) is
            [[physical_object:{*}@++](x)] ->
                (has_components) -> [Type_Of(physical_object){*}]
                (has_internal_structure) -> [Structure]
                ...]
```

This definition states that a substance is a set containing an uncountable number of physical objects of various types and that the members within the set have some internal structure. The missing element of the definitions are those relations and concepts associated with the behaviors of the objects within the regions of interest that have not been carried forth in this paper.

This definition has extremely interesting consequences, particularly when we observe that sets have a tendency to exhibit exactly the same troublesome behaviors as materials. Significantly, we know how to perform computations over sets and deal with most of the more troublesome behaviors.

3 Definition Consequences

Having achieved a definition for substances, we can now exploit that definition to explore what properties a substance can have. To do so, we just have to trace back to first principles the types of relations that can exist among the physical objects and see what effects that has on the set defining the substance. In this paper, we are interested in four very tough problems (composition, state, viscosity, and reactivity) to represent and reason about.

3.1 Composition.

This property refers to the types of physical objects that are members of the set - that is the components of the substance - and the physical relationships that exist between them. For example, a grain of salt is made of the substance salt which has two components: Sodium and Chlorine. In order for a substance that has sodium and chlorine as components to be considered the substance salt, two conditions must be met: (1) there are equal numbers of sodium atoms and chlorine atoms in the set and (2) a sodium atom must exist in a specific physical relationship with another chlorine atom - that relationship being (bound_to).

The act of pouring one substance into another creates a gallimaufry. A gallimaufry is an inhomogeneous substance of two or more substances and has several canonical properties that have significant consequences in reasoning about the composition of a substance. A gallimaufry that has one component that is a gallimaufry that has two component substances and second component that is a third substance is equivalent to a gallimaufry that has three component substances. A gallimaufry that contains only a single component substance is equivalent to that substance. A gallimaufry that has two instance components that are of identical substance is equivalent to a single substance instance.

A number of interesting cases result when we pour one substance into another. For example, if we pour two substances that exist in an oil-water relationship into a glass the result is that the glass will contain a single object that is made of a gallimaufry. As time passes, the object in the glass splits into three objects that we call layers with one layer existing in a between relationship with the other two.The layers will be oriented along the horizontal direction and form a vertical column. The top layer is composed of the oil-substance, the middle layer is composed of the gallimaufry, and the bottom layer is composed of the water-substance. When we examine what is going on at the instance set level, what we find is that the set of physical objects has divided into three subsets that have distinct boundaries between them. Each subset has a distinct internal structure, distribution, composition, and location relative to the other two. Because this has occurred, we can follow the derivation backwards taking the description from one that is dealing with one object to one that is dealing with an ensemble of three objects and ending at a conceptual ization that is dealing with three individual objects. The continued transferal of substance from the middle layer to the other two layers is actually a changing of set membership of the unspecified instances at the set level.

A second interesting case occurs when we pour two quantities of different non-reacting substances into a glass - for example water and alcohol. The result is a gallimaufry that eventually settles into a homogeneous mixture that is a distinct substance. What is interesting about this case is that it matters how we measure the quantities of the original substances and the quantity of the final substance. If we measure based on volume, then we will find that the quantity of final substance is less than the sum of the quantities of the initial substances. If we measure quantity based on mass, then we find that the quantity of final substance is identical to the sum of the quantities of the initial substances. The difference between the two measures can be explained at the set level. When we measure quantity by mass, we are summing over the masses of the individual instances that are in the set and mass won't change as a result of new interactions with other physical objects. When we measure by volume, we are measuring a property of the distribution within the ensemble and the distribution can easily change as a result of new interactions among the instances.

3.2 State.

There are three well recognized states that substances exist in, namely, solid, liquid, and gas. In order to understand state, we can fall back on the above discussion where the set of physical objects was still an ensemble. In a solid state, the physical objects comprising the ensemble exist in a well defined periodic distribution in which there is a zero distribution expansion rate (it is fixed). In a liquid, the distribution is statistically driven and the ensemble has a zero or negative distribution expansion acceleration (it holds together). In a gas, the distribution is statistically driven and the ensemble has a positive distribution expansion acceleration (it flies apart).

Transitions between states result because an ensemble has a distribution that is actually a composite of two competing distributions. Hence state changes can be described in terms of changes in the membership of individual instances in two competing distributions. Freezing is a net transferal of instances from a liquid distribution to a solid distribution. Boiling is a net transferal of instances from a liquid distribution to a gaseous. It is the actions of cooling and heating that forces the change in distribution memberships.

3.3 Viscosity.

Viscosity is related to the way in which a substance responds to non-uniform applied forces. Viscosity is usually considered in terms of fluid flow, but fluid flow is actually a result of non-uniform forces applied to a physical object, namely, the body of fluid. In order to understand viscosity, we can employ a description of substances that relies on the physical relationships that exists among the members of the ensemble. In particular, viscosity is understood in terms of changes in the positional relationships between the physical objects that result from an applied force to a few members of the ensemble. In a highly viscous substance like glass the application of a small force has no effect on the relative positions of the physical objects although it will affect the overall velocity of the ensemble. In a moderately viscous substance like molasses the application of a small force will affect of the relative positions of some members of the ensemble and will affect the overall velocity of the ensemble. Applying a force to a few physical objects of a substance with zero viscosity will affect the relative positions of only those objects to which it was applied.

Related to viscosity is the difference between motion of an object and motion of the substance that the object is made of. For example, when we speak of a running river we are not referring to a river that is translating across the country side. The river is essentially static being bounded by the surface and the river banks. What we mean is that the substance within the river is in motion. We measure the motion of the substance by examining individual instances (or sets of instances) of the physical objects that are members of the ensemble and measuring their velocity. When we say that the river is rising, what we mean is that the boundaries of the river are changing as the quantity of substances is increased. On the other hand, a falling drop of water is in motion while the substance within the object is not moving relative to the location of the object.

3.4 Reactivity.

Reactivity refers to the tendency for the composition of some substances to change over time resulting in radically different properties. In this case we are talking about physical objects that exist as compounds in reactive relationships with each other that force a change in the types of the physical objects within the ensemble. In terms of conceptual graphs this can be demonstrated for a case where two physical objects (molecules) react with each other,

```
[[[ROI: R] ->
            (contains) -> [molecule1: c1]
            (contains) -> [molecule2: c2] ->
                    (reactive) -> [molecule1: c1]
                    (adjacent_to) -> [molecule1: c1]] ->
    <time> ->
    [[ROI: R] ->
            (contains) -> [molecule3: c3]
            (contains) -> [molecule4: c4]]]
```
When dealing with substances rather than individual physical objects, the conceptual graph is more complex. It is,

```
[[[ROI: R] ->
        (establishes_time) -> [Time: t1]
        (contains) -> [physical_object: #1] ->
            (Made_Of) -> (Substance) -> [Substance0] ->
                (has_component) -> [Substance1] ->
                    (has_quantity) -> [Quantity: q1]
                (has_component) -> [Substance2] ->
                    (has_quantity) -> [Quantity: q2]
                    (reactive) -> [Substance1]] ->
    <time> ->
```

```
[[ROI: R] ->
    (establishes_time) -> [Time: t1 + t]
    (contains) -> [physical_object: #1] ->
        (Made_Of) -> (Substance) -> [Substance0] ->
            (has_component) -> [Substance1] ->
                (has_quantity) -> [Quantity: q1 - f(a • t) q1]
            (has_component) -> [Substance2] ->
                (has_quantity) -> [Quantity: q2 - f(b • t) q2]
                (reactive) -> [Substance1]]
            (has_component) -> [Substance3] ->
                (has_quantity) -> [Quantity: f(c • t)]
            (has_component) -> [Substance3] ->
                (has_quantity) -> [Quantity: f(d • t)]]].
```

In other words, as time passes the quantities of the two reactive substances decrease while the quantities of the product substances increase in very well defined manners given by the function $f()$ and the original quantities. The reaction can be halted by using the canonical graph that states:

```
[[physical_object: #1] ->
        (Made_Of) -> (Substance) -> [Substance0] ->
            (has_component) -> [Substance1] ->
                (has_quantity) -> [Quantity: q1]
            (has_component) -> [Substance2] ->
                (has_quantity) -> [Quantity: 0]] ->
    (equivalent_to) ->
    [[physical_object: #1] ->
            (Made_Of) -> (Substance) -> [Substance0] ->
                (has_component) -> [Substance1] ->
                    (has_quantity) -> [Quantity: q1]] ->
    (equivalent_to) ->
    [[physical_object: #1] ->
            (Made_Of) -> (Substance) -> [Substance1] ->
                (has_quantity) -> [Quantity: q1]] ->
```

to remove substances when their quantity is zero.

We can introduce additional conceptual graphs to handle other types of reactions. Of greater value is to introduce a canonical graph that can transform the conceptual graph for molecules reacting into a graph for the substances reacting. The conceptual graphs for the molecular reaction and the substance reaction points out the way. However, the complete graph required to accomplish this must be developed from first principles to establish how the parameters a,b,c, and d of the substance graph arise.

4 Conclusions

The derivation presented within this paper, although not carried out to full detail, provides a road map for an extremely detailed derivation of substances that incorporates all substance properties. The reason that such a detailed presentation was not provided here was one of page limit and not inability or inexperience. The conceptual graphs required to present this derivation easily consume more than ten pages of text.

Also not presented in this paper were the large number of canonical graphs that result from the different steps in the derivation. These canonical graphs describe the changes within substances that can take place over time. In many ways, these canonical graphs are the most important result of the derivation since we can generate an ad hoc definition for substance. However, without going through the derivation the canonical graphs are extremely difficult to identify and without understanding how these canonical graphs arise they are virtually impossible to justify. These graphs are the subject of a future paper.

References

[1]. F. Mish, ed., Webster's Ninth New Collegiate Dictionary, Merriam-Webster, Inc., Springfield, Mass (1986).

[2]. W. Tepfenhart and A. Lazzara, "The Situation Data Model", Proceedings of the Third Workshop on Conceptual Graphs, St. Paul, Minnesota, 1988 (pg 3.1.10-1)

[3]. W. Tepfenhart, "Using the Situation Data Model to Construct A Conceptual Basis", Proceedings of the Fourth Annual Workshop on Conceptual Structures, Detroit, Michigan, 1989 (pg.3.10).

[4]. W. Tepfenhart, "Problem Solving Using a Conceptual Graph Representation", Proceedings of the Fourth Annual Workshop on Conceptual Structures, Detroit, Michigan, 1989 (pg.4.13).

Representing Temporal Knowledge in Discourse: an Approach Extending the Conceptual Graph Theory

Bernard Moulin, Professor

Université Laval, Département d'informatique
Ste-Foy Quebec G1K 7P4 Canada

Abstract. This paper presents the main characteristics of an approach that can be used to model temporal information found in discourses. It extends the conceptual graph approach thanks to the introduction of the notion of time coordinate systems in the form of agents' perspectives and temporal localizations. The model may represent several types of temporal situations (events, processes, states and several sub-categories) which are related together and linked with temporal perspectives and localizations by means of temporal relations. We use an extension of Allen's temporal relations (augmented with parameters) as well as special relations used to represent multiple time intervals. This framework enables us to represent several conceptual issues such as iterativity, temporal metaknowledge, temporal operators.

1 Introduction

The model that we are proposing aims at representing most of the temporal characteristics that can be found in discourses. Linguists usually distinguish in discourses information corresponding to different levels of analysis, namely the syntactic, semantic and pragmatic levels. Temporal information in discourses may be highly context-dependent. For instance, we need to introduce a reference to the time when a sentence is uttered by a locutor (the so-called "speech-time") in order to explain adequately how verb tenses are used in that sentence. Hence, it seems relevant to study in the same framework both semantic and pragmatic properties of temporal characteristics in discourses. In our approach, the concepts contained in discourses are analysed according to an ontology which is based on a characterization of the worlds (real or hypothetical) that are described by discourses from a multi-agent perspective[1].

Through the discourse, the "narrator agent" communicates her knowledge of some world (real or hypothetical), this knowledge resulting from the agent's symbolic perception of some information from outside herself (physical perception with eyes, ears etc.) or within herself (internal visualization with "the mind's eye"). In a conversation, several agents play the narrator's role in turn. In the discourse content the narrator can evoke several situations related to the worlds she describes. These situations are temporally localized with respect to the narrator's perspective. In these situations one can describe a world containing objects and/or agents. The discourse may also contain a new perspective in which one reports another discourse taking place between other agents. In a discourse, perspectives and situations may also be temporally localized with respect to other perspectives and situations. The proposed ontology is based on a "naive model" [6] considering worlds as composed of temporal situations in which agents act physically, mentally or illocutionarily (through the performance of speech acts). The ontology is built on the basis of some main notions that will be described in the following sections:

- non-temporal world characteristics are described by *concepts and conceptual relations*;
- *temporal objects* (like day, year etc.) which are concepts associated with a time interval;
- *temporal situations* which describe states, processes, events which are also associated with time intervals;
- *temporal coordinates systems* that help in localizing temporally situations relatively to agents' perspectives, as well as to temporal references;
- temporal objects, situations and coordinate systems are related by *temporal relations*;
- *agents* who may be represented by concepts in the non-temporal description of worlds, or by their temporal perspectives which localize them relatively to temporal situations.

Non-temporal world characteristics are described by semantic networks composed of concepts and relations in the form of conceptual graphs [16]. All properties of simple conceptual graphs apply in our framework. They will not be detailed here.

In section 2, we introduce the notions of time interval, temporal object, temporal situation, and temporal relation. In section 3, the issues related to time coordinate systems are discussed: narrator's and agent's perspectives, temporal localizations. Section 4 presents the characteristics of multiple temporal objects and situations: iterativity, selectivity, subdivisions of temporal objects. In section 5, structures for representing temporal metaknowledge are suggested: temporal metasituations, temporal operators, causality relation. In section 6, we propose a linear notation for our conceptual representation of temporal information in discourse.

2. Time intervals, temporal objects and situations

2.1 Time intervals and temporal objects

All temporal structures are associated with a time interval. Let us consider an absolute time reference TR (called "time axis") composed of a set of elements called "time points" (this set is equivalent to the positive real numbers set) and a total order relation defined between these elements (denoted ">") which is called the precedence relation.

An *elementary time interval* is a continuous sub-set of TR. The elementary time interval is specified by a list of parameters :

- the begin-time BT which is the lower bound of the time interval on the TR axis;
- the end-time ET which is the upper bound of the time interval on the TR axis;
- the time scale TS indicates the unit which is used to measure the begin- and end-time on the TR axis;
- the duration DU indicates the duration of the time interval;
- the duration scale DS indicates the unit which is used to measure the duration.

The time interval duration is introduced here because a time interval may only be known by its duration. These parameters may be measured using different units (time scales) such as "minutes", "hours", "days", "months", "years" etc.

A *multiple time interval* is composed of a set of elementary time intervals, which may be contiguous or not.

A *temporal object* is a concept that is characterized by a time interval. "Day", "week", "month", "year" etc. are typical examples of temporal objects. A temporal object can be decomposed in other temporal objects (relation "part-of"). For instance "day" is usually decomposed in "morning", "noon", "afternoon", "evening", "night". Each of the component objects is associated with a sub-interval of the time interval of the decomposed temporal object. The corresponding time interval may only be known approximatively (the time interval begin- and end- times are not precisely known) as in the case of "the beginning", "the end" of temporal objects such as "afternoon" or "evening".

A *temporal object* is characterized by a triple <OD, OPC, OTI> where:

- the object description OD is a couple <object-descriptor, object-definition>; the object-descriptor is used for reference purposes; the object-definition corresponds to the concept that represents the temporal object; it is specified using the linear form of Sowa's conceptual graphs.

- the object propositional content OPC contains the description of other temporal objects or situations that semantically characterize the object.

- the object time interval OTI which aggregates the temporal information associated with the object.

Graphically we represent a temporal object with a rectangle decomposed in two parts. The rectangle represents the corresponding time interval. In the upper part of the rectangle we indicate the object description OD, as well as the relevant parameters of the object time interval OTI. In the lower part of the rectangle we represent the object propositional content in the form of temporal objects or situations related to the embedding rectangle by a relation ("part of" or a "temporal relation"). For example "the beginning of a stay of ten days" is represented by the diagram of figure 1a. Note that no time interval parameters are known for the temporal object "beginning", while duration parameters are known for the object "stay".

A simple temporal object is a temporal object associated with an elementary time interval. In a discourse we can consider a set of elementary temporal objects as a whole, that is called a "*multiple temporal object*". Its time interval is multiple and composed of the set of time intervals associated with the elementary objects. We anotate the time interval with the information 'multiple' as well as with the information 'continuous' (respectively 'discontinuous') if the resulting time interval is continuous (resp. 'discontinuous'). Figure 1b gives the representation of "the two first weeks of a three months holiday". Note that we use Sowa's notation of sets in the referent of concept "WEEK". Time interval annotations are represented after the semicolon mark. The relation "part-of" indicates that the resulting time interval is included in the time interval associated with "HOLIDAY". The temporal relation START indicates that the two time intervals start at the same time, and the annotation "first" indicate the linguistic formulation of this relation in the text. Multiple temporal objects are further discussed in section 4.

2.2 Temporal situations

In addition to temporal objects, a world may contain several temporal situations that correspond to events, processes, states of affairs etc., which take place in this world.
A temporal situation is a triple <SD, SPC, STI> where:

 - the situation description SD is a couple [situation-type : situation-descriptor] used to identify the temporal situation. The situation type is used to distinguish semantically different kinds of temporal situations : events, states, processes etc. The situation descriptor is used for referential purposes.

 - the situation propositional content SPC is a non-temporal knowledge structure described by a conceptual graph. It explicits some semantic characteristics of the situation;

 - The situation time interval STI is a structure which agregates the temporal information associated with the temporal situation.

Graphically we represent a temporal situation with a rectangle decomposed in two parts. The rectangle represents the corresponding time interval. In the upper part of the rectangle we indicate the situation description SD, as well as the relevant parameters of the object time interval STI. In the lower part of the rectangle we represent the situation propositional content SPC. For instance the sentence "John married Helen in

New York, January 23 1991" is represented by the following diagram:

```
┌─────────────────────────────────────────────────────────────────────┐
│ EVENT: ev1                              BT: January 23 1991, TS: date  │
├─────────────────────────────────────────────────────────────────────┤
│                                                                       │
│  [PERSON:John ] <- (AGNT) <- [MARRY] -> (PAT) -> [PERSON: Helen]       │
│                               └> (LOC) -> [CITY: New York]             │
│                                                                       │
└─────────────────────────────────────────────────────────────────────┘
```

Several authors have proposed different ways for categorizing temporal situations [17] [8]. We will use a taxonomy of temporal situations which is mainly based on Desclés' work [4] who applied the properties of topology for elaborating his classification. Here we present the main concepts introduced by Desclés.

"An interval is an oriented set of contiguous points (which are members of a continuous line) delimited to the left and to the right by two *boundary points* which separate the *interior* (the points between the boundary points) and the *exterior* (the points which are not between the boundary points) of the interval"... "An *open interval* is an interval where both boundaries to the left and to the right are opened. A *closed interval* is an interval where both boundaries are closed"...

"A *state (or static situation)* is characterized by absence of change of discontinuity; all phases of the static situation are the same. In a state we have neither a starting point (indicating change) nor an ending point (also indicating change). Each state is represented by an open interval"...

"An *event* marks discontinuity against the static background. Each occurrence of an event is a single whole viewed without regard to what happens before or after this occurrence. Each event is represented by a closed interval"...

"A *process* is a change from an initial static situation toward a final static situation. A process is always characterized by a beginning, that is the starting instant of the change (from the initial static situation). Each process is represented by an interval where the boundary point to the left is closed"...

"A *progressive process* at a *landmark point* T is a process not interrupted at T (or before T). It is represented by an interval closed to the left and opened to the right, where the landmark point T coincides with the boundary (last instant) to the right"...

"John was swimming" is an example of a progressive process. Such a sentence may usually be situated relatively to what Desclés calls a "landmark point" which provides a reference point for the progressive process such as: "When Mary called him, John was swimming". We will refine the concept of landmark point by introducing the notions of temporal perspectives and localizations (see section 3).

"When a process is interrupted in a flow, we may obtain two kinds of processes - *completed processes* or *non-completed processes* - depending on whether the landmark point where the process is interrupted coincides with the ending point of change involved in the process or not, that is depending on whether the final state is obtained or not" ... "In both cases, when the process is interrupted, we obtain a *non-*

progressive process" ... "Each *interrupted process* consists of :

a) a *resulting state* - a state which is in immediate contiguity with an interrupted process and is true after it;

b) an event, called the *event causing the interruption of the process* or *interrupting event*[2]...

The sentence "Now, John has written a letter" describes the resulting state of the process "John was writting a letter" which is now completed.

Figure 2 presents the taxonomy of temporal situations that we will be using in our approach. It is mainly based on Desclés proposal. We add a refinement for 'non-punctual events" in order to distinguish a simple event (associated with an elementary time interval) from a multiple event (associated with a multiple time interval). Multiple events are further refined in events which can be either continuous (the time intervals are contiguous) or discontinuous (the time intervals are not contiguous). We add also the notion of nucleus that is borrowed from Moens and Steedman [8], which describes the way an interrupted process is related to the interrupting event and the resulting state.

2.3 Temporal relations

Temporal situations and objects may be related together by *temporal relations*. In this section we will only consider temporal relations existing between elementary time intervals associated with temporal objects or situations. Relations between multiple time intervals will be examined in section 4. We use an extended form of Allen's temporal relations [1]. We consider two primitive relations called "BEFORE" and "DURING".

Given two intervals X and Y, the relation BEFORE(X, Y, Lap) holds if we have the following constraints between the begin- and end- times of X and Y compared on a time scale with the operators $\{>, <, =\}$: $BT(X) < ET(X)$; $BT(Y) < ET(Y)$; $BT(X) < BT(Y)$; $ET(X) < ET(Y)$; $BT(Y) - ET(X) = Lap$.

The Lap parameter measures the distance between the beginning of interval Y and the end of interval X on the time scale.

Given two intervals X and Y, the relation DURING(X, Y, DB, DE) holds if we have the following constraints between the begin- and end- times of X and Y compared on a time scale with the operators $\{>, <, =\}$: $BT(X) < ET(X)$; $BT(Y) < ET(Y)$; $BT(X) > BT(Y)$; $ET(X) < ET(Y)$; $BT(X) - BT(Y) = DB$; $ET(Y) - ET(X) = DE$.

The parameter DB indicates the distance between the beginning-time of X and the beginning-time of Y, and the parameter DE indicates the distance between the end-time of Y and the end-time of X, measured on a time scale. Using the DURING and

78

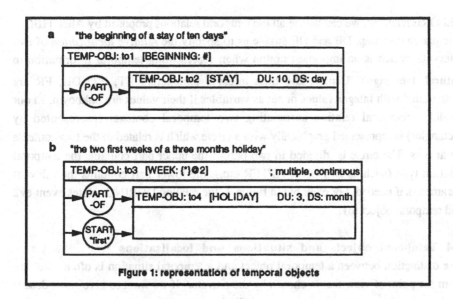

Figure 1: representation of temporal objects

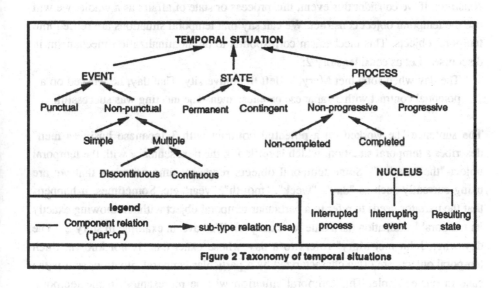

Figure 2 Taxonomy of temporal situations

BEFORE relations, we can define all the temporal relations proposed by Allen [10].
The parameters Lap, DB and DE enable us to specify the *relative localization* of two
intervals, which is an important notion when representing temporal information in
natural language[3]. Usually the time interval parameters BT, ET, DU, FR are
instantiated with integer values or set as variables if their values are unknown. In our
model a temporal relation associating two temporal contexts (represented by
rectangles) is represented graphically with a circle which is related to the two contexts
by arrows. The circle is divided in two parts : the upper part contains the temporal
relation type (such as BEFORE, AFTER etc.); the lower part contains the relevant
parameters if needed. For example in figure 3, the relation DURING relates event ev2
and temporal object to1.

2.4 Temporal objects and situations and localizations

The distinction between a temporal object and a temporal situation is often resulting
from the point of view that is chosen by the narrator. If we want to give some details
about an event, a process or a state of affairs, we will use the temporal situation
construct. If we consider this event, this process or state of affairs as a whole, we will
use the temporal object construct. We can say that temporal situations are reified into
temporal objects. This mechanism corresponds to the nominalization mechanism in
discourses. Let us consider story 2:

> The day when John met Mary, he left the university. That day, he worked on a
> potential contract with 3 Japanese business men. The meeting was successful.

The sentence "he worked on a potential contract with 3 Japanese business men"
describes a temporal situation, which is reified in the next sentence with the temporal
object "the meeting". Some temporal objects represent time concepts that we are
using currently such as "day", "week", "month", "year" etc. Sometimes, it happens
that the narrator needs to refer to a particular temporal object without knowing exactly
its temporal localization (as a date for example). We have an example in story 2: "The
day when John met Mary" refers to a day whose exact date is not known. Such
temporal objects are often characterized by a particular temporal situation, as it is the
case in our example. This temporal situation will be represented in the temporal
object propositional content as it is shown in figure3: the event ev1, "John met
Mary", is related to the temporal object time interval (embedding rectangle of to1)
with the temporal relation CONTAIN, which is anotated with the linguistic
formulation "when".

3. Time coordinate systems
3.1 Introduction
Since discourses result from the interactions taking place between a writer/speaker and

a reader/hearer, many temporal linguistic constructs are used to express directly or indirectly temporal relationships between the speaker's and hearer's temporal localization and the temporal situations that are described. Reichenbach (1947) proposed a model for interpreting semantically verb tenses. He localized on a time axis the point of speech (S), the point of reference (R) and the point of event (E) and gave a semantic interpretation of English tenses. Several researchers are using extensions of Reichenbach's model to explain and model temporal knowledge in natural language. See for instance Lo Cascio and Vet [7], Borillo et al. [2], Dorfmüller-Karpusa [5], Moulin and Côté [9].

In fact, Reichenbach's markers are a first approximation of a more general notion which is fundamental for explaining temporal phenomena in natural language : the notion of temporal coordinate systems. In physics, researchers use several temporal coordinate systems (absolute and relative) for explaining physical phenomena.

In [10] we claimed that several temporal linguistic phenomena can be explained by introducing explicitly temporal coordinate systems and by specifying the temporal relations which relate temporal situations to these temporal coordinate systems. When we analyse discourses, we can notice that several time coordinate systems are used such as:

- The *official time coordinate system* which provides dates, hours, years etc., that are used as time references for localizing time intervals on an absolute time scale.

- The *narrator's perspective*[4] which is the narrator's time coordinate system. It may be the same as the reader/hearer's time coordinate system if the reader and writer interact directly together or different if the interaction is delayed.

- The narrator may describe temporal situations taking place in a temporal localization which is different from her own temporal localization. A set of temporal situations can be localized according to a time coordinate system which is called a *temporal localization*. We can also represent temporal localizations which are embedded within another temporal localization.

- When a narrator reports in a discourse the words that have been said by a person, the temporal coordinate system changes and is localized at the time when the person spoke. We will call *agent's perspective*[4], the temporal coordinate system attached to a person whose words are reported in a discourse in a direct way [11].

These time coordinate systems fall into two broad categories: temporal perspectives and temporal localizations.

3.2 Temporal Perspectives

When producing a discourse, an agent considers her own spatio-temporal localization as the main coordinate system ("here and now"), which sets a reference for localizing other temporal situations, objects as well as other agents. Each agent acts physically,

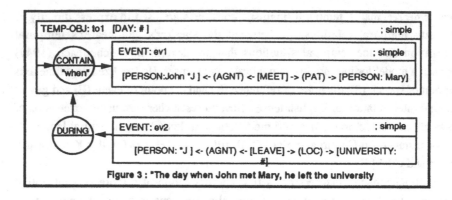

Figure 3 : "The day when John met Mary, he left the university

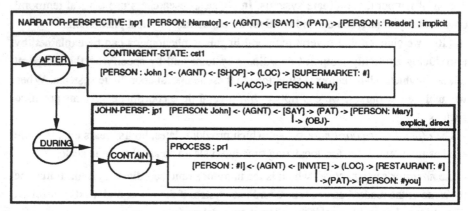

Figure 4: Narrator's and agent's perspectives

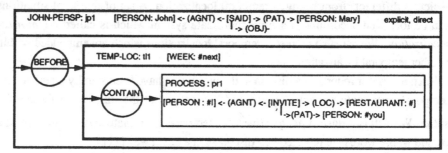

Figure 5: The temporal localization "next week"

mentally or illocutionarily from her own perspective. In a discourse the narrator's perspective sets the main time coordinate system. When reporting the words of another agent, the discourse perspective changes from the narrator's one to the agent's one. For instance, this perspective change is reflected in the text by the fact that verb tenses change in order to accomodate the new perspective. We distinguish the narrator's perspective[5] from other agents' perspectives. Let us consider story 3:

John was shopping with Mary at the supermarket.

John said to Mary: "I invite you to the restaurant".

Story 3 describes a scene that took place in the past relatively to the narrator's perspective: this explains the use of past tenses of verbs "was", "said". "John said Mary" introduces a new agent's perspective: the present tense "invite" is set relatively to John's perspective.

A *narrator's perspective* is a triple <NPD, NPS, NPTI> where NPD identifies the perspective description, NPS is called the perspective set and NPTI is the time interval associated with the origin of the time coordinate system that is represented by the narrator's perspective.

The *narrator's perspective description* NPD is a couple [perspective-descriptor, perspective description]. The perspective-descriptor is used for reference purposes. The perspective-description is specified by a conceptual graph that describes the narrator's action when generating the discourse :

[PERSON: narrator]<- (AGNT) <- [SAY] -> (PAT) ->[PERSON: reader].

Usually, these parameters are not indicated explicitly in a discourse. For example, in story 3, the narrator describes to the reader the situations that took place at the supermarket (see the upper rectangle in figure 4). The parameter 'implicit' indicates that the perspective description is not explicited in the discourse.

The *narrator's perspective set* of a narrator's perspective NPx is the set of all temporal situations, localizations and perspectives which are in the "temporal scope" of NPx. The narrator's perspective TPx sets the time coordinate system origin. The time coordinates of the temporal situations, localizations and perspectives within NPx temporal scope are evaluated according to NPx.

The *narrator's perspective time interval* NPTI has the same properties as the situation time interval (section 2.2). It specifies the characteristics of the time interval that is associated with the time coordinate system origin corresponding to the narrator's perspective. Usually it is implicit: that's the case of story 3. The interval NPTI sets what can be considered as the narrator's "now"[6].

The narrator's perspective is used in relation with the temporal situations and localizations which are in its perspective set, in order to determine the verb tenses which are used in sentences describing the corresponding situations. The notion of narrator's perspective is an extension of Reichenbach's "S marker" (speaker marker).

We represent graphically a narrator's perspective as a rectangle composed of two parts. In the upper rectangle, we indicate the narrator's perspective description and time interval parameters. In the lower rectangle we indicate the perspective set. The lower rectangle corresponds to the temporal scope of the perspective. The sides of the lower rectangle symbolize the time interval associated with the time coordinate system origin. The temporal relations link the side of the rectangle with the rectangles representing the temporal situations, localizations and perspectives which are elements of the narrator's perspective set. In figure 4 the narrator's perspective set contains a contingent state that describes the scene, and John's perspective.

An agent's perspective has the same characteristics as the narrator's perspective. It is specified by a triple <APD, APS, APTI> where APD identifies the perspective description, APS is called the perspective set and APTI is the time interval associated with the origin of the time coordinate system that is represented by the agent's perspective. The graphical conventions used to represent the agent's perspective are the same as those used for the narrator's perspective. The only difference relies in the perspective header which is characterized by the agent name, as JOHN-PERSP in figure 4. John's perspective contains a process that corresponds to his invitation. The indexicals #I and #you are discussed in [11].

The agent's perspective set may contain temporal situations, as well as temporal localizations and other agent's perspectives.

3.3 Temporal localizations

A temporal localization sets a secondary time coordinate system which is positioned relatively to a temporal perspective or to another temporal localization. A temporal localization can be seen as a temporal capsule within which a set of temporal situations is localized.

A temporal localization is a triple <TLD, TLS, TLTI> where TLD identifies the temporal localization description, TLS is called the temporal localization set and TLTI is the time interval associated with the temporal localization.

The *temporal localization description* TLD is usually implicit. It is a couple [localization-descriptor, localization-description]. The localization-descriptor is used for reference purposes. When explicited in the discourse, the localization-description is specified by a conceptual graph that usually specifies the temporal object describing the temporal localization.

The *temporal localization set* TLS of a temporal localization TLx is the set of all temporal situations, localizations and agent's perspectives which are in the "temporal scope" of TLx. The time coordinates of the temporal situations, localizations and agent's perspectives within TLx temporal scope are specified relatively to TLx.

The *temporal localization time interval* TLTI has the same properties as the situation time interval (section 2.2). It specifies the characteristics of the time interval that is associated with the temporal localization.

A temporal localization is an extension of Reichenbach's R-marker (reference marker). We represent graphically a temporal localization as a rectangle composed of two parts. In the upper rectangle, we indicate the temporal localization description and time interval parameters. In the lower rectangle we indicate the temporal localization set. The lower rectangle corresponds to the temporal scope of the localization. The sides of the lower rectangle symbolize the time interval associated with the time localization. The temporal relations link the side of the rectangle with the rectangles representing the temporal situations, localizations and agent's perspectives which are elements of the temporal localization set. For instance in story 3, let us slightly change John's words by : "Next week, I will invite you to the restaurant". The process of "inviting Mary to the restaurant" is localized in the future relatively to John's perspective. John's new perspective is represented in figure 4. It contains the temporal localization tl1 corresponding to "last week". The process pr1 is temporally localized relatively to tl1.

3.4 Temporal objects and localizations

Temporal objects may be used to "focus on" a new temporal localization. Let us consider for instance story 4:

John had never been ill until he arrived in Canada.

During the three first days after his arrival, he got a headache five times.

Twice he took an aspirine. Each time, he felt better.

"During the three first days after his arrival" sets a temporal localization which is used to localize the temporal situations "he got a headache five times", "twice, he took an aspirine" and "each time, he felt better" (the expressions "five times","twice" and "each time" are discussed in section 4).

Focusing is a cognitive process which enables the narrator to evoke a temporal object and to use it as a new temporal localization. Metaphorically, focusing is equivalent to looking inside the temporal object in order to describe the temporal localizations that are contained in it. Focusing is represented by a special relation which links the object to the corresponding temporal localization. This relation is denoted "FOCUS" and has the same properties as the temporal relation "EQUAL". Figure 6 presents the diagram representing story 4. The focus relation links temporal object to1 and temporal localization tl2.

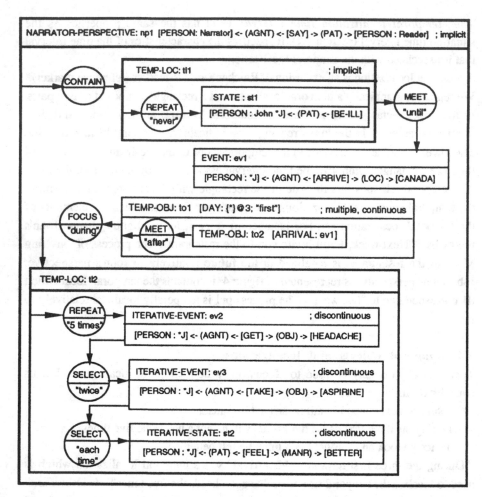

Figure 6: Representation of iterative events

4. Sets and sub-sets of temporal objects and situations
4.1 Multiple temporal objects

Most temporal logics are restricted to the study of temporal relations (such as Allen's relations) associating elementary time intervals. However we need a formalism and inference rules to reason on multiple intervals. Story 4 gives some examples of expressions corresponding to multiple events. Multiple temporal objects associated with continuous time intervals have been studied in section 2.1. In the present section we will only analyze multiple temporal objects with discontinuous time intervals. Two notions should be taken into account for explaining iterativity: the extent of the repetitive situation and the frequency of the repetitive phenomenon. In our model, we can have sub-categories of multiple events (states or processes) that can be characterized as "iterative" events (states or processes), "frequentative" events (states or processes), or "habitual" events (states or processes). In figure 6, we have occurrences of iterative events and states.

4.2 "REPEAT", the temporal relation of iterativity

Temporal quantifiers such as "never", "always" share a common ground with iterative temporal situations: the need to indicate the scope (or temporal extent) of the quantification. Typical examples are : "Until 1989, I never smoked. Since that year, I smoke from time to time"; "When he worked in the US, John never went in a Chinese restaurant". In fact, temporal quantifiers like "never" and "always" can be considered as special cases of iterativity. We introduce the special temporal relation REPEAT, which is used to express iterativity. We will adopt the notion of "cover of a time interval" which was proposed by Richards et al. [14 p. 43] for representing temporal quantifiers and iterativity.

A set of time intervals P is a *cover* for an interval i, if the elements of P are mutually distinct (non overlapping), if they are contiguous and if their union is identical to i^7.

Let X be a temporal object, situation or localization and Y be a temporal situation. Let i_X and i_Y be the time intervals respectively associated with X and Y. Let r be an integer, called the repetitive factor. The relation $REPEAT_r$ holds between X and Y (denoted [X] -> ($REPEAT_r$) -> [Y]) iff:

- there exists a cover P of i_X such that for exactly r elements $p_1, ..., p_r$ of P, the propositional content of Y is true, these elements compose the multiple interval i_Y associated with Y;

- for any element p_z, such that z œ [1,r], the propositional content of Y is false;

- for any other cover Q of i_X such that there exists an element q of Q for which the propositional content of Y is true, then there exists k Œ [1,r], such that q is a subinterval of p_k.

The temporal quantifier "never" is a special case. $[X] \to (\text{REPEAT}_{never}) \to [Y]$ holds iff there is no cover P of i_X such that for some p of P the propositional content of Y is true.

The temporal quantifier "always" is another special case. It is denoted $[X] \to (\text{REPEAT}_{always}) \to [Y]$ and holds iff there is a cover P of i_X such that for every p of P the propositional content of Y is true.

Graphically the relation $[X] \to (\text{REPEAT}_r) \to [Y]$ is represented by a circle containing the parameter r, and linking the rectangles representing the time intervals associated with X and Y. Figure 6 gives the representation of story 4.

The relation REPEAT_{never} links an implicit temporal localization tl1 and the event ev1: "John has never been ill". The temporal localization is not stated explicitly in the text, but it is needed in order to understand the extent of the temporal quantifier "never".

The relation $(\text{REPEAT}_5$; "five times') links the temporal localization tl2 and the repetitive event ev2 "John got a headache". The temporal localization tl2 results from an operation of focusing on the temporal object "during the three first days" (see section 3.4)

4.3 The relation SELECT

Another cognitive operation that can be used by the narrator consists in selecting some elements of a multiple temporal object or situation and in using the selected elements to localize other temporal situations. We have two examples in story 4, "*Twice* he took an aspirine" and "*Each time*, he felt better". These examples can be paraphrased in order to emphasize the selection operation: "he took an aspirine *twice* among the five times he got an headache" and "he felt better *each time* he took an aspirine". The selecting process is represented by a selection relation which links a multiple object or situation X with another situation (multiple or simple) Y. It is denoted by $[X] \to (\text{SELECT}_r) \to [Y]$.

Formally the selection operation consists in selecting a subset of the set of time intervals associated with the multiple temporal object or situation X. This subset of time intervals is associated with the situation Y. The parameter r indicates the way the selection among the set of time intervals is done.

Graphically, the selection $[X] \to (\text{SELECT}_r) \to [Y]$ is represented as a temporal relation linking the rectangle representing the time intervals corresponding to X and Y. The parameter r is represented in the lower part of the circle which represents the relation. In figure 8 we have the representations of "twice" and "each time".

4.4 Frequency of a multiple temporal object or situation

Some multiple temporal objects (respectively, situations) can have a known periodicity. If the temporal distance between two successive elementary objects (resp.

situations) contained in a multiple object (resp. situation) is a constant, we will call that distance the *frequency* of the multiple temporal object (resp. situation). In the sentence "Last month, John took an aspirine everyday", "*everyday*" indicates that we are dealing with a multiple temporal situation with a frequency of one day.

Since the frequency is a characteristic of a multiple time interval associated with a temporal object or situation, it will be specified by two additional parameters of the time interval :
- the frequency FR indicates the distance existing between two consecutive elementary time intervals contained in the multiple time interval;
- the frequency scale FS indicates the unit which is used to measure the frequency.

In some cases the frequency provides an additional information about the localization of the elementary time intervals contained in the corresponding multiple time interval. Let us consider the example : "Last week, John took an aspirine every morning". "*Every morning*" indicates that the frequency is "one day" and that each elementary event is localized "in the morning". We will still consider that this information represents the frequency of the temporal situation, but the parameter FR will be associated with a conceptual graph that represents the corresponding concept. The following graph represents this sentence.

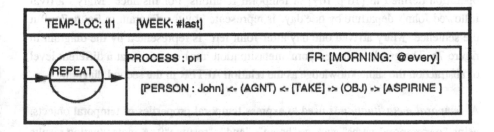

4.5 Subdivisions of temporal objects

Natural language enables agents to refer to sub-parts of temporal objects. Let us mention for instance the conventional divisions of the day : morning, noon, afternoon, evening, night. There are also subdivisions which apply to any kind of temporal object like: beginning, middle, end. In section 2.1, we indicated that these temporal object subdivisions are themselves considered as temporal objects. Figure 1 provides the representations of two examples: "The beginning of a stay of ten days" and "the two first weeks of a three months holiday". In each example the relation PART-OF specifies that one of the object (to1 and to3) is a part of another object (to2 and to3 respectively).

Let X and Y be two temporal objects. Let i_X and i_Y be the time intervals respectively associated with X and Y[7]. A relation PART-OF holds between X and Y, denoted [X]

-> (PART-OF) -> [Y], iff $i_Y \ldots i_X$.

The semantic properties of some concepts provide additional information with respect to their time intervals. For instance:

- from [BEGINNING: X] -> (PART-OF) -> [Y], we can deduce that
$$[i_X] \rightarrow (START) \rightarrow [i_Y];$$
- from [END: X] -> (PART-OF) -> [Y], we can deduce that $[i_X] \rightarrow (FINISH) \rightarrow [i_Y]$;
- from [MIDDLE: X] -> (PART-OF) -> [Y], we can deduce that
$$[i_X] \rightarrow (DURING) \rightarrow [i_Y].$$

Let us remark that the exact extent of the time intervals corresponding to the concepts of beginning, middle and end depends on the interpretation of the agent who uses these concepts.

5. Temporal meta-knowledge
5.1 Temporal metasituation
Natural language enables locutors to express knowledge at different levels of abstraction. That is the case for temporal knowledge [13]. Verbs like "begin", "start", "finish", "end", "resume", "follow", "occur", "precede" are used to denote temporal properties of temporal objects or situations. They can be considered as temporal metaknowledge. Several of them can be considered as expansions (in the sense of the operation defined in [16 p 109] of temporal relations. For instance "Mary's arrival followed John's departure by one day" is represented by the diagram in figure 7a. But the sentence "Mary arrived one day after John left" is represented by the diagram in figure 7b. The graph in the temporal metasituation tms1 expresses at a different level of abstraction the same knowledge as the relation AFTER in the second diagram.

A *temporal metasituation* is used to express temporal properties of temporal objects, using "pertemporal verbs" such as "begin", "end", "resume"[8]. A metasituation results from the performance of an illocutionary act by an agent, but it does not correspond to a new temporal situation with respect to the scene that is described by the discourse. As an illocutionary act, a metasituation bears all properties of temporal situations (time interval, linguistic parameters).

5.2 Temporal operators
Metaknowledge can apply also on temporal situations. Verbs such as "start", "finish", "end", "resume", "continue" can be used to express temporal properties of temporal situations. For instance, in the sentence "John began to work at the University in 1990" provides an information about the begin-time of the time interval associated with the temporal situation: the sentence can be paraphrased by "John worked at the University since 1990". We consider that these verbs express a temporal modality. We

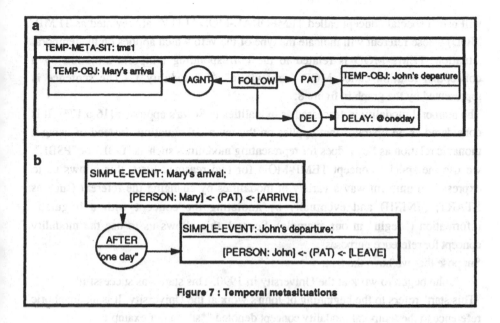

Figure 7 : Temporal metasituations

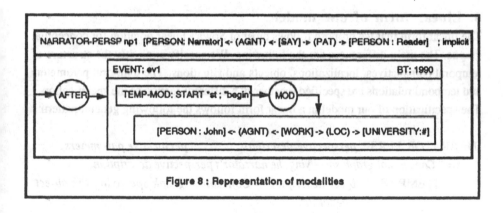

Figure 8 : Representation of modalities

will use a special concept called TEMPORAL-MODALITY (abbreviated as TEMP-MOD) whose referent will indicate the type of the verb which applies on the temporal situation. TEMP-MOD is related to the corresponding temporal situation by a conceptual relation called MOD (abbreviation of "modality"). The sentence is represented by the graph in figure 8.

The temporal modality is processed as modalities in Sowa's approach [16 p 175]. It is considered as an operator that applies on the temporal situation. Instead of using a monadic relation as Sowa does for representing modalities such as "OBL" or "PSBL", we use the special concept TEMP-MOD for two reasons. First, this allows us to express in a uniform way a variety of modalities by changing the referent (such as START, FINISH) and eventually by anotating the concept with a linguistic information ("begin" in our example). Second, this allows us to use the modality concept for reference purposes.

Suppose that we have the text:

"John began to work at the University in 1990. This start was successful".

"This start" refers to the beginning of John's work at the University. It is an anaphoric reference to the temporal modality concept denoted "*st" in our example.

The second sentence can be expressed by the following graph:

[START: *st] -> (CHRC) -> [SUCCESSFUL].

6. Linear form of our model

The linear form that we propose follows most of Sowa's linear notation for conceptual graphs. The main differences are the following. We specify explicitly the structure of temporal perspectives, localizations, objects and situations. The temporal parameters and temporal relations are specified together in a different section of the specification. The specification of our model in a linear form follows the following general pattern:

[NARRATOR-PERSP: *perspective-descriptor*   ; *perspective parameters*
    := *Conceptual graph specifying the narrator's perspective description*
    - - [TEMP-OBJ: *object-descriptor* := *Conceptual graph specifying the object*
                definition
            - - [*Temporal situation specification*
                ]
    ]
    - - [AGENT-PERSP: *perspective-descriptor*   ; *perspective parameters*
        := *Conceptual graph specifying the agent's perspective description*
        - - [*Temporal situation specification* and/or
                Temporal localization specification
            ]

            ....

```
          ]
- - [TEMP-LOC: localization-descriptor
          :=   Conceptual graph specifying the localization description
          - - [Temporal situation specification and/or
                    Temporal localization specification
                and/or Agent's perspective specification
                ]

                .............

          ]
- - [SITUATION-TYPE: situation-descriptor  :=
                Conceptual graph specifying the situation description
      ]
```

Specification of all the relevant temporal relations holding between temporal entities (perspectives, localizations, situations and objects):
[TINT: *temporal-entity-descriptor1*] -> (TEMPORAL-RELATION; temp.-relation-parameters) -> [TINT: *temporal-entity-descriptor2*]

....

The structures corresponding to temporal objects, localizations, situations and agent's perspectives can appear in any order .
As an example, we give here the linear form of the graphs contained in figure 4 where John's perspective is replaced by the content of figure 5.

```
[NARRATOR-PERSP: *np1                                ; implicit
   := [PERSON: writer]<- (AGNT) <- [SAY] -> (PAT) ->[PERSON: reader]
   - - [CONTINGENT-STATE: cst1   :=
            [PERSON: John]<- (AGNT) <- [SHOP] -
                        (LOC) ->[SUPERMARKET: #]
                        (ACC) ->[PERSON: Mary]
      ]
- - [JOHN-PERSP: jp1                                  ;event, explicit, direct
      :=   [PERSON: John *J]<- (AGNT) <- [SAY] -
                  (PAT) ->[PERSON: Mary]
                  (OBJ) -
      - - [TEMP-LOC: tl1  :=   [WEEK : #next]
          - - [PROCESS: pr1 :=
                  [PERSON: #I]<- (AGNT) <- [INVITE] -
                  (PAT) ->[PERSON: #you]
                  (LOC) ->[RESTAURANT: #]
              ]
```

```
                ]
          ]
    ]
```

[TINT: np1] -> (AFTER) -> [TINT: cst1]
[TINT: cst1] -> (DURING) -> [TINT: jp1]
[TINT: jp1] -> (BEFORE) -> [TINT: tl1]
[TINT: tl1] -> (CONTAIN) -> [TINT: pr1]

7. Conclusion

In this paper we have presented the main characteristics of an approach that can be used to model temporal information found in discourses. It extends the conceptual graph approach thanks to the introduction of the notions of time coordinate systems in the form of agents' perspectives and temporal localizations. The model may represent several types of temporal situations (events, processes, states and several sub-categories) which are related together and with temporal perspectives and localizations by means of temporal relations. We use an extension of Allen's temporal relations (augmented with parameters) as well as special relations used to represent multiple time intervals.

This framework enables us to represent several conceptual issues such as iterativity, temporal metaknowledge, temporal operators. Our framework also provides means for representing speech acts, anaphoric references, deictics, oblique contexts, sentence markers. These points are discussed in [12] along with a brief comparison of this framework with Sowa's approach.

In future papers, we will discuss how verb tenses can be automatically generated from our representation, how we can check the coherence of temporal information in discourses, and, how we can generate the logical content of knowledge bases as a result of the analysis of a discourse according to our concepual approach.

ACKNOWLEDGMENTS

This research is supported by the Natural Sciences and Engineering Research Council of Canada (grant OGP 0000 05518).

NOTES

1. We use the term "agent" in the same sense as it is used in distributed artificial intelligence : an agent who has various beliefs, makes decisions, elaborates plans and executes them autonomously. An agent may be human or artificial.

2. Desclés [4] calls this event "the stopped process". He admits that "this is a misleading name since it denotes an event". For clarity's sake, we chose to call that event "interrupting event".

3. In order to provide some variety for expressing temporal relations in the model, we can use the following relations :

AFTER (X,Y, Lap) with Lap > 0 which is equivalent to BEFORE (Y,X,Lap)

MEETS (X,Y) which is equivalent to BEFORE (X, Y ,0)

OVERLAPS (X, Y, Lap) with Lap > 0 which is equivalent to BEFORE (X, Y, -Lap)

CONTAINS (X, Y, DB, DE) with DB > 0 and DE > 0 which is equivalent to DURING (Y, X, DB, DE)

EQUAL (X, Y) which is equivalent to DURING (Y, X, 0, 0).

4. In the present paper the terms "narrator's perspective" and "agent's perspective" replace respectively the terms "temporal perspective" and "utterance perspective" that we used in [10,11] . In 1991 we indicated that these two notions were very similar. The new vocabulary is intended to reflect this similarity.

5. In fact, the discourse is the result of a communication taking place between the writer/speaker and the reader/hearer. It can be interpreted as a set of illocutionnary acts [15]. Here the narrator's perspective describes partially the context of occurrence of these illocutionary acts [10,11].

6. In other discourses the narrator's perspective origin time interval can be indicated explicitly as in the following sentence : "*We are in 1789, July 14. The Parisian People is marching towards La Bastille*". "July 14 1789" sets explicitly the date of the new narrator's perspective. Notice that in the second sentence the progressive present tense is used, reflecting the perspective change from the narrator's present to the "1789 present".

7. Note that all the relations defined between time intervals (such as contiguity, inclusion, equality etc.) can be defined as some combinations of elementary relations (<,>, =) between the begin- and end- time of these intervals.

8. Since these verbs express temporal properties of temporal objects or situations, we will call them "*pertemporal verbs*" by analogy with the *perlocutory verbs* that express explicitly the illocutionary acts they intend to perform (such as "*I request that* you come again").

BIBLIOGRAPHY

1. ALLEN J. F. (1983), Maintaining Knowledge about Temporal Intervals, *Communications of the ACM*, vol 26 n11.
2. BORILLO A. (1988) et Borillo M., Bras M., Une approche cognitive du raisonnement temporel, extrait des "Actes des journées nationales en intelligence artificielle", édité par Teknea, Toulouse, mars 1988.

3. COMP-LING (1988) *Computational Linguistics*, Special Issue on tense and Aspect, vol 14 n 2., pp 15-28.

4. DEsCLES J-P. (1989), State, Event, Process and Topology, in *General Linguistics*, vol 29 n3 pp 161-199, Pensylvania State University Press

5. DORFMULLER-KARPUSA K. (1988), Temporal and aspectual relations as text-constitutive elements, in *Text and Discourse Constitution*, J. S. Petöfi editor, Walter de Gruyter pub.

6. HAYES P. J. (1985) The second naive physics manifesto, in *Formal theories of the Commonsense World*, J. Hobbs and R. Moore edts, Ablex, Norwood.

7. LO CASCIO V. (1986) and Vet C. editors, *Temporal Structure in Sentence and Discourse*, Foris Publications, Dordrecht.

8. MOENS M. (1988), Steedman M., Temporal ontology and temporal reference, in [3] pp 15-28.

9. MOULIN B. (1990), Côté D. , Extending the conceptual graph model for differentiating temporal and non-temporal knowledge, in proceedings of the Fifth Annual Workshop on Conceptual Structures, AAAI conference, Boston July 1990. Appeared also in *Knowledge-Based Systems Journal*, vol4 n4 pp 197-208, December 1991.

10. MOULIN B. (1991a), A conceptual graph approach for representing temporal information in discourse, in proceedings of the Sixth Annual Workshop on Conceptual Structure, Binghamton, New York, July 1991, to appear in *Knowledge-Based Systems* journal.

11. MOULIN B. (1991b), D. Rousseau, D. Vanderveken, Speech acts in a connected discourse, a computational representation based on conceptual graph theory, in proceedings of the Sixth Annual Workshop on Conceptual Structures, Binghamton, New York, July 1991, to appear in the *Journal of Experimental and Theoritical Artificial Intelligence*.

12. MOULIN B. (1992), An approach for representing temporal knowledge in discourse: an extension of Sowa's conceptual graph theory", Research Report DIUL-RR-9202, Université Laval, Département d'informatique (50 pages).

13. PASSONNEAU R. J. (1988), A computational model of the semantics of tense and aspect, in [3], pp 44-60.

14. RICHARDS B. (1989), I. Bethke, J. van der Does, J. Oberlander, *Temporal Representation and Inference*, Academic Press.

15. SEARLE J. R. (1985), and D. Vanderveken, *Foundations of Illocutionary Logic*, New York, Cambridge University Press.

16. SOWA J.F. (1984), *Conceptual Structures : Information Processing in Mind and Machine*, Addison Wesley.

17. VENDLER Z. (1967) Verbs and times, in Z. Vendler editor, *Linguistics and Philosophy*, Ithaca, Cornell University Press.

Towards a Semantics of Inchoative and Causation Events in Conceptual Graphs

Pavel Kocura

Department of Computer Studies
Loughborough University of Technology
Loughborough, Leicestershire LE11 3TU, UK

Abstract. Jackendoff's TRH and Talmy's FD semantic theories can be analysed and further refined, in terms of deep knowledge, using Sowa's CGs. This is shown on the example of inchoative and causal events. Causality is divided into the thematic and action tiers. The action tier represents the roles of doer vs. undergoer, agonist vs. antagonist, beneficiary vs. 'negative' beneficiary. These are augmented by the aspect of voluntariness vs. involuntariness. The objective is to develop deep-knowledge models of such events which would become entries of the conceptual and relational catalogues. Such models will enable the system to produce correct logical and pragmatic inferences. These approaches extend to a number of semantic categories and cognitive domains. Although the motivation of the TRH and FD is primarily linguistic, the objective of the present investigation is the building of qualitative models which provide appropriate logical and pragmatic inferences.

1 Introduction

If the results of independent researchers converge or complement each other, it increases their validity. There appears to be such a parallel between Sowa's Conceptual Graphs (CGs) and Ray Jackendoff's theory of 'semantic structures', which is based on Gruber's Thematic Relations Hypothesis (TRH). The close correspondence of Sowa's and Jackendoff's notations was first shown by Kocura and the use of Jackendoff's semantics as the basis of CGs systems has been further explored [4, 5].

Although the original working semantics of Sowa's CGs [7] is based on case grammar, the CGs formalism and inference mechanism can readily accept other semantic and syntactic theories. For example, McHale and Myaeng [6] suggest a CGs-based NL system which incorporates Chomsky's Government-Binding Theory. McHale and Myaeng note the advantages of using GB grammar instead of the Augmented Phrase Structure Grammar. Their findings indirectly support Jackendoff's unified theory [3], which has its distant origins in GB grammar. Interestingly, although Jackendoff sees case grammar as direct opposition to the TRH, Cook [11] considers TRH-based grammar only another variant of case grammar.

The ability to generate semantic and pragmatic inferences is an essential property of any viable knowledge representation formalism. Inference should be possible across the whole range of expressions, from the surface statements to the deep-level structures. In the original case-grammar-based semantics of CGs inference appears to be based on a potentially large set of rules and schemata. These rules and schemata are extraneous to the lexical items and statements about which the inferences are made. Ideally, all logical and pragmatic inferences should be implicit from a statement. For this they have to be explicit in the detail of the deep-knowledge representation of individual lexical items, which is distributed amongst the various structures of the CGs system: the type and relation lattices and conceptual and relational catalogues and their linking in a proposition. Jackendoff's semantic system aims to map this explicit knowledge. However, Jackendoff's notation, as he freely admits, is too complex, difficult to read and non-computational. Thus, the CGs notation, because of its clarity and principled, processable nature, together with its inference and theorem proving system based on the open-world model, lends itself easily to the representation, validation and further development of Jackendoff's theory. The paper will introduce a small example of Jackendoff's semantic system and suggest a way to its reconstruction in terms of CGs, concentrating on the semantic and pragmatic modelling of inchoative and causal events.

2 Representing Inchoative Events

Jackendoff's semantics is partly based on Gruber's Thematic Relations Hypothesis (TRH), which abstracts the semantic structure of statements about the location and motion of objects in space [1, 9]. This abstraction is then adapted to represent other semantic categories. Inchoative verbs describe events in which Theme, having traversed a [PATH] in a semantic space, ends up in a new [STATE]. Jackendoff realised that his original representation of inchoative verbs by the event [GO] was not coherent, as shown in the sentence *The weathervane pointed north.* In the inchoative interpretation, using the original [GO], it would have been represented as shown in the following graph:

(1)   *The weathervane pointed north.*

              [WEATHERVANE]-
                 (E-THEME)-<-[GO_spc]-
                       (E-GOAL)-[PATH]-
                                (TOWARDS)->-[PLACE: North Pole],,. [1]

However, (1) would be normally interpreted to mean that the weathervane *bodily travelled* towards the North Pole, which is not what the sentence said. To accommodate situations in which [GO] produces incorrect interpretations, Jackendoff introduces a new concept, [INCH], which represents *inchoative* events. The elementary form of its use and its translation into a conceptual graph is shown in (2).

(2)   [_Event INCH ([STATE])] =   [INCH]->-(E-GOAL)->-[STATE].

The relation E-GOAL is rather general; it might be replaced with a more specific relation RESULT. Because of the infix notation of CGs, it is felt that [INCH] should be also linked to the Theme, as shown below:

---

[1] The relation label E-THEME = Event Theme, replaces the label SSUB = State Subject, and the label E-GOAL = Event Goal replaces the label SOBJ = State Object, which were tentatively used in previous papers on this topic [4, 5].

(3)   [THING]-<-(E-THEME)-<-[INCH]->-(E-GOAL)->-[STATE].

Then Jackendoff's original formula in (4) can be translated into the graph in (5)

(4)

$$[_{Event}INCH([_{State}BE_{Space}([THING], [PLACE])])]$$

(5)

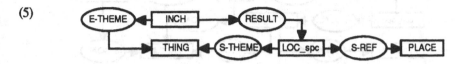

Having the advantage of recursive concept decomposition provided by CGs, we might not necessarily need to reject the general concept [GO] as a representation of the event in (1). One can envisage defining a subtype of GO, a 'GO_ROT', which would represent any rotational movement.

The reason why we do want to reject [GO] as a model of inchoative events is that it does not represent punctual events and does not provide unambiguous inferences about the resulting state of Theme. This situation is shown, using the original semantics lacking [INCH], in (6).

(6)   *x travelled from a, through b, to c* (incomplete representation)

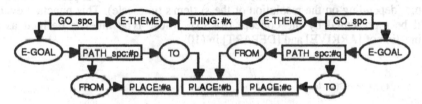

The sentence in (6) implies, pragmatically speaking, that *x* passed through *b* without stopping, and eventually stopped at *c* and stayed there. The graph in (6) represents only the *durative* events of *x* going from *a* to *b* , and from *b* to *c* . It does not represent the *punctual* events of passing and stopping, and the durative state of staying. This means that we cannot include temporal or other information about these events. Then we cannot make any inferences about the resulting location of *x*. An analysis of the adapted interpretation of [INCH], shown in (5), suggests that [INCH] can be further decomposed. It can be split into the punctual event [ARRIVE] which represents the arrival of Theme at a [PLACE], and into the creation of a new [LOC_spc], which completes Theme's travel. Thus, the label INCH can be defined from ARRIVE, as shown in (7).

(7)   **type INCH(x) is**

[ARRIVE: *x]->-(RESULT)->-[STATE].

Using [INCH], more of the intended meaning of the sentence in (6) may be represented:

(8)   *x travelled from a to b, where it stayed* (incomplete representation)

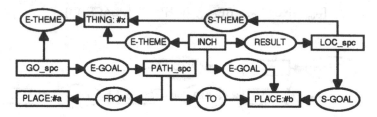

We may extend this analysis and try to represent the punctual events of Theme *departing* a [PLACE] and *terminating* its original state of [LOC_spc]. Let us represent this event by the concept [DEPART], as shown in (9).

(9)

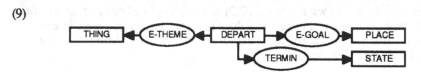

In (6) *x* passes through [PLACE:#b] without staying there for any appreciable length of time (depending on the resolution of the system's timescale). This punctual event could be represented by the concept [PASS], whose type label is defined as a combination of [ARRIVE] and [DEPART] in (10).

(10)

type PASS(x) is

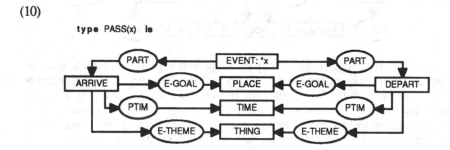

The punctual event [PASS] will be appropriately linked to the punctual version of the state [LOC_spc], as shown in the following tentative representation.

(11)

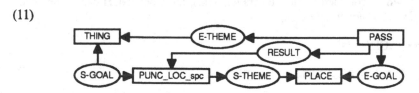

Now we can construct a graph, in (12), which will explicitly include much of the pragmatic knowledge underlying (6), enabling queries such as *When did x leave/arrive?*, *How long did it travel?*, *What places did x pass and when?*, *Is x at a?* or *Where is x now?*, etc.

(12)   *x travelled from a, through b, to c, where it stayed*

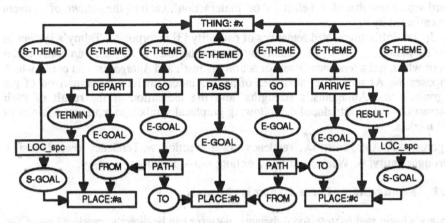

Although one may aesthetically appreciate its symmetry, the graph in (12) is by no means the definitive model of elementary movement in space. Outstanding tasks include determining and formalizing the relationship between the essentially durative event [GO_spc] and the punctual [PASS], [DEPART] and [ARRIVE], and between the punctual and durative versions of the state [LOC_spc]. There is also the present uncertainty whether the (S-GOAL) of the punctual events [ARRIVE] and [DEPART] is a [PLACE] or rather a semi-determined [PATH]. However, even in this form, the expanded graph will enable the system to make inferences about Theme.

There are a number of other loose ends concerning which part of this knowledge will be 'factored out' into the conceptual and relational catalogues. In the same way we do not include type lattice, definition and catalogue information in a single graph, all the parts of the definitive form of the graph in (12) will not be always displayed at the user interface. Those parts which represent general and pragmatic knowledge will be held in separate structures of the knowledge base and brought together only if a query or some other operation called for it. Such implicit knowledge will, for example, include the rule, not necessarily in the *if-then* form, that one object cannot be in several places, and that by departing from a place, Theme is no longer there. Eventually, the general structure and rules of the semantics of space will have to be adapted for the representation of the other semantic fields, e.g. possession, ascription of attributes and characteristics, causality, existence, etc.

3   Representing Causal Events

Jackendoff's original theory [1, 2] was criticized for its incomplete representation of a number of aspects of causality. One was volitionality, i.e. whether an event was caused by a volitional agent on purpose or inadvertently, or whether it was caused by an external instigator. Another was the degree of the success of a causal event. The third aspect was that of 'benefaction' or 'malefaction', i.e. how the *patient* of an event was affected by it.

Jackendoff's augmented semantics of causality [3] incorporates Talmy's system of *force-dynamics* (FD) [10]. Talmy introduces the roles of *Agonist* - an entity under focus which has a tendency either for action or 'rest', and *Antagonist* - an entity which opposes the Agonist. The attributes of an FD model include a comparison of the Agonist's and Antagonist's strengths, and the indication of the result of their interaction. Talmy introduced the following graphical symbols to represent a family of FD models:

Agonist: O, Antagonist: Ɫ, Tendency towards action: >, Tendency towards rest: •, Stronger entity: +, Weaker entity: -, Action: -->--, Inaction: --•--.

3.1   Steady-State Force-Dynamic Models

Talmy's basic *steady-state* force-dynamic patterns can be directly translated into CGs. This translation introduces a potential wealth of detail and logical dependencies amongst a pattern's constituent concepts and relations. A general, 'semi-formal' translation of the individual patterns of the steady-state force-dynamic patterns into CGs is shown below.

a)   A stronger Antagonist affecting a weaker Agonist with a tendency towards rest, resulting in action, e.g. *The ball kept rolling because of the wind blowing it.*

(13)

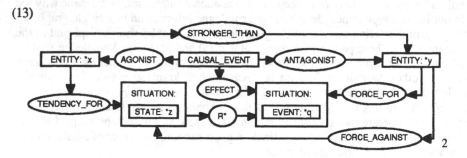

b)   A weaker Antagonist affecting a stronger Agonist with a tendency towards rest, resulting in inaction, e.g.*The shed kept standing despite the gale wind blowing against it.*

---

2 The relation (R*) is a 'place holder' for the logical interdependency of the Agonist's [STATE] and resulting [EVENT].

(14)

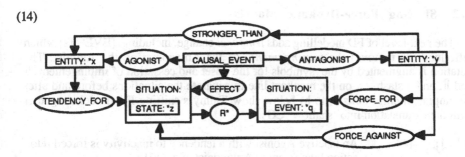

c) A weaker Antagonist affecting a stronger Agonist with a tendency to action, resulting in action, e.g. *The ball kept rolling despite the stiff grass.*

(15)

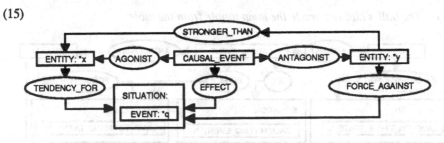

d) A stronger Antagonist affecting a weaker Agonist with a tendency to action, resulting in inaction, e.g. *The log kept lying on the incline because of the ridge there.*

(16)

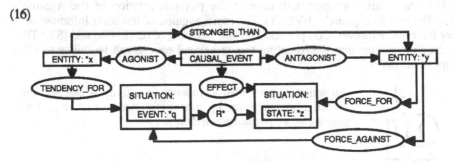

Translating Talmy's FD patterns into CGs introduces an automatic check on the consistency of the statement, e.g. the dependence of the resultant on the relative strengths of the Agonist and Antagonist and on their respective tendencies. The superficially described [SITUATIONs] can be expanded to describe their constituent [STATEs] and [EVENTs] in full detail. This will provide additional logical, semantic and pragmatic interactions not representable in the original model.

3.2 Shifting Force-Dynamic Models

The next level of FD modelling adds *temporal* change, including [EVENTs] which introduce (or remove) the *impingement* of the Antagonist on the Agonist. The notation is augmented by the symbols for the onset and cessation of impingement, ⇓ and ⇑, and a slash, / , on the resultant line to separate the situations before and after the impingement. Points e) to h) below show Talmy's shifting FD models, together with their translation into 'shallow' CGs.

e)   ⇓   An inactive Agonist with a tendency to inactivity is forced into action by a stronger Antagonist, e.g. (17)

(17)   *The ball's hitting it made the lamp topple from the table.*

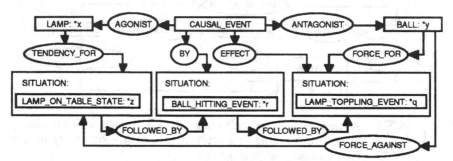

In (17) the relative strength will refer to the physical relation of the Agonist's [STATE] and Antagonist's [EVENT]. The representation of temporal information is very informal. However, CGs provide a precise mapping of [EVENTs] and [STATEs] on the time continuum. Representing this is beyond easy graph legibility and the scope of the paper.

f)   ⇓   An active Agonist with a tendency to activity is stopped and forced into inactivity by a stronger Antagonist, e.g. *The pouring of water extinguished the fire.*

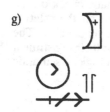

g)   The removal of the impingement of a stronger Antagonist enables an inactive Agonist with a tendency to activity to become active, e.g.*The plug's coming loose let the water flow from the tank.*

h) 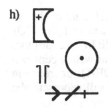 The removal of the impingement of a stronger Antagonist enables an active Agonist with a tendency to inactivity to become inactive, e.g. *The stirring rod's breaking let the particles settle.*

The general format of the CGs translations of the patterns in f) to h) will be the same as that of the graph in 17). The intrinsic differences will appear in the detailed representation of the constituent [STATEs] and [EVENTs].

It is evident that Talmy's semantic analysis of force-dynamics can be directly translated into the formal, computational representation of CGs. This requires a level of detail and rigour 'orders of magnitude' higher than that of Talmy's notation, as a CGs model explicitly shows the internal logical, semantic and pragmatic consistency of each pattern.

Apart from analysing basic causal patterns, Talmy shows convincingly that the force dynamics extends to the modelling of internal psychological motivations and social interactions and pressures. This includes situations of internal struggle, such as in the sentence *I refrained from being impolite* , or situations such as in *He's under a lot of pressure to keep silent*, or *Getting job security relieved the pressure on her to perform.*

3.3 Combining Force-Dynamic Semantics with the Thematic Relations Hypothesis

As briefly shown by examples in (13) to (17), although Talmy's analysis provides an excellent methodology for the general format of the CG representation, it does not address the internal detail of the individual [STATEs] and [EVENTs]. Jackendoff's enhanced treatment of causality provides this detail by, amongst other things, dividing causation into two tiers: a) the *thematic tier* representing Theme's motion in a semantic field, and b) the *action tier* representing the *Doer/Undergoer* or *Actor/Patient* relations. An example of Jackendoff's notation, involving both tiers, is shown in (18).

(18)   The car hit the tree.

$$\begin{bmatrix} & \text{INCH [BE}_c \text{ ([CAR], [AT}_c \text{ [TREE]])]} \\ \text{Event} & \text{AFF ([CAR], [TREE])} \end{bmatrix}$$

To represent the action tier, Jackendoff uses a complex concept/binary relation [AFF(..,..)]. Its meaning can be further specialized by notational variation and parameters. The [AFF(..,..)] relation can appear in the following forms:

a)   [AFF([X],)]   =   Actor only,
b)   [AFF(,[Y])]   =   Patient only,
c)   [AFF([ ], [Y])]   =   implicit Actor,
d)   [AFF([X], [ ])]   =   implicit Patient,
e)   [AFF([X], [Y])]   =   both Actor and Patient explicit.

In CGs notation, the equivalent of the relation [AFF(..,..)] cannot be so ambiguous about the number of its compulsory links. Let us connect the CGs version, [AFF], to the relevant event in the thematic tier and specialize it into two forms: a) one which will accept *only* one argument, and b) one that will accept *two* arguments. In a) it will be either the relation (ACTOR/DOER) or relation (PATIENT/UNDERGOER), in b) the use of the generic referent will represent the implicit Actor or Patient.

Jackendoff further specializes [AFF(..,..)] by information abut the *volitionality* of the Actor/Doer. Using the subscript *vol*, the following example shows three possible interpretations for the sentence *Bill rolled down the hill* .

(19)   Bill rolled down the hill.

$$\begin{bmatrix} \text{GO([BILL], [DOWN ([HILL])])} \\ \left\{ \begin{array}{l} \text{a. AFF}_{+vol}(\text{[BILL], })\\ \text{b. AFF}_{-vol}(\text{[BILL], })\\ \text{c. AFF (,[BILL])} \end{array} \right\} \\ {}_{\text{Event}} \end{bmatrix}$$

The meaning of the options in the curly brackets is: a. Bill = willful doer/volitional Actor, 'he wanted to do it', b. Bill=unwilling doer, 'he did not want to do it but had to', and c. Bill = undergoer, 'he was rolled by somebody or something'.

There are several ways how to deal with the suffix *vol* in CGs. On the one hand, [AFF] could be linked to the attribute [VOL], which would take the values 'positive' and 'negative'. In fact, we could define a much richer, possibly continuous scale of volitionality on which we could represent adjectives such as 'keenly', 'rather reluctantly', etc. On the other hand, we could put all this information into specialized relations, such as (WILLFUL_DOER), (NONWILLFUL_DOER), etc. Then we could eliminate [AFF] altogether and attach the specialized relations directly to the appropriate verb at the thematic tier. Alternatively, the degree of volition could be inferred from the attached representation of the motivational patterns of the Doer. Deciding which is best will require more investigation. Obviously, the value of [AFF] or of any relations that we may substitute for it will have to agree with the type of the concept that is attached to the (DOER) and (UNDERGOER) relations. Example (20) shows a CGs version of the graph in (18).

(20)   The car hit the tree.

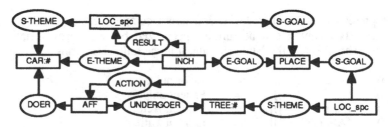

The graph in (21.a) shows a CGs version of the example in (19).

(21.a) Bill rolled down the hill.

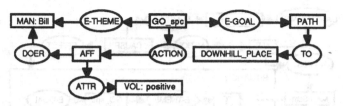

The following graph shows a version of (21.a), with [AFF] and its relations contracted into the relation (WILLFUL_DOER).

(21.b)

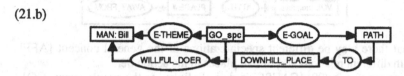

In (21.b) the concept [MAN: Bill] is linked to [GO_spc] by two relations, each of them representing a view of a different tier. The thematic tier will answer queries such as *Where did Bill go?*, *Where is Bill now?*, etc. The action tier will answer queries such as *Who/what has caused Bill's going down the hill?* However, the contracted representation in (21.b) makes it rather difficult to specialize the incorporated concept [AFF] with further information.

When representing situations in which an *extrinsic instigator* causally affects a volitional agent, we have to include one more action tier to express their relationship. This tier is represented by [CS], a high-level concept of causality, and by its arguments *agonist*, who is motivated to maintain or achieve a certain situation, and *antagonist*, who is motivated to prevent it. The example in (22), using Jackendoff's original notation, shows the first [AFF(..,...)] belonging to the [GO] of the thematic tier, the second [AFF(..,...)] belonging to the [CS] or [CAUSE] at the action tier.

(22) *Harry forced Sam to go away.*

$$\left[\begin{array}{l} \text{CAUSE([HARRY], } \left[\begin{array}{l} \text{GO([SAM], [AWAY])} \\ \text{AFF([SAM],} \quad) \end{array} \right]) \\ \text{AFF}^-(\text{[HARRY], [SAM])} \end{array} \right]$$

In (22) the concept [AFF(..,...)] appears to behave in the same way at both levels of the action tier. However, the CGs notation can, and by definition has to, express much more detail. This makes it necessary to introduce a number of explicit relations which were only implicit in Jackendoff's notation. Then occurrences of [AFF] at different levels will have different relations linked to them, as seen in (23). However, this seems to indicate that the two different occurrences of [AFF] do in fact represent different concepts.

(23)   Harry forced Sam to go away.

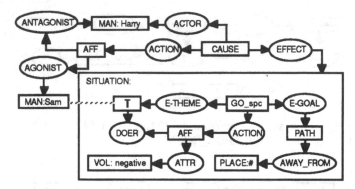

It follows that there may be different specializations of the general concept [AFF] appearing with different relations at different places in the action tier.

One may notice that in (22) [CAUSE] is directly linked to the simple event [GO], whereas in (23) the corresponding concept [CS] is connected, by the relation (EFFECT), to a [SITUATION], not to a single [EVENT]. This is because a single Actor may cause multiple or negated events, which may need to be logically clustered. On the other hand, by separating the individual [EVENTs] of a complex [SITUATION] we may want to show the direct causal links. The ideal solution to this problem will show the mapping between the 'factored out' and encapsulated cause of a [SITUATION] and the 'distributed' causes of its individual components.

The following example shows a successful prevention, i.e. a causation of an event *not* happening:

(24)   Harry prevented Sam from going away.

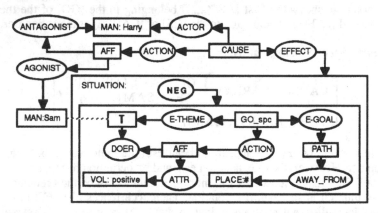

The concept [CS] can be further specialized with information about the antagonist's success in influencing the agonist. This is represented as CS$^+$='successful', CS$^-$='unsuccessful' and CS^u='success undetermined'. The examples in (23) and (24) show successful causations. As with the degrees of volitionality specializing [AFF], we may specialize [CS] with the attribute [SUCCESS] with values from a continuum, say, of 0%-100%.

When translating Jackendoff's formulae into CGs, because of their greater explicit detail, some redundancy becomes apparent. As seen in examples (18) to (22), the success of a causal event is given a) by the value of the attribute [SUCCESS] and b) is evident from the overall structure of the graph, namely, from the relations (EFFECT), (TENDENCY_FOR) and (FORCE_FOR). This means that the success of a causation may be inferred by comparing directly the outcome of the causation with a description of the intention or motivation of the Actor or Doer. Then the value [SUCCESS: undetermined] of a causation may be indicated directly by the omission of the graph describing the outcome. Thus, there is no need to attach the [CS] parameter [SUCCESS]. This, and other similar observations, suggests an important problem for KR by CGs, namely, the necessity to differentiate between the object level and 'meta' levels of a CGs model. Thus, while at the object level information about the success of a causal event is implicit from the structure of the graph, at the meta-level it is explicitly stated, e.g. [SUCCESS: %100]. One assumption of this investigation, which has yet to be implemented and thoroughly tested, is that a deep knowledge representation by CGs should essentially use the object-level; the meta-level structures would be automatically inferred by the system. The example in (25) shows an unsuccessful causation.

(25)   Harry unsuccessfully pressured Sam to go away.

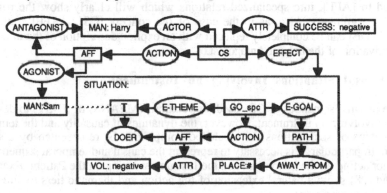

By negating the whole [SITUATION] and by varying the value of the concept [SUCCESS], we could represent the sentence *Harry (successfully/ unsuccessfully) impeded/hindered Sam's going away.* By giving a negative value to the attribute concept [SUCCESS], we can express sentences indicating failure, e.g. *Harry failed to go away.* By putting Harry into the double role of DOER and AGONIST, we may represent the sentence *Harry tried to go away.* The negative value of the attribute [SUCCESS] would mean *Harry tried to go away but did not succeed.*

3.4 Letting and Helping

The two protagonists of Talmy's action tier do not have to oppose each other. In the original notation [AFF(.....)] is modified with superscripts to indicate the 'attitude' of the antagonist towards the agonist's goal. Thus, verbs of *helping* are represented by

AFF⁺, verbs of *letting* by AFF⁰. We may tentatively translate this into CGs by specializing the concept [AFF] attached to the concept [CS] with the attribute [ATTITUDE]. (21) is an example of non-opposition.

(26)   Harry let Sam leave/allowed Sam to leave (and so Sam left).

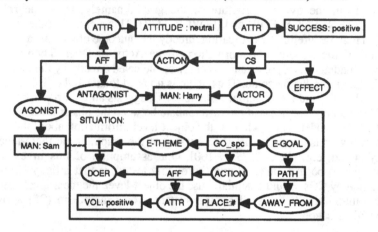

However, in the long run it may prove preferable to put all the information, now attached to [AFF], into specialized relations which will clearly show the roles of [AFF]'s arguments. Also, as in the case of the concept [CS], [AFF]'s attribute [ATTITUDE] can be completely replaced with an explicit description of the intentions and motivation of the helpful Antagonist.

3.5   Causal Situations Involving an Instrument

The original semantics of CGs used the relations (MEANS) or (INST) to describe actions involving an instrument. However, the dynamics of causality and the temporal relationships of such actions are much too complex to be represented by a single relation. In particular, it is necessary to represent the causal and temporal sequences of the Actor acting on the Instrument and the Instrument acting on the Patient. Examples (27) and (28) show a natural extension of the action and thematic tiers to situations involving an instrument.

(27.a) Sue hit Fred with a stick.

$$\begin{bmatrix} \text{CS}^+(\ [\text{SUE}], [\text{INCH} [\text{BE}_c([\text{STICK}], [\text{AT}_c[\text{FRED}]])]]) \\ \text{AFF}^-(\ [\text{SUE}], [\text{FRED}]) \\ [\text{BY} \begin{bmatrix} \text{CS}^+(\ [\text{SUE}], [\text{AFF}^-(\ [\text{STICK}], [\text{FRED}])]) \\ \text{AFF}^-(\ [\text{SUE}], [\text{STICK}]) \end{bmatrix}] \end{bmatrix}$$

(27.b)

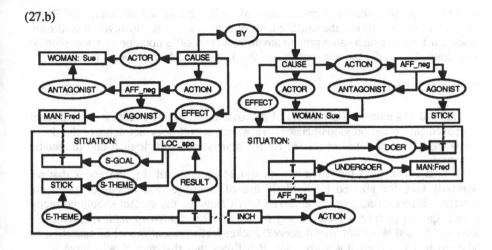

(28)   Pete hit the ball into the field with a stick.

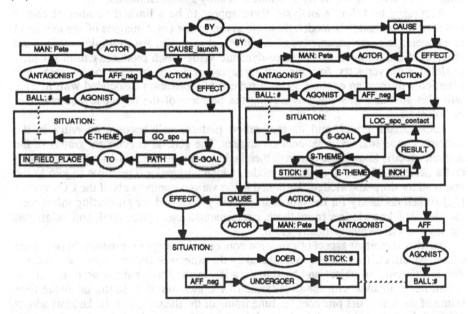

The graphs in (27) and (28) are tentative, more or less direct translations of Jackendoff's examples. Example (27) shows the linking of the two actions of Sue affecting the stick and the stick affecting Fred and so Sue affecting Fred. (28) shows that Pete affects the ball by hitting it with the stick, which is the doer of the dynamic contact with the ball.

A CGs representation provides, and requires, a level of detail qualitatively much higher than that of Jackendoff's notation, let alone Talmy's original patterns. Thus, the deep-knowledge structure will have to be carefully analysed and the individual patterns and models rationally reconstructed in terms of CGs. Only then will it be possible to include the temporal relations of individual situations, their [EVENTs] and [STATEs], and model the causal chains of complex events. However, it is already evident that CGs which can represent anything Jackendoff's notation can, and provide an unlimited scope for the incorporation of the deeper models mentioned above.

4   Discussion

Jackendoff's enhanced semantics of inchoative and causal events is a significant improvement on the previous versions. However, it still does not achieve the deep-knowledge level of detail necessary for the generation of all logical and pragmatic inferences.

A comprehensive CGs-based system should generate all the inferences that we normally take for granted in the daily use of language. For example, given the sentence *Harry unsuccessfully hindered Sam's leaving,* the system should correctly answer queries such as *What did Harry want?* or What *did Harry achieve?, What did Sam want?* and *What did Sam achieve?* Jackendoff's present model of causality, as adapted to CGs, does not allow that. It follows that this model will have to be elaborated. This representation will have to include an explicit representation of motivation, which is implicitly available in Talmy's original model.

According to Talmy's analysis, there appear to be a limited number of causal patterns. In Jackendoff's model, these are given by the combinations of the low-level components, e.g. the parameters of the different relations. An example of this is the analysis of the combinations of force-dynamic verbs which produces causal situation models for the verbs *try, fail, succeed, force* and *resist.* In the CGs model, each such situation will have an explicit, general pattern of motivation associated with it. This will also go some way towards solving the problem of the object level/meta-level dichotomy.

Each causal model, and its motivation pattern, will become an entry in the system's conceptual and relational catalogues. The general conceptual graphs of the domain-specific knowledge base will then include only the domain-specific knowledge and a skeleton of the full semantic model, which will make it possible to join to the graph all the deep-knowledge distributed in the various components of the CGs system [12]. When necessary for the processing of the graphs and for generating inferences, the detailed knowledge from these components, e.g. conceptual and relational catalogues, will be automatically added.

Another important aspect of causation concerns the temporal relations between the cause and effect. This may be exemplified by the difference between causing and event through moving an object and by throwing it. In the former the time frame of the action tier coincides with the time frame of the thematic tier. In the latter the time frame of the action tier precedes the time frame of the thematic tier. Jackendoff adopts Michotte's terms of 'entraining' for dragging and 'launching' for throwing. This difference can be generalised to other types of events. In Jackendoff's notation, the temporal information can be only indirectly inferred from the attributes of the [CS] concept. In the CGs notation, temporal information can be attached to the individual concepts. A dedicated structure and a set of procedures for the representation and processing of temporal knowledge, based on the TRH, will provide correct temporal inferences.

The semantics of force dynamics is uniquely compatible with that of the TRH. The TRH has been convincingly shown by Jacknedoff to apply to a comprehensive range of semantic categories or cognitive domains. The application of the TRH to CGs lends itself directly to the qualitative modelling of specialized dynamic systems, e.g. physical, biological and industrial processes, robotics, design, etc.

Similarly, FD semantics extends to semantic categories other than that of simple causality. It appears to embody the underlying principles necessary for the representation of psychosocial domains, including, as Talmy shows, concepts of emotions, drives, psychological repression and resistance. Thus FD will make possible the qualitative modelling of psychological and sociological theories, as well as the simulation of general social processes.

Neither Jackendoff's nor Talmy's approach is fully formalized, let alone computational. The role of these theories in their application to CGs is that of powerful informal methodologies which are being substantially expanded by their formalization and simultaneously logically and pragmatically validated.

References:

1.   R. Jackendoff: Semantics and Cognition, The MIT Press 1983.
2.   R. Jackendoff: Consciousness and the Computational Mind, The MIT Press 1987.
3.   R. Jackendoff: Semantic Structures, The MIT Press 1990.
4.   P. Kocura: Deep Knowledge Semantics for Conceptual Graphs", in Proc. 5th Annual Workshop on Conceptual Structures, Stockholm, pp. 297-306 (1990)
5.   P.Kocura: Conceptual and Semantic Structures, in Knowledge-Based Systems, Butterworth Scientific Ltd, (to be published September 1992)
6.   M.L. McHale, Sung H.Myaeng: The Integration of Conceptual Graphs and Government Binding Theory, in Proc. 6th Annual Workshop on Conceptual Graphs, SUNY Binghamton, July, pp. 223-234 (1991).
7.   J.F. Sowa: Conceptual Structures: Information Processing in Mind and Machine, Adison-Wesley 1984.
9.   J.S.Gruber: Studies in Lexical Relations. Doctoral dissertation MIT, Cambridge 1965; Indiana University Linguistics Club, Bloomington, Ind. Reprinted as part of Lexical Structures in Syntax and Semantics, North-Holland, Amsterdam 1976.
10. L. Talmy: Force Dynamics in Language and Thought, in Cognitive Science 12, 49-100 (1988)
11. W.A.Cook: Case Grammar Theory, Georgetown University Press, Washington, D.C., 1989.
12. P.Kocura et al.: Aspects of Conceptual Graphs Processor Design, in Proc. 6th Annual Workshop on Conceptual Structures, SUNY Binghamton, July, pp.317-329 (1991)

Does Every Difference Make Difference?

Jonathan C. Oh and Steven Graham

Computer Science
University of Missouri at Kansas City

Abstract It is argued that natural language syntactic differences must be used a useful cues in determining the contents of input sentences to be inserted into a knowledge base. To demonstrate how some apparently less important syntactic details might turn out to be useful, relative clauses have been analyzed in the context of conceptual graph theory.

1 Introduction

Pat Hayes raises in a very stimulating discussion with John Sowa a question of whether every surface syntactic difference should be mirrored in the semantic formalism. It would seem uncontroversial that not EVERY difference in word or structural choice need and even should be reflected on in a conceptual representation for EVERY application. On the other hand, it seems quite reasonable to expect to find important guides to significant semantic cues in the syntactic structure of that expression in question. Therefore, one can take the question to be one of policy. But I feel it IS important to have facilities available to express any semantic finitudes when an application calls for them. How much syntax is to be considered sugar is clearly application-dependent.

In this paper we will illustrate and support the present thesis, namely that one should not be dogmatic and insist that certain semantic features should or should not be covered in a semantic theory without any regard to an actual application. We will consider some of the utilities of Lambda expressions in the conceptual graph theory.

2 Synonymy and Conceptual Graphs

The conceptual graph below can be read in any one of the following and possibly some more:

```
[MONKEY] <- (AGNT) <- [EAT] -> (OBJ) -> [WALNUT]
                        |                  |
                      (INST)             (PART)
                        |                  |
                        V                  V
                     [SPOON] -> (MATR) -> [SHELL]
```

1. *A monkey eats a walnut with a spoon made out of the walnut's shell.*
2. *A monkey eating a walnut with a spoon made out of the walnut's shell.*
3. *A spoon that a monkey eats a walnut with which is made out of the walnut's shell.*
4. *A walnut which a monkey is eating with a spoon made out of the walnut's shell.*
5. *A walnut's shell out of which a spoon is made which a monkey is eating a walnut with*
6. *A spoon that a monkey eats a walnut with is made out of the walnut's shell.*
7. *A monkey easts a walnut out of whose shell a spoon he eats it with is made*

Should all these readings be taken to be synonymous with each other in all applications? In conceptual analysis, inferencing, and natural language generation, would there be a situation where these syntactic differences need to be marked on the conceptual level? Should the differences among these English translations be merely pragmatic in nature?

One might insist that the mapping between natural language and conceptual graph is not a matter of semantics. Presumably, WHAT needs to be represented on the conceptual level for an application in question is semantics, but HOW that representation is arrived at, for example, from natural language is quite a different matter. But in all fairness, the mapping involves more than just the methodology of translation. In view of the fact that any significant semantic distinction would HAVE TO be expressed in natural language, it would be only a prudent thing to give attention to linguistic structural differences in deciding what to include in the semantics.

3 Lambda Construction for Relative Clauses

Consider the following English sentences and how their meaning might be represented:

8. *The book that the lady wrote won the Pulitzer price.*
9. *The lady did not write the book that she wrote.*
10. *The lady did write the book that she wrote.*
11. *The man mentioned that the book that the lady wrote won the Pulitzer price.*
12. *The man asked whether the lady did (actually) write the book that she wrote.*
13. *The man claimed that the lady did not write the book that she wrote.*

8'. [WRITE] -
      (AGNT) -> [LADY: #]
      (OBJ) -> [BOOK: #] <- (INST) <- [WIN] -
                            -> (OBJ) -> [PRIZE: #Pulitzer]

10'. [WRITE] -
      (AGNT) -> [LADY: #12345]
      (OBJ) -> [BOOK: #12346] <- (OBJ) <- [WRITE] -
                            -> (AGNT) -> [LADY:#12345]

Sentence 9 is a contradiction and Sentence 10 is repetitive and redundant. Conceptual graphs for these two sentences must and do show these abnormalities. But Sentences 12 and 13 which contain them are not necessarily contradictory nor redundant. We will consider what this implies for conceptual analysis and resultant conceptual representation. I propose that Sentence 13 be translated as below The Lambda expression here is a type definition which does not use a label. This corresponds to a lambda expression in Lisp and an unnamed block structure in an imperative language. It is not part of what is claimed but just a means to specify a type, hence no contradiction.:

```
[CLAIM] -
        (AGNT) -> [MAN: #]
        (OBJ) -> [PROPOSITION:
            (NEG) -> [[WRITE] -
                    (AGNT) <- [LADY:#12345]
                    (OBJ) -> [[WRITE] -
                        (OBJ)<-[BOOK:Lambda]
                        (AGNT)->[LADY: #12345]: #] ] ]
```

It does not seem too far-fetched to consider the lambda construction a semantic counterpart of linguistic transformation known as Relativization. Equipped with this rule, lambda expressions for relative clauses, we will go back to sentences 1-5. Sentence 1 may be represented as below:

1. *A monkey eats a walnut with a spoon made out of the walnut's shell.*

```
1'. [EAT] -
        (AGNT) -> [MONKEY]
        (INST) -> [SPOON] -> (MATR) -> [SHELL] <- (PART) <- [WALNUT:*x]
        (OBJ) -> [WALNUT: *x]
```

The conceptual graphs for sentences 2-5 will be the same as the above except that lambda is attached in the referent field for different concepts; on [monkey] for sentence 2, on [spoon] for sentence 3, on [walnut] for sentence 4, and finally on [shell] for sentence 5. The conceptual graph for sentence 1 is a proposition, that for sentence 2 a description of a particular subtype of the type monkey, etc. Remember that the conceptual structures are a semantic language which itself needs to be interpreted.

Level conflation may be allowed under certain conditions, thus explaining synonymy relation among certain expressions in spite of their structural differences. Sentences 1, 6, and 7 are all synonymous which fact is represented in the conceptual graph theory by the conceptual graph of any of these being merely a lambda converted form of another.

6. *A spoon that a monkey eats a walnut with is made out of the walnut's shell.*

```
[[EAT] -
        (agnt) -> [MONKEY]
        (inst) -> [SPOON:lambda]
        (obj) -> [WALNUT:*x]] -> (MATR) -> [SHELL] <- (PART) <-
                                        [WALNUT:*x]
```

7. *A monkey eats a walnut out of whose shell a spoon he eats it with*
    is made

```
[EAT] -
        (AGNT) -> [MONKEY:*m]
        (obj) ->
                [[EAT] -
                        (AGNT) -> [*m]
                        (obj) -> [WALNUT: lambda *w]
                        (INST) -> [SPOON] -> (MATR) ->
                        [SHELL] <- (PART) <- [WALNUT:*w]: *x]
```

Similar ideas are discussed in Sowa 1991. Sowa discusses the representation of a quantification; the quantifier in the referent field, the restrictor as a lambda expression in the type field, and the scope being the entire context in which the quantifier occurs. The restrictor is functionally similar to relative clauses. Sentence 14 gets translated into a conceptual graph and by phi further translated into a predicate logic expression as below

14. *Every donkey-farmer (or farmer who owns a donkey) beats it.*

```
[BEAT] -
    (agnt) -> [[FARMER: lambda] -> (stat) -> [OWN] ->(obj) -> [DONKEY:x]: ALL]
    (obj) -> [T: #x]
```

```
IF [[FARMER: lambda] -> (stat) -> [OWN] ->(obj) -> [DONKEY:x]: *y]
THEN [y*] <- (agnt) <- [BEAT] -> (obj) -> [T: #x]
```

The familiar lambda conversion will reduce the above into

```
IF [FARMER: *y] -> (stat) -> [OWN] ->(obj) -> [DONKEY:x]
THEN [y*] <- (agnt) <- [BEAT] -> (obj) -> [T: #x]
```

4 Implications

This suggests a two-step type expansion. Instead of doing a direct maximal join between the differentia u of a type definition 'TYPE t = lambda a u' and a conceptual graph containing the type name t, one can first simply replace t by the corresponding lambda expression and only when certain conditions are satisfied does one apply type conflation. One such condition would be that the proposition that contains the lambda expression is not in the scope of an intensional predicate. The propositional contents of a lambda expression used to characterize a type is in one context and an intensional predicate might introduce a new context distinct from the first, in which case the conflation is not always safe. Type contraction is applied when a lambda expression is used in the type field for some concept node and in addition there is a type definition which has that same lambda expression as its right hand side.

Linguists have long recognized that part of a sentence is to express new information and' the rest to provide a syntactic frame for that expression. That part of the sentence that is used to express new information has been variously called 'higher predicate' or 'assertive force'. Various tests have been used such as negation test, question test, and test based on other illocutionary acts. In a conceptual graph, those predicates which are not the highest and thus are not negated, questioned, etc. may be hidden away in lambda constructions.

In evaluating a conceptual graph containing a lambda expression in the type field of a node, the PROJECT move will have to be charged with the task of checking non-emptiness of the type denoted by the lambda expression. It is also conceivable that a conceptual graph theory represents presupposition and assertion separately, so that 'John stopped beating his wife' be represented as [PRESUPPOSITION: <John used to beat his wife>] and [PROPOSITION: <John does not beat his wife>] where <x> is the conceptual graph representation of some linguistic expression x. Within such a theory, the checking of validity of the lambda expression might be done in a projection of the lambda body into the presupposition set instead of by a bolstered PROJECT move. A closed world in such a theory might be defined as a triple of <T,P,I> where P is the pressuposition set used in the evaluation. A similar idea has been explored in Guha 1991.

While it is perfectly understandable that the first batch of sentences -- Sentences 1 - 5-- be represented by a single conceptual graph, thus the linguistic difference not making difference on the conceptual level, the similar syntactic distinction in the second batch of sentences --Sentences 6-11--is shown to represent some significant semantic element. We do not necessarily insisit that all relative constructions everywhere should be realized semantically as a lambda expression. But it is not a bad policy to use such syntactic cues in alerting a conceptual analyzer for probable semantic features.

5   Additional Observations

Wilensky's analysis [Wilensky 91] of situation reveals something quite noteworthy in the context of this paper. In analyzing a quantified sentence such as 'Everyone looked at Mary', Wilensky finds it necessary to extend the usual ontology by adding in this case an entity he calls *Complex Event*. He characterizes the meaning of this sentence as below:

$$
\begin{aligned}
&\text{Exist c1 (AIO(c1, Complex-event) \&} \\
&\qquad \text{All x (AIO(x, Person) -> Exists l (AIO(l, Looking-at) \&} \\
&\qquad\qquad \text{Actor(x,l) \&} \\
&\qquad\qquad \text{Patient(mary,l)) \&} \\
&\qquad\qquad \text{Sub-event(l,cl))))}
\end{aligned}
$$

The motivation for proposing this somewhat odd ontology is so that not only the total event of everyone looking at Mary but component looking events may also be expressed. If we continue the sentence with 'and someone looked at her derisively', the latter event clearly is not unrelated to the first. The above predicate logic expression indicates both the total and component events. But compare this with the conceptual graph assigned to this sentence:

$$
\begin{aligned}
&\text{-[[PERSON:*x] -[[PERSON:*x] <- (agnt) <- [LOOK] -} \\
&\qquad\qquad\qquad\qquad \text{(obj) -> [PERSON:mary]]]}
\end{aligned}
$$

The Phi translation function will provide us with what we need, ie.

- (E x)(E y) [Person (x) & -(Look(y) & Person(mary) &
agent(x,y) & obj(mary,y))]

which is equivalent to

(All x)(All y) [Person(x) & Look(y) & Person(mary) ->
agent(x,y) & obj(mary,y)]

in this all the component events as well as the total event is given. The conceptual
graph for the sentence is considerably more natural and less contrived. If we apply
Sowa's quantification analysis discussed above, this point is probably clearer. Both
quantifier and restrictor are attached to the nominal expression, not on the verb of the
sentence, thus keeping us from Wilensky's style analysis.

6 Conclusions

The key issue here is what or how much needs to be represented for a sentence within
a knowledge base. It is pointless to argue whether all syntactic features must be
exploited in obtaining relevant semantics. How much or what needs to be
accommodated depends on the application in question. But one thing that is important
is that the knowledge representation can provide all that is needed when called for.
Human language is an intricate and highly versatile system that has been used for all
sorts of human reasoning. As we search for a cognitively satisfactory semantic
language, natural language in its rich syntax has much to offer.

References

1. Brachman, R.J., H. J. Levesque, and R. Reiter. (eds) 1989.   *Proceedings of
    the first international conference on Principles of Knowledge
    Representation and Reasoning.*   Morgan Kaufmann.

2. Guha, R.V. 1991. Contexts: a formalization and some applications. MCC
    Technical Report Number ACT-CYC-423-91 Microelectronics and
    Computer Technology Corporation

3. Sowa, John 1984. *Conceptual Structures.* Addison-Wesley.

4. Sowa, John. (ed) 1991 *Principles of Semantic Networks.* Morgan Kaufmann
    Wilensky, Robert. 1991. "Sentences, Situations, and Propositions", in Sowa
    (ed) 1991 *Principles of Semantic Networks..*

An Exploration Into Semantic Distance

Harry S. Delugach

Department of Computer Science
Computer Science Building – Rm 109
University of Alabama in Huntsville
Huntsville, AL 35899 U. S. A.
Electronic mail: delugach@cs.uah.edu

1. Introduction

The issue of semantic distance has been discussed in several previous papers. Various strategies have been proposed for obtaining some quantitative measure of the similarity between things. In conceptual graph terms, the various approaches may often be classified according to operations involving the type hierarchy, canonical graph definitions and the context in which particular concepts are found. This chapter will discuss the following topics:

- The meaning of *semantic distance*.
- Issues regarding semantic distance measures
- Taxonomy of semantic distance measurement schemes
- How to conduct an empirical study of intuitive semantic distance

Readers are advised that this work is intentionally speculative. It is not an exhaustive attempt to cover all of the subject; such would be beyond the scope of a brief paper. The purpose of this paper is to suggest an approach for incorporating the notion of semantic distance into the conceptual graph theory, while preserving its meaning as already studied by psychologists, philosophers and cognitive scientists.

2. The Meaning of Semantic Distance

The issue of whether two things are similar may be re–cast as the question: Can we identify one or more categories in which the two things belong? As a trivial statement, we could say that if two things do not belong to any identifiable category, then they are completely dissimilar. In practice, however, *some* category may always be identified, although in some cases we may have to extend our imagination to do so. For example, one might make the claim that JUSTICE and TELEPHONE have no similarity at all, but I can put them both in the category *concepts appearing in this sentence*.

At this point, the reader may well argue that not all classifications are real categories; that there are some valid classes we hardly ever use to organize our thoughts about something. Psychologists, such as Smith, agree; e.g., consider the class of objects that weigh an even number of grams. While identifying members of such a class is straightforward, something seems to be missing from its usefulness as a category. As Smith says in [5]: in general, we divide the world so as to maximize within–category similarity while minimizing between–category similarity.

Smith [5] identifies three characteristics of categories:

1. Category comprises a class of things that belong together.
2. Category constitutes a coding of experience, reducing the demand
      on perceptual processes.
3. Categorization of an object enables inductive inferences about that
      object.

Efforts to develop a single ontology in conceptual graphs have much to offer in using the first characteristic and the third. Indeed, Sowa's conceptual catalog ([6], Appendix B) is an early attempt in this regard.

Little discussion has been focused on the second characteristic, namely that categorization represents a coding of experience (in the words of one psychologist, we are all "cognitive misers") to reduce thinking time. Once we begin talking about perception and experience, we must take into account individual differences among people. The theory does not always address issues in the same way as in the real world.

As an illustration: in the conceptual graph type hierarchy, an object is a not only an instance of its type, it is also an instance of each of its type's supertypes. Conceptual graph theory is interested, for example, in minimal supertypes in order to characterize minimal generalizations. For most things, however, there is a basic level of categorization adopted by most people that is considerd most natural. If I hold up an apple and ask "What is this?" nearly everyone will say, "an apple". Other categorizations, while clearly valid, don't come to mind. Few would say, "a McIntosh Apple," or "a red apple" (subordinate levels of categorization) or "a fruit," "a food," or "an entity," (superordinate levels of categorization), even though all of the other categorizations are valid.

3. Issues Regarding Semantic Distance Measures

In order to evaluate particular measurement schemes, we require some set of properties that we would expect of any reasonable measure of semantic distance. A measuring scheme should have at least the following properties.

1. Insertion of new types into some hierarchy does not affect previous
      types' similarity measures.
2. Symmetry – semantic distance from a to b is the same as the
      semantic distance from b to a.
3. Computational complexity should be no worse than subgraph
      isomorphism problem.
4. Relative similarity must correspond to some intuitive notion of
      similarity, even if restricted to some particular domain.

Criteria 2 has been shown to be false in real–life. Given pairs of objects, the order in which they are presented affects one's perception of similarity. For example, the poet did not say: "Shall I compare a summer's day to thee?" The difficulty with the criteria 4 is that intuitive notions tend to be rather variable. Two effects are most important: context effects and perceiver effects. To quote Tversky: "Like other judgments, similarity depends on context and frame of reference."

A context effect is illustrated as follows. Suppose we want to compare two black birds. The context in which the birds appear has a large effect on whether they are perceived as similar. For example, a black bird that is stuffed on display in a museum may seem completely different from a bird of even the same species that I see perched on a tree by my kitchen window. Psychological experiments (summarized in [5]) confirm this. When subjects were given an object and asked which was most similar of three other objects given, the responses were markedly different when just one of the three items was changed.

A perceiver effect is as follows. I may see two black birds flying side by side as quite similar, while an experienced bird watcher may recognize the two birds as being of two completely different varieties, and therefore different in their habitat, food source, call sound, etc. Perceiver effects seem to have several aspects, among them: relevance to the perceiver, experience of the perceiver and vocabulary of the prerceiver.

To further complicate matters, the two effects are not independent of each other. Relevance and experience may influence what parts of an object's context are even visible to the perceiver!

A semantic distance measure will therefore need to take into account both these effects if the notion of similarity is to have any value to graph analysis. Determining the effects in a given situation is difficult; expressing the effect in some usable way is even more difficult.

Context effects lend themselves more naturally to being expressed by conceptual graphs. While a conceptual graph *context* is a more formal notion than similarity context, we can still envision a similarity context of concept A defined as the neighbors to A in the graph(s) in which it appears.

The effect of the context may be modeled in one of several ways. In one approach, each neighboring concept be assigned a relative weight of importance in the overall context, depending only upon their distance from A. Thus in Fig. 1, concepts C, E, F, H, and J are nearest neighbors and should therefore be more important in the similarity determination than B, D, and G.

One drawback to the first approach is that all neighboring concepts are treated as being equally important with respect to concept A. A second approach is to assign an individual weight to each neighbor, based perhaps on the definition of A. For example, an automobile may have attributes of 4 WHEELs and a COLOR, but most comparisons would give more importance to the wheels than the color. A third approach is to develop a more complicated formula for each definition, e.g., COLOR is an optional attribute for comparing to AUTO, but WHEEL is not.

Another issue has to with whether we intend to measure similarity or difference; i.e., measuring common features vs. measuring distinctive features. As Tversky notes [7]: In the assessment of similarity between objects the subject may attend more to their common features, whereas in the assessment of differentce between objects the subject may attend more to their distinctive features. Thus the relative weight of the common features will be greater in the former task than in the latter task.

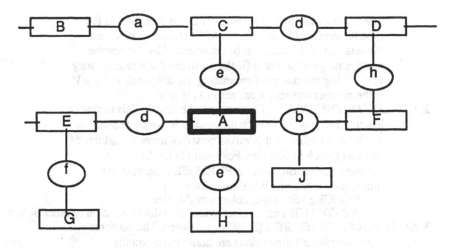

Fig. 1. A similarity context.

Perceiver effects will be much harder to capture. In theory, a similarity measurement might depend on having a complete characterization of the perceiver: his/her experience, relevent concepts, and personal definitions. We consider obtaining such a characterization to be nearly impossible: even if a person could describe himself completely (an enormous task!), there might still be underlying unstated assumptions that affect his perception of things.

A more tractable approach is to develop a characterization of a perceiver with respect to a particular domain of interest. This means essentially building a "box" around the perceiver and considering only those characteristics inside the box. There are two aspects to the problem: first, deciding the domain boundaries, and second, deciding what characteristics to put in the box.

For most applications, deciding the domain will not be an insurmountable problem. Previous uses of conceptual graphs have fairly well–defined domains: e.g., software requirements specifications, financial planning, product features for delivery to customers, etc. In practice, most characterizations of perceivers are based on well–defined domains. For example, when deciding whether to loan someone money, a bank will look at a few well–defined pieces of information: one's credit history, current income, and current assets. Other information about the person is therefore considered irrelevant to the bank (as perceiver), such as physical appearance, address, etc.

4. Taxonomy of Measurement Schemes

We now propose a taxonomy of various strategies for measuring semantic distance. This scheme should be considered preliminary, as a point of departure for discussion and eventual refinement of the taxonomy.

1. *NUMBER OF DIMENSIONS*: The notion of *dimension* refers to any single aspect of being along which objects can be

placed. For example, COLOR, TYPE (as referring to the type hierarchy), PURPOSE, SHAPE, etc. are considered dimensions. This taxon can be represented by the number itself, a range of numbers (if the number of dimensions may vary during the course of measuring the distance), or by a V if the number can vary from zero to infinity.

2. *QUALITY OF DIMENSION* (Concrete vs. Abstract): This notion has to do with whether each dimension is purely featural in character, or whether it involves possibly unseen qualities of an object, such as function. For example COLOR is a concrete dimension, whereas PURPOSE is an abstract dimension. Some possible dimensions are:
   CONCRETE: Shape, color, weight, size
   ABSTRACT: Purpose, behavior, transform, relation to other objects.

3. *SALIENCE OF DIMENSION* (Equal or Skewed): The taxonomy notes whether all dimensions are considered equally important in measuring semantic distance.

4. *SALIENCE VARIABILITY BY DIMENSION* (Fixed or Variable by Factor): For each dimension, the taxonomy records whether that dimension's salience (contribution to the measure) is fixed, or may vary. Importance may often be measured by some real number, or it may be denoted by whether a dimension is included or not. In either case, the salience taxon would be called variable for that dimension. If salience is variable for some dimension, the taxonomy must record what affects its variability; e.g., context, definitions, etc.

For example, in earlier work involving matching graphs, I used the number of hops in a hierarchy between two types as their semantic distance from one another. Since type closeness is a single dimension, this approach would be categorized as a 1–dimensional, concrete quality, equal, fixed semantic distance measure.

Tversky's original contrast model was an n–dimensional, concrete, equal, fixed measure of semantic distance.

Features of a taxonomy must be orthogonal (i.e., independent) of each other. It might be argued that taxons 3 and 4 are related. If salience (taxon 3) is Skewed, then each dimension's importance may be either fixed (at unequal levels) or else variable. If salience is Equal, then perhaps Variable by Factor is not relevant.

In geometric or featural approaches, objects or items are represented as points in some multidimensional space such that distance corresponds to dissimilarity between them. These are n–dimensional, abstract, equal, fixed approaches.

5. Empirical study of intuitive semantic distance

Since conceptual graphs originated from a desire to capture ideas expressed in natural language, then the notion of semantic distance, if it has any validity at all, ought to reflect some intuitive understanding about similarity between things expressed in natural language. The dictionary provides several words that mean something akin to semantic distance; e.g., *similarity, likeness, resemblance, correspondence,* etc. More studies need to performed with human subjects in order to discover whether semantic distance actually has any meaning independent of a particular person, and how to use semantic distance in a meaningful way.

6. Conclusion

The notion of semantic distance or similarity measurement is closely related to the process of categorization. Any theory of measurement must take into account some real–world results from cognitive science: it is not possible to make *a priori* judgments about how similar are two random concepts.

The taxonomy presented herein is not unbiased. It clearly implies that the more complex is the semantic distance measure, the more accurate and in line with experience it will be. That is not a particularly comforting result, but the purpose of the taxonomy was to organize the search for accurate semantic distance measures.

References

[1] Aronson, Jerrold L., "Conceptual Graphs and Modality," *Proc. 5th Wkshop on Conceptual Structures*, AAAI-90, Detroit. MI.

[2] Delugach, Harry S., "A Multiple–Viewed Approach to Software Requirements," Ph.D. Dissertation, Department of Computer Science, University of Virginia, Charlottesville, VA, May, 1991.

[3] Gentner, Dedre, "Structure Mapping: A Theoretical Framework For Analogy," *Cognitive Science*, vol. 7, pp. 155–170, 1983.

[4] Smith, Edward E., "Concepts and Induction," in *Foundations of Cognitive Science*, Posner, Michael, ed., Bradford Books (MIT Press), 1989.

[5] Smith, Edward E., "Categorization," in *An Invitation To Cognitive Science*, vol. 3, *Thinking*, Osterson, Daniel and Smith, Edward, eds., MIT Press, 1990.

[6] Sowa, John F., *Conceptual Structures: Information Processing in Mind and Machine*, Addison–Wesley Publ. Co., Reading, MA, 1984

[7] Tversky, Amos, "Features of similarity," *Psychological Review*, vol. 84, pp. 327–352, 1977.

Appendix – Semantic Distance Survey

This enclosed survey is meant to enhance the reader's appreciation for the rich complexity involved in making judgments of semantic distance. It is suggested the reader fill out the survey, while trying to be consciously aware of the multiplicity of mental decisions being made on each question.

III.   Reasoning

A Reconstruction of Conceptual Graphs on Top of a Production System*

Jacques Bouaud and Pierre Zweigenbaum

DIAM, INSERM U194 & SIM AP-HP
91, bd de l'Hpital, F-75634 Paris cedex 13 – France
{bouaud,zweig}@frsim51.bitnet

Abstract. In this paper, we study how several aspects of the Conceptual Graph theory can be implemented using the pattern-matching mechanisms of production systems. Usually, standard pattern matching applies to arbitrary data that, unlike CGs, do not rely on a particular theory. Reconstructions of Conceptual Graphs in terms of basic graphs have been proposed in the literature. We show that K, a graph representation language with "high-level" (rule-based) graph manipulation facilities, allows an elegant implementation of these proposals. We show how the CG projection is reconstructed from standard pattern matching. Such a mechanism provides the user with graph retrieval facilities. Moreover, K's inherent features, such as forward reasoning rules, are gracefully transferred to the resulting CG implementation with no further effort. The result is a production system that operates within the CG theory thus providing the basis for a flexible CG processor.

1 Introduction

Graph matching operations constitute an important issue in the context of Sowa's Conceptual Graphs (CGs) [1] for they allow associative graph access. Production Systems [2] are also known to provide, through the process of pattern matching, an associative access to arbitrary data that unlike CGs do not rely on a particular theory. This last point has often been criticized. A reconstruction of CGs in terms of more basic "Knowledge Graphs" has been proposed by Willems [3]. Furthermore, Ellis and Willems [4] showed how fundamental operations on CGs can be described using graph grammar production rules.

In this paper, we explore how production systems and their standard pattern matching (SPM) techniques provide an elegant framework to implement easily and explicitly several aspects of the CG theory. We use the production system of K [5], a graph representation language, and show that it is convenient for implementing such ideas. An important point is that the fundamental CG projection, which determines the specialization/generalization relations between CGs, is reconstructed in terms of SPM. Moreover, the production system features are gracefully transferred to the resulting CG implementation with no further effort. The result is a production system that operates on CGs within the CG theory.

* This work has been partly supported by program PRC-IA of the French Ministry of Research and by the European Community project MENELAS (AIM 2023).

The paper is organized as follows. The second section introduces the production system terminology and presents our CG implementation. The third section describes how CG projection is turned into SPM. The fourth section presents potential pattern-directed applications with CGs. Efficiency issues about this approach are mainly discussed in the fifth section. The last section concludes.

2 The Production System Framework

2.1 Standard Pattern Matching

SPM is the semi-unification between data abstractions or patterns, and actual data or facts. A fact, or working memory element, is a piece of data usually represented as a data structure whose positions contain symbols. A simple pattern is a data abstraction. It has the same structure as a fact but may contain variables.

A fact d matches a pattern p, and conversely, if there exists a mapping m from the set of variables of p to the set of symbols of d such that d is identical to the pattern p where each variable occurrence has been substituted by its value in m. The mapping m corresponds to the set of variable bindings of the match.

This pattern-to-fact match is extended for a conjunctive sets of patterns, or multi-pattern, possibly sharing common variables. A multi-pattern matches a set of facts if each pattern matches a fact with consistent variable bindings: each occurrence of a variable must be bound to the same value. This ensures that if two patterns refer to the same variable, their corresponding facts will refer to the same symbol. A set of facts correctly matching a multi-pattern is called an instanciation.

2.2 Production Systems

A production system [2] is characterized by a set of If-Then statements or rules, a database or working memory (WM), and an inference engine. Rules are pieces of problem solving knowledge. The WM contains facts which are temporary assertions describing a particular problem to be solved. A rule describes in its condition part a generic configuration of data, and in its action part actions to be taken when a specific instanciation of the condition part is found in WM. A rule is said to be fireable for each instanciation of its condition part. During its Recognize-Select-Act cycle, the production system interpreter searches for the valid rule instanciations, then selects one and finally executes the corresponding actions. As the rule condition parts are usually multi-patterns, SPM algorithms for production systems have to find all possible instanciations of a set of multi-patterns.

2.3 K

The production system we use in this work is K, whose purpose is to serve as an experimentation tool for implementing knowledge representation formalisms. The elementary factual structure is very simple and reduced to a triple of symbols. A triple $[n_1 \ r \ n_2]$ can be assumed to correspond to an arc labeled r linking two nodes labeled n_1 and n_2. As a result, WM can be viewed as a labeled directed graph.

K includes basic graph construction and manipulation primitives. Basic patterns can be combined with connectors to form more complex pattern expressions. The conjunctive operator, **and**, is used to express multi-patterns. These expressions are used for static data retrieval in WM and in the condition parts of K's forward rule based system.

3 Conceptual Graphs and Production Systems

3.1 Implementing Conceptual Graphs

A reconstruction of CGs in terms of basic "Knowledge Graphs" has been proposed by Willems [3] where CGs are turned into simpler graphs. We use this work as a starting point. A concept is naturally characterized by its type and its referent; a conceptual relation by its label and its n related arcs. Each concept or relation is associated to a unique identifier or token:

1. A concept token is related to its type by a *typ* link and to its referent by a *ref* link. The type must be in the type lattice and the referent must be either an individual marker or $\star$.
2. A relation token is related to its label by a *lab* link and to its ith token by a -i-link.

This implementation specification is naturally operationalized in K. Tokens, types and markers are assumed to be symbols, the links to be triples. A given CG is represented by a set of triples which constitutes the corresponding D-graph (Fig. 1). Let us call D-transformation this transformation from a CG to its data representation. With these basic data structures for, algorithms for the four canonical operators *copy*, *restrict*, *join*, and *simplify* can be programmed in a standard way.

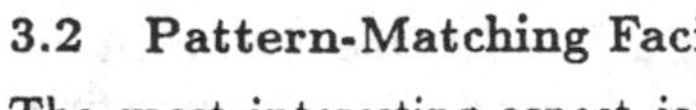

$$!1\boxed{\text{girl:sue}}\xleftarrow{\ !2\ }(\text{agnt})\xleftarrow{\ !3\ }\boxed{\text{eat}} \iff \begin{array}{l} \{\ [!1\ \textit{typ}\ \text{girl}]\ [!1\ \textit{ref}\ \text{sue}] \\ \ [!3\ \textit{typ}\ \text{eat}]\ [!3\ \textit{ref}\ \star] \\ \ [!2\ \textit{lab}\ \text{agnt}][!2\ \textit{-1-}\ !3]\ \ [!2\ \textit{-2-}\ !1]\ \} \end{array}$$

Fig. 1. The D-graph representation of a conceptual graph

3.2 Pattern-Matching Facilities

The most interesting aspect is that once CGs are represented in K's WM as D-graphs, SPM can apply to them. It consequently allows an associative access to every part of the CG implementation. This access does not necessarily correspond to selection operations permitted by the CG theory. For instance, the multi-pattern: (**and** [?c *ref* sue] [?r *-2-* ?c]) matches all pairs of facts corresponding to a token ?c with a *ref* link to sue and which is the end of a *-2-* link beginning by a token ?r. In other words, it specifies every couple of a concept and a relation (?c,?r) such that the concept referent is sue and the concept is the destination of the relation's second link. This multi-pattern matched against the D-graph in Fig. 1 would yield the variable bindings {(?c,!1);(?r,!2)}.

4 From Standard Pattern Matching to Projection

Projection is a fundamental notion in the CG theory for it determines the special-ization/generalization relations between CGs. It is the basis of CG retrieval. SPM is a kind of projection from a pattern to a fact. We examine here how to obtain CG projection from SPM. We first reproduce subgraph morphism, then projection.

4.1 Subgraph Morphism

According to the previous notation, let us represent a CG as a set of triples, but instead of symbol token identifiers, use distinct variables as tokens. As a result, fact specifications are turned into patterns. This new representation is a multi-pattern. Let us call it an M-graph.

In this situation, if a CG, G, is represented as an M-graph, M, then an instan-ciation I of M is a set of facts. This set of facts necessarily satisfies the constraints of M. Each variable token of M is bound to a symbol token in I. SPM ensures that the links between symbol tokens are the same as those between variable tokens. As a result, I is a D-graph representing a CG G', possibly part of a larger CG. The variable bindings of I are the mapping between the tokens of G and those of G'. This mapping correspond to a graph morphism between G and G' — i.e. it preserves concept types and referents, conceptual relation labels, and their connections.

4.2 Conceptual Graph Projection

CG projection is a kind of morphism that preserves conceptual relation labels but may restrict concepts. A concept can be restricted by replacing its type by a sub-type, replacing the $\star$ referent of a generic concept by an individual marker, or both. Projection requires a type hierarchy. We suggest to represent type subsumption re-lations explicitly within K. Each type is a symbol and a subtype relation between a type and a supertype is represented by an *inf* link. We assume here that each type is linked to each of its supertypes including itself. Subtype relationships are now available for SPM.

We introduce now another CG transformation, or P-transformation, in order to simulate projection. From a given CG, we build a new multi-pattern, or P-graph. This P-graph is close to the M-graph but has several differences in order to take into account the problem of concept restriction. The P-transformation is the following:

1. A generic concept $^{?c}\boxed{\text{typ}}$ is turned into (and [?c *typ* ?t_1][?t_1 *inf* typ]) where ?t_1 is a new free variable.
2. An individual concept $^{?i}\boxed{\text{typ:ref}}$ is turned into (and [?i *ref* ref] [?i *typ* ?t_2] [?t_2 *inf* typ]) where ?t_2 is a new free variable.
3. Conceptual relations have the same representation as in M-graphs. A binary relation ?c1 $\xrightarrow{?r}$ (lab) $\longrightarrow$?c2 is turned into (and [?r *lab* lab] [?r *-1-* ?c1] [?r *-2-* ?c2]).

Let G be a CG, P its corresponding P-graph, and I an instanciation of P. The variable bindings of I include a mapping between variable tokens and symbol tokens. These three points ensure that, (i) if ?c is a generic concept bound to !c, then the type of !c ($?t_1$) is a subtype of the type of ?c (typ) whatever the referent of !c, $\star$ or marker; (*ii*) if ?i is an individual concept bound to !i, then they have the same referent and the type of !i is a subtype of the type of ?i, and finally, (*iii*) connections between tokens and relation labels are preserved.

In this situation, I corresponds to a D-graph D corresponding itself to a CG G' such that there exists a projection from G to G'. Figure 2 illustrates how CG projection is obtained from SPM in our framework. An analogous approach could be used to take into account a relation lattice [6] and adapt the projection mechanism.

$$G \xrightarrow{\quad projection \quad} G'$$

$$\text{P-transformation} \downarrow \qquad\qquad\qquad \downarrow \text{ D-transformation}$$

$$P \xrightarrow{\quad SPM \quad} I \Leftrightarrow D$$

Fig. 2. Conceptual graph projection and standard pattern-matching

Figure 3 exemplifies a CG and its corresponding P-graph. As this CG projects into the CG in Fig. 1, this P-graph matches the previous D-graph yielding the following variable bindings: $\{(?a,!1);(?t_1,\text{girl});(?b,!2);(?c,!3);(?t_2,\text{eat})\}$. The mapping for projection would only include the bindings between tokens.

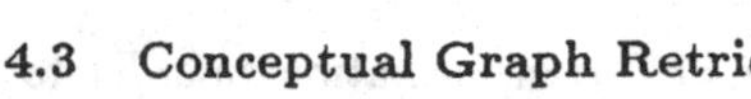

$?a\ \boxed{\text{person:sue}} \xleftarrow{?b} (\text{agnt}) \xleftarrow{?c} \boxed{\text{eat}} \iff$

(and [?a *typ* $?t_1$] [?a *ref* sue] [$?t_1$ *inf* person]
[?c *typ* $?t_2$] [$?t_2$ *inf* eat]
[?b *lab* agnt][?b *-1-* ?c] [?b *-2-* ?a])

Fig. 3. The P-graph representation of a conceptual graph

4.3 Conceptual Graph Retrieval

The multi-pattern representation of CGs enables the selection of arbitrary CGs at different levels of abstraction (graph morphism and projection). With projection, CG queries can be performed, hence enabling the retrieval of more specific CGs [7, 8]. A CG query Q is compiled into a P-graph, and as the matcher returns every instanciation of this P-graph, we obtain every D-graph whose corresponding asserted CG is a subgraph of a larger CG A such that $A < Q$. The pattern-matcher then computes the projective extent of Q in WM.

5 Pattern-Directed Applications with Conceptual Graphs

As CGs are asserted in WM and can be matched by SPM, K's forward production system becomes available for CGs with no further effort. This enables the application of production rules dealing with CGs.

These rules can apply either at the implementation level, when they address the internal structure (triples) of CGs, or at the CG level, when they directly address CGs using the P-transformation. At the implementation level we illustrate the computation of the *inf* closure and the way actors can be triggered. The same type of matching could be used to check the conformity of asserted CGs with some canonical CGs. At the CG level, we show how a user can write CG production rules.

5.1 At the Implementation Level

Computing the *inf* Closure. Proper subtype relationships are represented in K with a *sub* link between two types, *e.g.* GIRL<PERSON is represented by the triple [girl *sub* person]. The closure is actually computed with the two following rules:

$$\text{If } [?t_1 \ sub \ ?t_2] \text{ Then } [?t_1 \ inf \ ?t_1]$$
$$\text{If (and } [?t_1 \ sub \ ?t_2] \ [?t_2 \ inf \ ?t_3]) \text{ Then } [?t_1 \ inf \ ?t_3]$$

They respectively deal with reflexivity and transitivity. The process is initialized with [T *inf* T] where T is the universal type. The result is that all *inf* links are built.

Actors. The notion of actor is easily represented in our framework. Close to a conceptual relation, a token is associated to an actor. It is related to the function by a *fun* link, to the ith input concept by an inp_i link, and to the jth output concept by an out_j link. An actor's behavior can be implemented as a rule. The referents of the generic output concepts can be computed as soon as the input concepts have individual referents. Figure 4 exemplifies this for the square-root actor: $?c_i \xrightarrow{?a} \text{<sqrt>} \longrightarrow ?c_o$. Such rules used with a data-driven firing strategy provide a demon-like forward activation of actors. This should be possible without conflict in the case of functional dataflow graphs.

If (and [?a *fun* sqrt] [?a *inp* $?c_i$] [?a *out* $?c_o$] [$?c_o$ *ref* *] [$?c_i$ *ref* ?i $\neq$ *])
Then Restrict-ref($?c_o$,sqrt(?i))

Fig. 4. The square root actor's behavior as a production rule

5.2 At the Conceptual Graph Level

The pattern representations for CGs (M-graphs and D-graphs) can be combined using the other pattern matching operators of K (not, or, forall, exists...). Sophisticated CG configurations can be described, dealing with unconnected CGs that exceed the usual CG framework. These configurations can be used as queries as well as for writing graph production rules. One may choose for instance to systematically apply type expansion, or to join the concepts that have a given referent as soon as they are created.

Moreover, the user can define his/her own application dependent graph production rules that correspond to CG inferences going beyond single basic operations. In the development of an application, the system behavior can consequently be described by a declarative rule base instead of a program.

The following example is borrowed from a medical natural language processing system [9] ; the rule stating that *the presence of a high dose of radioactive iodine in the body causes the death of all thyroid cells,* can be easily written. The representation of this statement as a production rule ensures that the relevant conclusion will be added into memory every time the body contains radioactive iodine in high quantity and there exists some thyroid cell that is not dead. The rule is stated in a CG format as in Fig. 5. Each CG is expanded by P-transformation. The expansion results in 30 patterns. We solve the problem of coreference between the concepts $\boxed{\text{cell:} \ast \text{x}}$ which appear twice in the rule condition by using the same variable token.

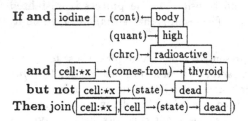

Fig. 5. A production rule at the conceptual graph level

6 Discussion: Efficiency Issues

Manipulating CGs with rules does not eliminate the drawbacks of the rule-based approach. However, it must be noticed that if the action parts of rules make an exclusive use of the four canonical CG operators instead of K primitives, then the system is ensured to evolve within the CG framework.

An issue that cannot be evinced is the problem of efficiency as our implementation relies on SPM which is known to be a costly process. The determination of an optimal SPM algorithm is NP-complete and SPM algorithms are exponential. In order to determine the instanciations of a multi-pattern, matchers explore a search tree and build a solution tree. This process is in the worst case in $O(n \times wm^c)$, where n is the number of rules, wm the size of WM, and c the mean number of patterns per rule. For a rule, wm corresponds to the branching factor of the search tree, and c to its depth. RETE [10] and TREAT [11] are currently the most efficient production system matchers but do not provide a complexity reduction in the general case. Fortunately, the upper exponential bound is not reached in practice because the constraints due to variable sharing prune the search. Nevertheless, there are several ways to gain efficiency for the production match: ordering patterns, restricting expressiveness, indexing techniques, and sharing patterns. We briefly show why our CG implementation satisfies most of the requirements for using these techniques.

Ordering Multi-Patterns. Matchers usually treat a multi-pattern from left to right. The aim of pattern ordering is to reduce the branching factor of the solution tree [12]. Constraints due to variable sharing are used to limit the possible matches and avoid the costly cross-products during joins.

Ordering the multi-pattern representations of CGs is very important because the number of patterns is high, even for a simple CG and patterns are poorly discriminant — their cardinalities can be high. In this respect, the square-root actor rule in Fig. 4 has a good ordering because every pattern is constrained by a bound variable. Conversely, the P-graph in Fig. 3 generates three cross-products. As a CG is a connected graph, its connections are reflected in its P-graph through common variables shared by its patterns. We could show that each pattern is constrained. As a result, cross-products can be avoided in the P-transformation, even if finding the best ordering is not possible.

Table 1 illustrates a better ordering for the P-graph of Fig. 3. For each ranked pattern, we give the constraints due to bound variables[2], the branching factor of the solution tree, and a comment. Numbers with a /0 indicate places where the search can be cut. We can see that more than half of the patterns have a branching factor of 1, which reduces the match complexity. This ordering is probably good but is not necessarily the best one.

Table 1. An ordering for the P-graph of [person:sue]←(agnt)←[eat]

Rank	Pattern	Branching factor	Comment
1	[?a *ref* sue]	n/0	the number of concepts whose referent is sue
2	[•a *typ* ?t_1]	1	the unique type of •a
3	[•t_1 *inf* person]	1/0	•$t_1 \leq$ person, true or false
4	[?b -*2*- •a]	k/0	the number of relations reaching •a
5	[•b *lab* agnt]	1/0	the label of •b = agnt, true or false
6	[•b -*1*- ?c]	1	the unique incoming concept of •b
7	[•c *typ* ?t_2]	1	the unique type of •c
8	[•t_2 *inf* eat]	1/0	•$t_2 \leq$ eat, true or false

The Unique-Attribute Representation. Tambe and Rosenbloom [13] proposed a linear match in $O(wm \times c)$ instead of the exponential $O(wm^c)$ by acting on the branching factor of the solution tree. This result is obtained with the unique attribute representation: the attribute of an object must have a unique value. The fact *(obj attr val)* is unique with *obj* and *attr* known. When this situation occurs during

---

[2] When bound, the question mark of a variable is changed into a bullet: ?x becomes •x.

match, the branching factor is at most 1. Imposing unique attributes constrains the expressiveness of the representation.

However, we can observe that in our framework all the links (except *inf*) are equivalent to unique attributes. In the P-transformation, we locally and naturally benefit from this result when the first element of a triple is bound which is a frequent situation. This property can be used in ordering (see Table 1). For instance, it is easy to see that the square-root actor rule perfectly fits the linear match.

Indexing. The aim of indexing is to reduce the branching factor of the search tree at each join to the one of the solution tree. In most standard matchers, like RETE and TREAT, facts are indexed according to the patterns they match. During joins, they scan the "matching extent" of the considered patterns.

The idea for indexing is to take advantage of the equality tests resulting ¿from variable bindings. Hashing techniques are often used. The hash code that indexes a fact depends on the pattern it matches, its rule, its rank in the multi-pattern, and the symbol to be tested. The matching algorithm of K is TREE [14], a RETE-like matcher, suited for matching connected facts. Each symbol S indexes both the triples leaving and reaching S: $[S \ ? \ ?]$, resp. $[? \ ? \ S]$. TREE's heuristic join strategy exploits these connections between facts for treating connected patterns.

For instance, the join with the constrained pattern $[\bullet a \ typ \ ?t_1]$ is treated using the following indexes and tests:

1. $[?x \ typ \ ?t_1]$ and $?x = \bullet a$ with standard RETE;
2. $[\bullet a \ ?x \ ?t_1]$ and $?x = typ$ with TREE;
3. $\mathrm{Hash}([?x \ typ \ ?t_1], P, 2, \bullet a)$ and $?x = \bullet a$ with hashed RETE.

The cardinalities of these sets are respectively, the total number of concepts in WM, the number of links leaving a concept token, *i.e.* 2, and the number of facts having the same hash code, probably less than 10. TREE is then adapted for such joins which are, unlike the others, independent of WM size. Moreover, experimental measurements [14] on other production system applications showed that TREE's overall performance is similar to hashed RETE in other respects.

Multi-Pattern Sharing. The fact that rules share common parts is exploited in RETE-like matchers to avoid performing the same matching sequence several times.

This property can be used in our CG framework. Let two CGs U, V be such that $U < V$. Then V projects into U and the P-graph of U intersects with the P-graph of V. We can make a kind of difference (P-graph(U) − P-graph(V)) which is a set of patterns corresponding to the additional constraints a D-graph should satisfy to produce an instanciation of P-graph(U), knowing that it already produced an instanciation of P-graph(V). This kind of sharing is close to the one used in the compilation of a CG hierarchy [8].

7   Conclusion

In this work, the production system framework has been explored in order to implement CGs. CG projection is reconstructed from SPM. This leads to several types

of representation for graphs: facts and multi-patterns. Some CGs are asserted in
WM. The multi-pattern representation for CGs is used for stating queries as well as
production rules. These rules are used for internal needs and for user-defined applications. The result is a running production system operating in the framework of CGs.
As much of the implementation detail is made explicit, such a CG processor can be
easily modified and/or extended. This approach will be used for the CG processor
in the natural language component of the European project "MENELAS".

References

1. J. F. Sowa. *Conceptual Structures: Information Processing in Mind and Machine.*
   Addison-Wesley, London, 1984.
2. D. A. Waterman and F. Hayes-Roth, editors. *Pattern-Directed Inference Systems.* Academic Press, New York, 1978.
3. M. Willems. Generalization of conceptual graphs. In *Proceedings of the 6th Annual
   Workshop on Conceptual Graphs*, pages 27–37, July 11–13, 1991.
4. G. Ellis and M. Willems. Analysis of semantic networks using graph grammars. In
   AAAI Spring Symposium on Propositional Knowledge Representation, 1992.
5. J. Bouaud. K, un langage pour l'implémentation d'outils de représentation des connaissances. PhD thesis, Université Paris 7, 22 juin 1989.
6. P. Kocura, K.K. Ho, D. Moorhouse, and G. Sharpe. Aspects of conceptual graphs
   processor design. In *Proceedings of the 6th Annual Workshop on Conceptual Graphs*,
   pages 317–319, July 11–13, 1991.
7. R. Levinson and G. Ellis. Multi-level hierarchical retrieval. In *Proceedings of the 6th
   Annual Workshop on Conceptual Graphs*, pages 67–81, July 11–13, 1991.
8. G. Ellis. Compiled hierarchical retrieval. In *Proceedings of the 6th Annual Workshop
   on Conceptual Graphs*, pages 187–207, July 11–13, 1991.
9. M. Cavazza and P. Zweigenbaum. A semantic analyzer for natural language understanding in an expert domain. *Applied Artificial Intelligence*, 1992. *(To appear).*
10. C. L. Forgy. Rete: A fast algorithm for the many pattern/many object pattern match
    problem. *Artificial Intelligence*, 19:17–37, 1982.
11. D. P. Miranker. TREAT - a better match algoritm for AI production systems. In
    Proceedings of the 6th National Conference on Artificial Intelligence - AAAI-87, pages
    42–47, Seattle, WA, July 13–17, 1987.
12. D. E. Smith and M. R. Genesereth. Ordering conjunctive queries. *Artificial Intelli-
    gence*, 26:171–215, 1985.
13. M. Tambe and P. S. Rosenbloom. A framework for investigating production system
    formulations with polynomially bounded match. In *Proceedings of the 8th National
    Conference on Artificial Intelligence - AAAI-90*, pages 693–700, Boston, MA, July 29
    - August 3, 1990.
14. J. Bouaud. TREE : une stratégie de jointure pour système de production fondée sur
    la propagation de contraintes. In *Proceedings of the 8th Conference RFIA-AFCET*,
    Lyon - Villeurbanne, France, November 27–29, 1991.

On Uncertainty Handling in Plausible Reasoning with Conceptual Graphs

Sung H. Myaeng & Christopher Khoo

School of Information Studies, Syracuse University

Syracuse, NY 14244

shmyaeng@mailbox.syr.edu; cskhoo@suvm.bitnet

Abstract. Plausible reasoning with conceptual graphs (CG) presents special problems. This paper treats plausible reasoning as a problem of matching a query CG with a target CG and computing the likelihood of relevance of the target CG from the degree of match. While the emphasis is on information retrieval modeled as plausible reasoning, the approach is applicable to schema selection, case-based reasoning and inferencing with production rules which have a CG as their premise. Two levels of uncertainty handling are suggested. Micro-level analysis combines evidence from individual concept and relation node matches between the query CG and a target CG. Macro-level analysis combines evidence from multiple, possibly overlapping, fragments of the query CG that occur in different parts of the target CG. The use of the Dempster-Shafer theory of evidence and Bayesian inference for combining and propagating evidence is examined.

1. Introduction

The notion of uncertainty has been much discussed in the literature of artificial intelligence, especially in the context of modeling the uncertain nature of decision making processes to develop expert systems. In this paper, we discuss our effort to develop a scheme for handling uncertainty within the Conceptual Graph (CG) framework (Sowa, 1984) with a particular application in mind, namely, information retrieval. We argue that such a scheme can be applied to the general problem of schema selection as well as the retrieval of cases in case-based reasoning.

The CG theory has been known for its versatility in allowing both for precise interpretation of its meaning representation of natural language sentences and for informal reasoning. While its logic formalism is strongly based on the model theory and is thus appropriate for deductive inference, the graph manipulation operators together with the notion of schemata and prototypes facilitate informal or plausible reasoning. This flexibility is in line with the view that reasoning should be treated as distinct from logic itself (Israel, 1983). Nonetheless, the CG theory has not been developed and/or extended enough to fully exploit its strength on plausibility representation and manipulation.

Practical problem-solving situations involve different kinds of uncertainty. A number of methods for handling uncertainty have been developed based on differing formal theories, each with its assumptions about the nature of uncertain knowledge and information, and the nature of the decision making process.

Rule-based systems have to handle two distinct but related tasks involving uncertainty. Given production rules of the following form:

$$P \Rightarrow Q$$

where P is a conjunction of atomic propositions and Q is an atomic proposition, the system must be able to 1) combine the truth values (or probability values) associated with individual atomic propositions in P to determine the overall truth value for P, and 2) calculate the truth value for the conclusion Q via a mechanism by which the truth value associated with the production rule is propagated when inference is performed.

Since the CG theory is basically a system of logic, this type of uncertainty handling in deductive reasoning can be incorporated into a CG system in a straightforward way. With the φ operator that maps CGs into formulas in the first-order logic (FOL) form, a general CG production rule

$$CG1 \Rightarrow CG2$$

with uncertainty values attached to individual connected graphs can be translated into an equivalent in FOL with truth values associated with formulas. Unlike many rule-based systems which allow a rule to fire only when the entire premise of the rule exactly matches with facts, we want to allow both partial and inexact matching between CG1 (the premise of the rule) and a fact CG.

Another aspect of CG theory that is germane to plausible reasoning is the idea of the schema, exemplified in the theory of scripts (Schank & Abelson, 1977). A schema describes a typical situation, lists default attributes for a concept, or presents a perspective on a way a concept may be used (Sowa, 1984). Using a schema, a new conclusion can be drawn based on plausible reasoning rather than deductive reasoning. When selecting an appropriate schema for a particular situation, we are interested in the degree to which a schema CG and the CG representing the situation match and share conceptual structures. Viewing CG matching as an important element of plausible reasoning (Myaeng, 1990), we feel that it is essential to devise a method by which the best or most plausible script or schema for a particular situation can be chosen in an intuitively correct, theoretically justifiable, and computationally feasible way.

Our present work is motivated by our DR-LINK project (Myaeng & Liddy, 1991) where we use CGs as the underlying representation framework for information retrieval. We believe that the problem of retrieving documents relevant to the information need stated in natural language can be seen as a form of plausible reasoning since there are uncertainties involved both in document and query representations. We view information retrieval as a process of determining the degree to which a query CG and a document CG share a conceptual structure (Myaeng, 1990). We believe that the matching process can be modeled as a form of plausible reasoning that involves handling uncertain information and gathering evidence from multiple sources of evidence.

2. Problem Characterization: Two Levels of Uncertainty Handling

There are at least two levels of uncertainty handling we are concerned with: micro and macro. Consider, for example, a fragment of a CG, representing a script for a situation, a document, or the premise of a rule:

CG1:   [cat] ->(on)-> [mat]->(in)->[city-hall].

We shall refer to this as the target CG. Suppose we have the following CG (called the query CG) representing a situation, a query or a fact:

CG2:   [cat] -> (sitting) -> [mat]->(in)->[building]

Each node in the query CG is associated with weights w1, w2, w3, w4, and w5 representing the relative importance of the nodes.

For a query CG (e.g. CG2), we want to retrieve target CGs (e.g. CG1) that are relevant or appropriate to the situation represented by the query CG. For each target CG, a decision has to be made whether it is sufficiently close to the query CG to be selected for whatever processing (e.g. inferencing) necessary. In other words, we want to know to what extent the target CG satisfies or covers the query CG.

There may be several unconnected fragments of the target CG that together partially or fully cover the query CG. The micro-level analysis addresses our interest in combining evidence from individual nodes forming a subgraph in a target CG that match with nodes in the query CG. The macro-level analysis deals with the problem of combining evidence from multiple subgraphs of the target CG, each of which has a counterpart in the query CG. The counterpart subgraphs in the query CG may overlap.

Using our example above, suppose we try to match CG1 with CG2. If we decide that the relations *(on)* and *(sitting)* are not close enough to match but that *[city-hall]* is a subtype of *[building]*, we can identify two subgraphs of CG1 that partially cover the query CG: [cat] and [mat]->(in)->[city-hall]. Calculating the evidence for each of the subgraphs requires micro-level analysis, and combining the evidence from the two subgraphs to determine the total evidence for the target CG is referred to as the macro-level analysis. It should be evident that our interest is not in truth-preserving inference.

3. Handling Uncertainty at the Micro Level

Our approach at the micro level is to view the evidence from a partial match as an accumulation of evidence from the components (concepts and relations) of the target CG that has counterparts in the query CG. Since each concept and relation in the query CG that finds a match contributes to the system's confidence that the target is relevant, we need a scheme to combine evidence from individual concept and relation matches.

In general, the amount of evidence that the target CG is relevant depends on several factors:

-   *The semantic distance* between corresponding matching nodes. A query concept and a target concept may not match exactly. An inexact match between two concepts that are semantically distant should contribute less to the evidence.

- *The importance weights* associated with each query node. Not all concepts and relations are equally important in the query. Important concepts will contribute more to the evidence. It is also possible to have weights associated with nodes in the target CGs.

- *The relative importance* between concept nodes and relation nodes. Since we use domain-independent case (or thematic) relations in our representation, we assume that matching concepts contribute more to the overall evidence for relevancy than matching relations.

We explore two formalisms for handling uncertainty at the micro level: belief functions using Dempster-Shafer's theory of evidence (Shafer, 1976) and Bayesian inference.

3.1. Dempster-Shafer's Theory

A set of hypotheses that is exhaustive and mutually exclusive is called a frame of discernment and is denoted by Θ. In our case, we have only 2 hypotheses, *relevant* and *not-relevant* denoted by {REL, NREL}. Let $P(\Theta)$ denote the power set of Θ. In the theory, any subset of Θ can be treated as a hypothesis, not just singletons, resulting in 2^n hypotheses. In our situation, however, {REL}, {NREL}, and Θ itself are the only hypotheses to which we can assign belief, making the computational burden manageable. The empty set $\emptyset$ by definition is assigned 0 belief.

Each piece of evidence is associated with a function m: $P(\Theta) \rightarrow [0,1]$ such that

$$m(\emptyset) = 0,$$

$$\sum_{A \subseteq \Theta} m(A) = 1$$

This function is called a *basic probability assignment* (*bpa*). The basic probability assignment distributes the total amount of belief among all the subsets of Θ. In our context, a *bpa* assigns a number in the interval [0,1] to each of the subsets of Θ, {REL}, {NREL} and {REL, NREL}. The value assigned to the subset {REL, NREL} represents the amount of belief not committed to either {REL} or {NREL}. Associated with each *bpa*, m, is a belief measure, *Bel*, defined by

$$Bel(A) = \sum_{B \subseteq A} m(B)$$

Where there are only two hypotheses, the distinction between m and *Bel* is not significant and we shall use them interchangeably.

When there are several independent pieces of evidence, each associated with a *bpa*, we can use Dempster's combination rule to combine the *bpa*'s for all the pieces of evidence into a new *bpa*, which represents the pooling of evidence from independent sources. A new *bpa*, m, from two *bpa*'s m1 and m2 is computed as the orthogonal sum:

$$m1 \oplus m2\,(A) = \sum_{X \cap Y = A} m1\,(X) \cdot m2\,(Y)$$

If applying the combination rule results in m(∅) > 0, the new *bpa* has to be normalized so that m(∅) = 0, and $\sum$ m(A) = 1.

3.2. Gathering Evidence from Concept and Relation Nodes using Belief Functions

To represent the importance of each concept and relation in a query CG, we can extend the CG notation in such a way that both concept and relation nodes contain a number in the range [0,1]. The meaning of this importance weight will differ from application to application. In the context of information retrieval, the importance weight indicates how important the concept or relation is in representing the information need of a user. We are currently developing heuristics for assigning weights to concept and relation nodes from a natural language statement but a discussion of this is beyond the scope of this paper.

This importance weight can, then, become the basis for a *bpa* for each node in the query CG. If a concept in the query CG occurs in a target CG, then the *bpa* associated with this concept assigns a quantity of belief to the hypothesis that the target CG is relevant. If the target CG has a projection on the query CG involving more than one concept node, Dempster's combination rule can be used to combine the *bpa*'s of the concept nodes, pooling the evidence from several concept matches.

As an example, suppose we have a target CG and a query CG as follows:

target CG: [C1] <- (R1) <- [C2] -> (R2) -> [C3]

query CG: [C2 | 0.6] -> (R2 | 0.1) -> [C3 | 0.3]

Each concept and relation in the query CG is assigned a *bpa*. The *bpa* for C2 is

$m_{C2}\,(\{REL\}) = 0.6, m_{C2}\,(\{NREL\}) = 0, m_{C2}\,(\{REL, NREL\}) = 0.4$

Similarly, the bpa for C3 is

$m_{C3}\,(\{REL\}) = 0.3, m_{C3}\,(\{NREL\}) = 0, m_{C3}\,(\{REL, NREL)\} = 0.7$

The combined evidence from these two concepts that the target CG is relevant to the query CG becomes

$m_{C2} \oplus m_{C3}\,(\{REL\}) = 0.6 \cdot 0.3 + 0.6 \cdot 0.7 + 0.4 \cdot 0.3 = 0.72$

It should be noted that it is possible for $m_i\,(\{NREL\})$ to have a non-zero value, indicating that the absence of a concept in a target CG has a negative impact on the relevancy.

It may appear, at first glance, that the same combination rule can be applied repeatedly to combine all pieces of evidence coming from all the nodes. Closer examination reveals a problem associated with relation nodes. The assumption of Dempster's combination rule is that the pieces of evidence being combined are independent of each other. While it can be argued that observing a concept in a document or a schema does not necessarily imply the occurrence of another concept (hence independence), the same cannot be said for relation nodes since, in the CG framework, a relation node cannot exist by itself. One way of handling this problem is to allow relation-based matching. Relation-based matching, which is meant to simulate analogical reasoning (Myaeng, 1990), seeks an exact match between a relation node in the query CG with a relation node in the target CG, while allowing the concept nodes linked by the relation to match any concept node. The case relations we are currently using are, however, so general or primitive that relation-based matching is meaningless.

This problem is made more explicit in the following scenario. Suppose a query CG contains two concepts linked by a relation. If the two concepts occur in a target CG, then a relation match between the two concepts provides additional evidence that the document is relevant. If, however, one of the concepts does not occur in the target CG, then the relation match does not add anything to the likelihood of relevance since the relation can occur with almost any other concept.

We have not found a completely satisfactory method within the belief function framework for handling evidence from relation node matches. However, we shall describe two approaches that we are investigating. The first approach is to divide the belief interval, [0,1], into two parts that reflect the relative contribution of concept matching and relation matching. For example, we can stipulate that concept matching without any relations can account for up to 0.7 of the belief interval. Relation matching will then account for 0.3 of the belief interval. The amount of belief from concept match calculated using Dempster's rule is rescaled to the interval [0, 0.7]. The total amount of belief is then obtained by simply adding together the amount of belief from concept matches and relation matches.

The second approach is to simply use Dempster's combination rule to combine evidence from both concepts and relations but imposing an order on what is combined first. Evidence from concept matches is gathered first, and then evidence from relation matches is added. This reflects the perceived difference between concepts and relations in their relative contributions to the final belief. A relation contributes evidence only if the two concepts that it links already occur in the target CG. A relation match evidence is thus additional evidence that is added to concept match evidence.

3.3 The Bayesian Approach for Handling Concept and Relation Matches

The Bayesian approach has a more satisfactory way of handling relation node matches but requires additional assumptions for concept node matches. We shall base our analysis on Bayes' updating rule, a modified version of Bayes' theorem that is amenable to recursive and incremental computation of multiple pieces of evidence.

Let $e_n = e^1, e^2, ..., e^n$ denote a sequence of data (e.g. the existence of certain query concepts and relations in the target CG) observed in the past. Let e denote a new fact (e.g. a concept or relation in the query that we are considering). Then the impact of the new datum on the likelihood of relevance can be computed by the formula:

$$P(REL|e_n, e) = \frac{P(REL|e_n) \cdot P(e|e_n, REL)}{P(REL|e_n) \cdot P(e|e_n, REL) + P(NREL|e_n) \cdot P(e|e_n, NREL)}$$

For the above formula to be useful, we have to assume that e and e_n are conditionally independent given REL so that we can use $P(e|REL)$ as an approximation for $P(e|e_n, REL)$. Otherwise, we would need to have a value for $P(e|e_n, REL)$ for each piece of evidence e and every combination of facts that comprises e_n.

While the Dempster's rule requires only that the occurrence of the query concepts be independent in the database or knowledge base of target CGs, the Bayesian approach requires that the concepts be independent in the subset of relevant target CGs. In an information retrieval situation, the occurrence of query concepts are generally not independent in the set of relevant documents.

The advantage of the Bayesian approach is that it provides a natural way of handling relation node matches. Suppose that a relation R_i links concepts C_j and C_k in the query CG, and that e_n includes all the possible concept matches (including C_j and C_k) between the query CG and the target CG plus an arbitrary number of relation matches. In an information retrieval situation, we can take $P(r_i/c_j, c_k, REL)$ as an approximation for $P(r_i/e_n, REL)$, where r_i represents the occurrence of R_i in the target CG, and c_j and c_k represents the occurrence of C_j and C_k. In other words, we can reasonably assume that the probability of R_i occurring in a relevant target CG depends only on the occurrence of C_j and C_k, and is independent of other concepts.

3.4. Uncertainty from Semantic Distance and from Abstraction

Another source of uncertainty is inexact matching between concepts and relations in the query CG with concepts and relations in the target CG. A concept C_1 in the query CG may match, inexactly, a concept C_2 in the target CG. The degree of match between C_1 and C_2 depends on the semantic distance between them.

The uncertainty resulting from this inexact match can be modeled as an inference rule. If C1 occurs in the target CG with a 100% certainty (i.e. exact match), then we have a belief b_1 that the target CG is relevant. This can be represented as the following inference rule:

(1) C1 --> Relevant (with belief b_1)

For each related concept, C_i, we can then have an inference rule

(2) C_i --> C1 (with belief b_i)

where b_i is a function of the semantic distance between C_i and C1. When the target CG contains C_i instead of C1, the degree of belief that C_i contributes to the relevancy can be calculated by chaining (1) and (2), and taking the product of the two belief measures (Pearl, 1988, p. 444; Mantaras, 1990, p. 54):

(3) C_i --> Relevant (with belief $b_1 \cdot b_i$)

In the Bayesian formalism, propagation of evidence in a chain is given by the formula (Pearl, 1988, p. 154):

$$P(REL|C_i) = \frac{P(REL)}{P(C_i)} \cdot [P(C_i|C1, REL) \cdot P(C1|REL) + P(C_i|\neg C1, REL) \cdot P(\neg C1|REL)]$$

If we approximate $P(C_i|C1,REL)$ with $P(C_i|C1)$, and if $P(C_i|\neg C1,REL)=0$, then the formula simplifies to:

$$P(REL|C_i) = \frac{P(REL)}{P(C_i)} \cdot [P(C_i|C1) \cdot P(C1|REL)]$$

A related issue in uncertainty handling stems from abstraction. A query concept C may be defined by a lambda abstraction, CG'. While an exact match between CG' and a target CG gives us 100% certainty that the query concept C occurs in the target CG, a partial match between the lambda abstraction (i.e. CG') and the target CG gives us less confidence that concept C exists in the target CG. We, thus, have to compute the degree of belief we have in the existence of C from the concepts and relations in CG' that matches the target CG. Since the process of determining the degree of belief from the partial match between the lambda abstraction and the corresponding target CG is identical to that of determining the same for a query and a target CG, the same evidence gathering process can apply. Nonetheless, this type of recursive CG matching should be avoided since it is computationally expensive.

3.5. Handling Negation

Negation can appear in the query CG or the target CG. Both needs careful handling.

If the query CG matches a subgraph of a target CG, then there is a perfect match and we have maximum belief that the target CG is relevant. This is true, however, only if the matching subgraph in the target CG is not negated. If in fact the matching subgraph in the target CG is negated, then our belief that the target CG is relevant is less than if the subgraph was not negated. So during matching, we have to check whether a matching subgraph in the target CG appears within a negative context.

Negation can also appear in a query, as exemplified in the following query statement:

I'm interested in antitrust cases that are NOT a result of a routine review.

The following examples of target CGs indicate the kinds of problems that a matching scheme has to handle:

Target CG1: [antitrust case: x]

(not)-> [[antitrust case: x] <-(result)- [routine review]]

Target CG2: [antitrust case]

Target CG3: [antitrust case] <-(result)- [routine review]

Target CG1 is the most likely to be relevant. Target CG2 is likely to be relevant but slightly less so than CG1. Target CG3 is not relevant to the query.

4. Handling Uncertainty at the Macro Level

There are two related problems involved in handling uncertainty at the macro level. The first problem is how to combine evidence from several fragments of the query CG occurring in the target CG. If the fragments are non-overlapping subgraphs of the query CG, then Dempster's rule can be used to combine their separate *bpa*'s. It is possible, however, for the matching subgraphs to overlap in the query CG.

A related problem is how to handle multiple matches between the query CG and the target CG. A concept in the query CG may occur several times in the target CG. Similarly, a subgraph in the query CG may occur several times in a target CG. In the context of information retrieval, multiple occurrence of a concept in a document suggests that the concept is a central topic of the document. A document that has multiple occurrence of query concepts is more likely to be relevant than a document in which query concepts appear only once.

We would like to develop a scheme for increasing the belief of relevance when subgraphs of the query CG occur multiple times in the target CG. This problem is intertwined with the first problem of overlapping fragments because a query concept may occur multiple times in a target CG as part of subgraphs that match with overlapping fragments of the query CG.

The following example illustrates both problems. Suppose we have the following query CG:

Query CG: [C1] --(R1)--> [C2] --(R2)--> [C3]

and we have the following target CGs:

Target CG1: [C1'] --(R1')--> [C2']

Target CG2: [C1'] --(R1')--> [C2'] [C1"] --(R1")--> [C2"]

Target CG3: [C1'] --(R1')--> [C2'] [C2"] --(R2')--> [C3']

Target CG4: [C1'] --(R1')--> [C2'] --(R2')--> [C3']

where C1' represents an instance of the concept C1, and C1' and C1" are different instances of the concept C1.

An appropriate scheme would assign a higher belief of relevance to target CG2 than to target CG1 because the matching subgraph occurs twice in target CG2. However, the belief assigned to target CG2 should be less than the belief assigned to target CG3, which should, in turn, be less than that assigned to target CG4. The belief of relevance for multiple partial matches should not exceed the belief of relevance for a complete match.

5. Conclusion

The capability of retrieving potentially relevant CGs from a store of many CGs is essential to many applications, including information and fact retrieval, case-based reasoning, and retrieval of appropriate scripts for story understanding. We have formulated the CG retrieval problem using the notion of uncertainty handling and have identified areas in which appropriate uncertainty handling capabilities are required.

We explored the use of the Dempster-Shafer theory of evidence and the Bayesian formalism, and identified some of the strengths and weaknesses of both in our application.

For some of the problems (e.g. macro-level handling of uncertainty), we have not been able to find satisfactory solutions and have only identified conditions or criteria that a suitable uncertainty handling scheme should meet.

To handle some of these problems in our DR-LINK project, we plan to devise alternative ad hoc procedures, implement them on top of the base CG matching algorithm we have developed (Myaeng & Lopez-Lopez, in press), and conduct experiments to determine the best one for information retrieval applications, at the same time searching for theoretically more elegant methods.

Acknowledgement

This work is supported by the Defense Advanced Research Projects Agency (DARPA) under Contract 91-F136100-000.

References

Israel, D. (1983). "The role of logic in knowledge representation." *IEEE Computer*, 16 (10): 37-42.

Mantaras, R.L. de (1990). *Approximate Reasoning Models*. Chichester: Ellis Horwood.

Myaeng, S. H. (1990) "Conceptual graph matching as a plausible inferencing technique for text retrieval." In *Proceedings of the 5th Annual Workshop on Conceptual Graphs*, held in conjunction with AAAI-90, July 1990.

Myaeng, S. H. & Liddy, E. D. (1991). "Information retrieval using linguistic knowledge and conceptual representation." *Technical Paper, School of Information Studies, Syracuse University.*

Myaeng, S. H. & Lopez-Lopez, A. (in press). "Conceptual graph matching: a flexible algorithm and experiments." *Journal of Experimental and Theoretical Artificial Intelligence.*

Pearl, J. (1988). *Probabilistic Reasoning in Intelligent Systems*. San Mateo, CA: Morgan Kaufmann.

Schank, R. C. & Abelson, R. P.(1977). *Scripts, Plans, Goals and Understanding*. New York: Lawrence Erlbaum Associates.

Shafer, G. (1976). *A Mathematical Theory of Evidence*. Princeton, NJ: Princeton University Press.

Sowa, J. F. (1984). *Conceptual Structures: Information Processing in Mind and Machine*. Reading, MA: Addison-Wesley.

Expert Humans and Expert Systems: Toward a Unity of Uncertain Reasoning

Weldon Whipple

IBM Corporation
Rochester, MN 55901

Abstract. Expert systems use a variety of techniques to reason about uncertain, incomplete, or unclear knowledge. This paper considers some methods expert systems have used to attempt to match human reasoning with and about uncertainty. It discusses MYCIN certainty factors, probability and Bayes' Theorem, Dempster-Shafer belief functions, and fuzzy logic. For each method, the paper suggests how conceptual graphs can be used to implement those representations, using concepts, relations, schemata, and actors.

1 Introduction

Among the problem domains of expert systems are disciplines whose knowledge is fuzzy, uncertain, or incomplete. Expert systems have dealt with these domains using techniques such as certainty factors, probability theory, belief functions, and fuzzy logic. Humans, by contrast, have been dealing intuitively with uncertainty for thousands of years. One of artificial intelligence's biggest challenges is to capture the essence of human reasoning and duplicate it in expert systems.

The flexibility and extensibility of conceptual graphs have attracted researchers in a variety of knowledge related fields. This paper proposes methods of dealing with uncertainty, and illustrates possible representations in conceptual graphs.

2 MYCIN

2.1 Representing Certainty Factors in Conceptual Graphs

In MYCIN, each fact stored in the knowledge base has an associated certainty factor lying in the interval $[-1.0, +1.0]$, ranging from definitely false to definitely true. One possible method for representing certainty factors in conceptual graphs would be to expand the notation of concepts to include certainty factors. Unfortunately, placing certainty factors in individual concepts prevents the knowledge base from being easily expanded or modified to support different or multiple certainty representations.

Another approach would be to use separate concepts and relations to represent certainty factors. With a [CF] concept and a (CFAC) relation, for any concept c in the knowledge base, a [CF] concept could be related to c by using the (CFAC) relation (Figure 1). This method would allow a preexisting MYCIN implementation to be more easily modified to support other representations.

```
[CF: 0.6] <- (CFAC) <- [PROPOSITION:
[TATTOO: #]-
  ->(OWNR)->[PERSON: Wilson]
  ->(IMAG)->[FISH]-
    ->(PART)->[SCALE: {*}]-
      ->(ATTR)->[COLOR: pink],,,].
```

Figure 1. Associating certainty factors with facts

2.2 Actors to Combine Certainty Factors

When two rules fire, both concluding the same concept, MYCIN uses a
commutative formula to combine the certainty factors (Harmon & King [1985], p.
51):

$$((1.0 - cf1) \times cf2) + cf1 \tag{1}$$

If **cf1** = 0.6 and **cf2** = 0.4, then the combined certainty factor is $((1.0 - 0.6) \times 0.4)$
+ 0.6 = 0.76. The <COMBINE-CF> actor (Figure 2) implements formula (1).

```
actor COMBINE-CF(in cf1, cf2; out cf) IS
  [MINUEND: 1.0]-><SUBTRACT>-
    <-[SUBTRAHEND: *cf1]
    ->[DIFFERENCE]-><IDENT>-
      ->[MULTIPLICAND]-><MULTIPLY>-
        <-[MULTIPLIER: *cf2]
        ->[PRODUCT]-><IDENT>->[AUGEND]-><ADD>-
          <-[ADDEND: *cf1]
          ->[SUM: *cf],,,,
  [SUBTRAHEND: *cf1]-><IDENT>->[ADDEND:   *cf1].
```

Figure 2. Actor to combine certainty factors

2.3 Natural Language and Schemata

MYCIN associates English descriptions with certainty factors (Harmon & King
[1985], p. 42). These associations can be effectively represented in conceptual
graphs using schemata. Figure 3 shows mappings from English to certainty factors.

```
schema for DEFINITE(x) IS
  [DEFINITE: *x]<-(CHRC)<-[PROPOSITION: *x]->(CFAC)->[CF: 1.0].
schema for SUGGESTIVE-EVIDENCE(x) IS
  [SUGGESTIVE-EVIDENCE:*x]<-(CHRC)<-[PROPOSITION: *x]-
    ->(CFAC)->[[CF: 0.45 ≤ *y] [CF: *y < 0.7]],.
schema for WEAKLY-SUGGESTIVE-EVIDENCE(x)IS
  [WEAKLY-SUGGESTIVE-EVIDENCE:*x]<-(CHRC)<-[PROPOSITION: *x]-
    ->(CFAC)->[[CF: 0.2 < *y] [CF: *y < 0.45]],.
```

Figure 3. Schemata provide default English values for certainty factors.

2.4 Certainty Factors in Rules

Certainty Factors in Antecedents. Rules fire only when their antecedent has certainty greater than 0.2. (Harmon & King [1985], p. 52). The schema in Figure 4 determines whether or not a rule can fire.

```
schema for PREMISE(x) IS
   [PREMISE: *x]->(CFAC)->[CF]->( > )->[NUMBER: 0.2].
```

Figure 4. Schema for premise consisting of a single concept

When the antecedent consists of a conjunction of concepts having different certainty factors, the *minimum* certainty factor must be greater than 0.2. If the antecedent is a disjunction, the *maximum* certainty factor must be greater than 0.2 (Figure 5).

```
schema for PREMISE(x,y) IS              schema for PREMISE(x,y) IS
   [PREMISE: *x]->(CFAC)->[CF: *z]         [PREMISE: *x]-
   [PREMISE: *y]->(CFAC)->[CF: *w]           ->(CFAC)->[CF: *z]
   [CF: *z]-><MIN>-                          ->(OR)->[PREMISE: *y]->(CFAC)->[CF: *w],
     <-[CF: *w]                            [CF: *z]-><MAX>-
     ->[CF]->( > )->[NUMBER: 0.2],.          <-[CF: *w]
                                             ->[CF]->( > )->[NUMBER: 0.2],.
```

Figure 5. Schemata for premise consisting of two ANDed and two ORed concepts.

Certainty Factors in Consequents. When a rule fires, the certainty factor of the antecedent is computed (using a PREMISE schema) and multiplied by the consequent's certainty factor to yield a certainty factor for the consequent (see Figure 6).

```
actor for NEW-CF(in x,y; out z) IS
   <MULTIPLY>-
     <-[CF: *x]<-(CFAC)<-[PREMISE]
     <-[CF: *y]<-(CFAC)<-[CONSEQUENT: *o]
     ->[CF: *z]<-(CFAC)<-[NEW-FACT: *c],
   [CONSEQUENT: *c]-><IDENT>->[NEW-FACT: *c].
```

Figure 6. Actor for deriving the certainty factor of the consequent

If the resulting fact does not already exist in the knowledge base, it is added to the knowledge base. If the fact has been previously asserted, the new and the old are merged using the <COMBINE-CF> actor (Figure 2 above).

3 Bayes' Theorem

3.1 Representing Probabilities in Conceptual Graphs

Probabilities are represented by the (PROB) relation and [PROBABILITY] concept. Figure 7 shows how to associate a probability of .8 with the statement "Wilson's tattoo is a fish with pink scales."

```
[PROBABILITY: 0.8]<-(PROB)<-[PROPOSITION:
  [TATTOO: #]-
   ->(OWNR)->[PERSON: Wilson]
   ->(IMAG)->[FISH]-
    ->(PART)->[SCALE: {*}]-
     ->(ATTR)->[COLOR: pink]],,,.
```

Figure 7. Conceptual graph illustrating probability

3.2 Combining Probabilities Using Bayes' Theorem

Formulas for Bayes' Theorem. Bayes' Theorem is shown in formula (2).

$$Pr(A_i|B) = \frac{Pr(B|A_i)\ Pr(A_i)}{\sum_{j=1}^{k} Pr(B|A_j)\ Pr(A_j)} \tag{2}$$

Formula (2) includes conditional probabilities, which are defined as

$$Pr(A|B) = \frac{Pr(AB)}{Pr(B)}, \quad [Pr(B)\neq 0] \tag{3}$$

The conjunction of probabilities of two independent events (used in figure (3)) is defined as

$$Pr(AB) = Pr(A)\ Pr(B) \tag{4}$$

Actors for Bayes' Theorem. We now define actors to realize those formulas. Figure 8 shows the <IND-PR-CONJ> (independent probability conjunction) actor.

```
actor IND-PR-CONJ(in pr1, pr2; out pr) IS
  [MULTIPLIER: *pr1]-><MULTIPLY>-
   <-[MULTIPLICAND: *pr2]
   ->[PRODUCT: *pr],.
```

Figure 8. The conjunction of two independent probabilities

The <COND-PR> (conditional probability) actor incorporates the <IND-PR-CONJ> actor:

```
actor COND-PR(in prA, prGiven; out pr) IS
  [MULTIPLIER: *prA]-><MULTIPLY>-
   <-[MULTIPLICAND: *prGiven]
   ->[PRODUCT]-><IDENT>->[DIVIDEND]-><DIVIDE>-
    <-[DIVISOR: *prGiven]
    ->[QUOTIENT: *pr],,.
```

Figure 9. Conditional probability actor

Figure 10 shows the <POST-PR> actor, which produces the output of Bayes' Theorem for $k = 2$.

```
actor POST-PR(in prA, prGiven; out pr) IS
  [MULTIPLIER: *prGiven]-><MULTIPLY>-
   <-[MULTIPLICAND: *prA]
   ->[PRODUCT]-><IDENT>->[DIVIDEND]-><DIVIDE>-
    <-[DIVISOR: *prA]
    ->[QUOTIENT]-><IDENT>->[MULTIPLICAND]-><MULTIPLY>-
     <-[MULTIPLIER: *prA]
     ->[PRODUCT]-><IDENT>-
      ->[ADDEND]-><ADD>-
       <-[AUGEND]<-<IDENT><-[PRODUCT]<-<MULTIPLY>-
        <-[MULTIPLICAND]<-<IDENT><-[QUOTIENT]<-<DIVIDE>-
         <-[DIVISOR: *x]<-<IDENT><-[MULTIPLICAND]-
          <-<IDENT><-[DIFFERENCE]<-<SUBTRACT>-
           <-[MINUEND: 1]
           <-[SUBTRAHEND: *prA],
           -><MULTIPLY>-
            <-[MULTIPLIER: *prGiven]
            ->[PRODUCT]-><IDENT>->[DIVIDEND:  *y],,
           <-[DIVIDEND: *y],
          <-[MULTIPLIER]<-<IDENT><-[DIVISOR:  *x],
         ->[SUM]-><IDENT>->[DIVISOR]-><DIVIDE>-
          <-[DIVIDEND: *z]
          ->[QUOTIENT: *pr],,
        ->[DIVIDEND: *z],,,,,
```

Figure 10. Actor that implements posterior probability for k = 2

4 Dempster-Shafer Belief Functions

Glenn Shafer introduced belief functions in his book entitled *A Mathematical Theory of Evidence* (Shafer, 1976). Belief functions address the inability of probabilistic methods to deal with ignorance.

4.1 Overview of Belief Functions

Belief functions are based on the mass function m. The theory assumes that the universe of discourse is a set of mutually exclusive and exhaustive elements, termed the environment, Θ. The degree of belief in any of the elements of the power set $P(\Theta)$ lies in the interval $[0,1]$. The mass function allows for ignorance (termed

153

nonbelief): the sum of the masses of belief in all the elements of P(Θ) can be less than 1.

The belief function **Bel** is the total mass of belief in a set (in the power set) and *all* its subsets. It is the inclusion of the phrase "all its subsets" that differentiates **Bel** from *m*. (Recall that the mass, in contrast, is assigned to a *single* set—a member of the power set P(Θ).) **Bel** is defined in terms of the mass function:

$$Bel(X) = \sum_{Y \subset X} m(Y) \qquad (5)$$

Dempster's Rule of Combination is used to combine masses of belief from different sources in support the same element of P(Θ):

$$m_1 \oplus m_2(Z) = \sum_{X \cap Y = Z} m_1(X) m_2(Y) \qquad (6)$$

Belief functions from different sources in support of a common set are combined by using a formula called an orthogonal sum. The orthogonal sum is defined in terms of Dempster's Rule of Combination, by summing the combined masses of a set and all its subsets. (See Giarratano & Riley [1989], pp. 276-290, for a further overview of belief functions.)

4.2 Representing Belief Functions in Conceptual Graphs

We follow the pattern established in earlier sections. The representation includes [DSBEL] and [DSMASS] concepts and (DSBEL) and (DSMASS) relations. Figure 11 illustrates the DSBEL concept and relation.

```
[DSBEL: 0.1]<-(DSBEL)<-[PROPOSITION:
  [TATTOO: #]-
    ->(OWNR)->[PERSON: Wilson]
    ->(IMAG)->[FISH]-
      ->(PART)->[SCALE: {*}]-
      ->(ATTR)->[COLOR: pink] ],,,.
```

Figure 11. Conceptual graph illustrating belief function

Schemata are used to map evidential intervals to English. The evidential interval is the part of the interval [0,1] in which the actual belief would lie, if we could eliminate all nonbelief. If there is no belief for or against a proposition—if ignorance is complete—the evidential interval is [0,1]. If we are 100% certain of the truth of an argument, the evidential interval is [1,1]. If we are totally certain of the falsity of a proposition, the evidential interval is [0,0]. In the interval, the lower bound is called belief (**Bel**—identical with the belief function), and the upper bound is called plausibility (**Pls**). Schemata for these ideas are shown in Figure 12.

```
schema for COMPLETELY-TRUE(x)IS
  [COMPLETELY-TRUE: *x]<-(CHRC)<-[PROPOSITION: *x]-
    ->(DSBEL)->[DSBEL: 1.0]
    ->(DSPLS)->[DSPLS: 1.0],.

schema for COMPLETELY-FALSE(x)IS
  [COMPLETELY-FALSE: *x]<-(CHRC)<-[PROPOSITION: *x]-
    ->(DSBEL)->[DSBEL: 0.0]
    ->(DSPLS)->[DSPLS: 0.0],.

schema for COMPLETELY-IGNORANT(x)IS
  [COMPLETELY-TRUE: *x]<-(CHRC)<-[PROPOSITION: *x]-
    ->(DSBEL)->[DSBEL: 0.0]
    ->(DSPLS)->[DSPLS: 1.0],.
```

Figure 12. Schemata for evidential intervals

Actors for Mass and Belief Functions. Actors for orthogonal summation, normalization, and other Dempster-Shafer operations can be implemented in conceptual graphs. They are not shown in this paper.

5 Fuzzy Logic

In 1965, Lotfi A. Zadeh first proposed fuzzy sets as a way of describing vague systems. Their popularity grew rather slowly until fairly recently. Fuzzy logic is now the focus of significant interest in Japan (Togai [1991]). This section discusses ways of representing fuzzy logic in conceptual graphs.

5.1 Set Membership

In traditional set theory, set membership is discrete: elements are either members of a set or they are not. The membership function μ is defined as

$$\mu_A(x) = \begin{cases} 1 & \text{if } x \in A \\ 0 & \text{if } x \notin A \end{cases} \tag{7}$$

The possible outputs of μ are in the *set* {0,1}. With fuzzy sets, on the other hand, the value of μ can be anywhere in the *interval* [0,1].

We illustrate with a fuzzy set named TALL. Figure 13 shows μ_{TALL} as it applies to humans. The value of $\mu_{TALL}(6)$ is 0.5. This is called the *crossover point*: 50% of all people agree that people at least 6 feet in height are tall. 90% of all people agree that humans who are at least 6.5 feet in height are tall. 100% of all people agree that anyone who is 7 feet in height is tall.

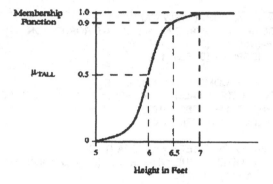

Figure 13. A membership function for the TALL

The *S-function* $S(x; \alpha,\beta,\gamma)$ is used to compute the value of μ (see Figure 14). For μ_{TALL}, the values of α, β, and γ are 5, 6, and 7 respectively.

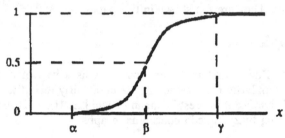

Figure 14. The S-function

The S-function is defined mathematically as follows:

$$S(x; \alpha,\beta,\gamma) = \begin{cases} 0 & \text{for } x \leq \alpha \\ 2(\dfrac{x-\alpha}{\gamma-\alpha})^2 & \text{for } \alpha \leq x \leq \beta \\ .5 & \text{for } x=\beta \\ 1-2(\dfrac{x-\gamma}{\gamma-\alpha})^2 & \text{for } \beta \leq x \leq \gamma \\ 1 & \text{for } x \geq \gamma \end{cases} \qquad (8)$$

We can compute μ only if parameters α, β, and γ are supplied to the S-function.

5.2 Representing Fuzzy Set Membership in Conceptual Graphs

Fuzzy set representation is straightforward using conceptual graphs. Before a system can reason with fuzzy sets, however, it must convert the fuzzy set name (e.g., TALL) to concrete values. The S-function is one method of making this conversion, but to compute μ, the S-function needs the values α, β, and γ.

Schemata can supply values of α, β, and γ. If the expert system must reason about TALL, we can provide schemata for TALL. Depending on the actual concepts involved, different schemata will provide default values for the S-function. Figure 15 shows a schema for tall people. Figure 16 shows schemata for tall buildings.

```
schema for TALL(x) IS
  [PERSON: *x]-
   ->(MBR)->[TALL: *x]-
    ->(CROSSOVER)->[HEIGHT: '6 feet']
    ->(ALPHA-VAL)->[HEIGHT: '5 feet']
    ->(GAMMA-VAL)->[HEIGHT: '7 feet'],..
```

Figure 15. Schema representing TALL people

```
schema for TALL(x) IS
  [BUILDING: *x]-
   ->(MBR)->[TALL: *x]-
    ->(CROSSOVER)->[HEIGHT: '50 floors']
    ->(ALPHA-VAL)->[HEIGHT: '25 floors']
    ->(GAMMA-VAL)->[HEIGHT: '75 floors'],
   ->(LOC)->[CITY: 'New York, N.Y.'],.
```
```
schema for TALL(x) IS
  [BUILDING: *x]-
   ->(MBR)->[TALL: *x]-
    ->(CROSSOVER)->[HEIGHT: '3 floors']
    ->(ALPHA-VAL)->[HEIGHT: '7 floors']
    ->(GAMMA-VAL)->[HEIGHT: '15 floors'],
   ->(LOC)->[CITY: 'Pierre, S.D.'],.
```

Figure 16. Two schemata for tall buildings

To promote determinism, we could provide more generic schemata for TALL, to provide a match of last resort. Figure 17 shows such a schema.

```
schema for TALL(x) IS
  [ENTITY: *x]-
   ->(MBR)->[TALL: *x]-
    ->(CROSSOVER)->[HEIGHT: '8 feet']
    ->(ALPHA-VAL)->[HEIGHT: '4 feet']
    ->(GAMMA-VAL)->[HEIGHT: '10 feet'],..
```

Figure 17. General-purpose schema for tall entities

The actual heights chosen for the generic TALL schema may vary, depending on the domain of the expert system.

5.3 Actors for Fuzzy Sets

Actors are needed for a variety of purposes when dealing with fuzzy sets. The S-function has already been mentioned above. An alternate method for computing μ is the Π-function (Giarratano & Riley [1989], 297 ff.). Besides these, actors must be provided to combine fuzzy values during reasoning. A few of the set operations are given in the following table. Other fuzzy set operations can be implemented as needed.

Actor	Operation	Mathematical Function
F-EQUALITY	A = B	$\mu_A(x) = \mu_B(x)$, $\forall x \in X$
F-SUBSET	A ⊆ B	$\mu_A(x) \leq \mu_B(x)$, $\forall x \in X$
F-PROPER-SUBSET	A ⊂ B	$\mu_A(x) \leq \mu_B(x)$ & $\mu_A(x) < \mu_B(x)$, for at least 1 x $\in$ X
F-UNION	A ∪ B	$\max(\mu_A(x), \mu_B(x))$ $\forall x \in X$
F-INTERSECT	A ∩ B	$\min(\mu_A(x), \mu_B(x))$ $\forall x \in X$

6 Summary

The goal of expert systems is to duplicate the behavior of human experts working under optimal conditions. We have demonstrated the ability of conceptual graphs to represent uncertainty using several methods found in existing knowledge-based systems. Conceptual graphs, by their flexibility of representations, show promise as a vehicle to achieve a unity of uncertain reasoning.

References

Giarratano, Joseph, & Gary Riley (1989) *Expert Systems: Principles and Programming*, PWS-KENT, Boston.

Harmon, Paul, & David King (1985) *Expert Systems: Artificial Intelligence in Business*, Wiley, New York.

Shafer, Glenn (1976) *A Mathematical Theory of Evidence*, Princeton University Press, Princeton, NJ.

Togai, Masaki, Lotfi A. Zadeh, & Piero P. Bonissone (1991) "Fuzzy Logic: Applications and Perspectives" IEEE Videoconferences, Seminars via Satellite, April 25, 1991.

Zadeh, Lotfi, A. (1965) "Fuzzy Sets," *Information and Control* 8 338-353.

Knowledge Fusion

Brian John Garner and Dickson Lukose

Deakin University, Geelong 3217, Australia

Abstract. The integration of conflicting views of experts remains one of the most serious impediments to expanded use by business and government of Expert Systems. Clark (1990) has suggested separation of the risk component (certainty sets) from Expert System rules for judgemental problem solving, and fuzzy rules extracted through simulation studies [2] have also been proposed.

In this paper, a new approach is suggested, based on knowledge fusion, which is achieved through expansion of the goal space to accommodate divergent views.

1   Introduction

An Expert Advisor was recently developed utilising goal interpretation methods [3] as a prototype for a new class of Executable Knowledge Structure. Subsequent recognition, however, of the need to accommodate the expertise of several experts focuses attention on how to expand the goal space to include divergent views.

As described in Lukose (1992), assume that the Goal Space available to the Knowledge Acquisition System is depicted in Figure 1. By utilising the Goal Interpretation Technique, the domain expertise of a particular domain expert can be elicited. In this paper, the authors describe how domain expertise of several domain experts can be elicited and represented as different Problem Maps, The authors also illustrate how the expertise of several such experts could be merged into a Problem Map.

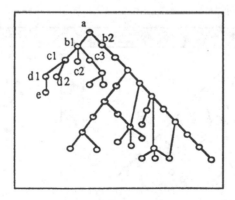

Figure 1. System's Goal Space

2 Eliciting Problem Solving Knowledge from Multiple Domain Experts

If **G** is the final goal state that the domain expert wants to achieve, then the Knowledge Acquisition System will attempt to explicate the planning and strategic knowledge required in order to achieve the final goal state [7]. Figure 2 depicts a Problem Map representing planning utilised by one (i.e., say the first) domain expert.

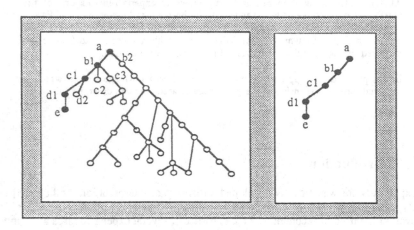

Figure 2. System's Goal Space and a Problem Map

The execution sequence of the above problem map is as follows:

$$e \rightarrow d_1 \rightarrow c_1 \rightarrow b_1 \rightarrow a$$

The second domain expert could utilise the same Knowledge Acquisition System and the same Goal Space, and his/her expertise could also be acquired to achieve the same goal **G** . His/Her problem solving techniques might be different from the techniques of the first domain expert, as depicted in Figure 3.

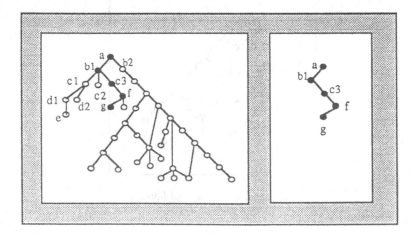

Figure 3. System's Goal Space and another Problem Map

Execution sequence for the above Problem Map is as follows:

$$g \rightarrow f \rightarrow c_3 \rightarrow b_1 \rightarrow a$$

Since the two Problem Maps (i.e., PM_1 and PM_2) are of level one complexity, each will have only a single Problem Map definition. In physical terms, there are two complex conceptual graphs for each of these Problem Maps. Each of these respective definitions will describe the same final goal state, but different execution sequences, and also different initial states. Say for example, the initial state requirement for PM_1 is described as $ISRPM_1$, and for PM_2 is described as $ISRPM_2$, then the Problem Map definitions for these Problem Maps will be as follows:

pmindex(PM_1, CG_1, CG_1id, A, Aid, [_], [$e \rightarrow d_1 \rightarrow c_1 \rightarrow b_1 \rightarrow a$], [G],[$ISRPM_1$],$PM_1$id).
pmindex(PM_2, CG_2, CG_2 id, A, Aid, [_], [$g \rightarrow f \rightarrow c_3 \rightarrow b_1 \rightarrow a$], [G],[$ISRPM_2$],PM_2id).

When the domain user communicates to the Problem Map Execution System that the desired goal state is G, it will select these two Problem Maps. It will then have to disambiguate these two Problem Maps by testing their initial state conditions against the current state of the Knowledge Base System.

Complications arise when the initial state requirements of PM_1 and PM_2 are identical. In this situation, there is no way for the Knowledge Base System to differentiate between these two Problem Maps. Thus, it will choose one first and will proceed to execute it. If it fails, then it will choose the alternative Problem Map and execute it. As one can observe from the Goal Space and the two Problem Maps, the sub-actions that could achieve the precondition of action b is the two actions c_1 or c_3, as determined by both the domain experts, respectively. Since the two Problem Maps are elicited and constructed at different times, there is no strategic knowledge that the execution module could utilise to disambiguate actions c_1 and c_3. In the next sub-section, the authors describe how two Problem Maps can be merged together to form one Problem Map, with a number of different execution sequences resulting from the merger.

3   Merging Domain Expertise

As described above, the Knowledge Acquisition System has the capability to elicit and represent domain expertise from multiple domain experts. Contemporary knowledge acquisition techniques, like "interview", and "thinking aloud", as described in Lukose (1992), can also be utilised (and sometimes very efficiently) to acquire expertise from multiple domain experts. What has been the difficult part, so far, is the ability to merge the problem solving knowledge from multiple experts. In the knowledge acquisition paradigm described in this thesis, the problem of merging domain expertise has been tremendously simplified, due to the utilisation of a semantically rich knowledge representation formalism (i.e., Conceptual Graph), which has a well-defined set of canonical rules and associated operators to manipulate conceptual graphs at both the intra-graph level as well as at the inter-graph level [8].

The two conceptual graphs corresponding to the Problem Maps depicted in Figure 2 and Figure 3, are, in fact, highly nested conceptual graphs, also known as complex graphs. Corresponding to each of the action nodes in a Problem Map is the Actor Graphs (i.e., the declarative structure and the Actor, which represents the procedural component of the "action" that is represented by the action node). Collectively, these action nodes are called Actor Graphs, and these Actor Graphs respond to messages. Each of the Actor Graphs in a Problem Map is activated by sending appropriate messages in a sequence, thus realising the problem solving activity that is implicit within the Problem Map [5].

With respect to the Problem Maps PM_1 and PM_2 described above, the final goal state that both of these Problem Maps arrive at is identical (i.e., G). The initial state for the Problem Map depicted in Figure 2 is the pre-condition of action e while the initial state for the Problem Map depicted in Figure 3 is the pre-condition of action g. Since both these Problem Maps are complex conceptual structures, the merging of these two conceptual graphs is quite simple. We could utilise the "Maximal Join" operation defined by Sowa (1984). In Lukose (1988), the author has described the Maximal Join operation applied to complex graphs, which has sets of other graphs as referents of its concepts. However, canonical constraints have to be imposed by the person who is performing the Join or Maximal Join operation between these types of graphs.

Even though the Problem Maps are complex conceptual graphs, they are unique types. They represent plans, goals, actions, preconditions, postconditions, and at times causal rules representing strategic knowledge. This planning knowledge is also organised in a certain manner to reflect the problem solving techniques peculiar to a certain domain expert. Also described in the Problem Map are the roles involved in the problem solving process. Therefore, constraints need to be imposed when merging the Problem Maps in order to maintain the consistency of the planning knowledge, as well as the roles described in each of the Problem Maps. After several trial and error activities, the author has identified a set of unique constraints under which Problem Maps can be merged. These constraints are described below:

(a)   The final goal state achievable by both the Problem Maps have to be identical;
(b)   The roles played by each of the AGNTs in both the Problem Maps have to be the same; and
(c)   The OBJECTS described in each of the Problem Maps also have to be identical.

The control script that would perform the merging of Problem Maps is quite a complex process. It will firstly have to check the above constraints. Once these constraints are satisfied, it will have to identify identical Actor Graphs, and reorganise the Problem Map by adding the appropriate Actor Graphs into the same nodes, based on the information stored in the appropriate Goal Specification Graphs. For example, the Goal Specification Graph that represents action b will point to three other sub-actions (i.e., c_1, c_2 and c_3). In Problem Map PM_1, c_1 is utilised, while in PM_2, c_3 is utilised. Thus, the control script will add these two actions as alternative actions that could satisfy the preconditions of action b. After merging two Problem Maps, the result is shown in Figure 4.

As each of the Problem Maps represent planning and strategic knowledge, just performing the maximal join operation and reorganising the Actor Graphs are quite insufficient to merge the domain expertise of multiple domain experts. The Problem Map depicted in Figure 4 has two different initial states but has one final state. Thus, the two execution sequences are as follows:

$$g \rightarrow f \rightarrow c_3 \rightarrow b_1 \rightarrow a$$
$$e \rightarrow d_1 \rightarrow c_1 \rightarrow b_1 \rightarrow a$$

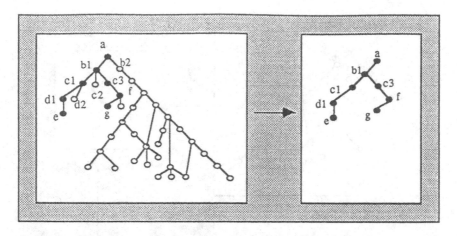

Figure 4. System's Goal Space and the Merged Problem Map

When the control script for merging the Problem Maps is executed between the Problem Maps in Figure 2 and Figure 3, we obtain the Problem Map in Figure 4, but there is one further step involved. From the Goal Specification graph for action b, we have already identified that currently, there are two alternative sub-actions (i.e., c_1 or c_3). To disambiguate these two sub-actions, we have to obtain from the domain experts the strategic knowledge relevant to determine in what situation c_1 would be adopted, and in what situation c_3 would be adopted. This knowledge is then transformed into strategic rules and stored as strategic knowledge. This is the knowledge that the preprocessor would utilise to determine the right sequence of problem map execution. On completion of the merging operation and the strategic knowledge acquisition process, the resultant Problem Map will be depicted as Figure 5.

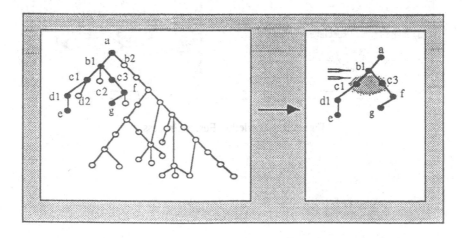

Figure 5. System's Goal Space and the resultant Problem Map

163

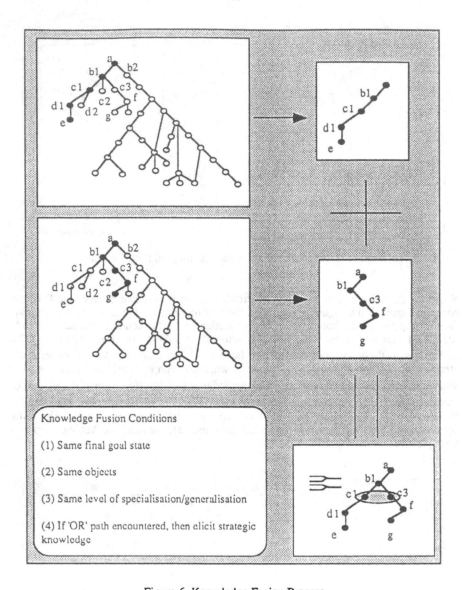

Figure 6. Knowledge Fusion Process

Figure 6 depicts in an elegant manner the whole process of merging the Problem Maps, the elicitation of appropriate strategic knowledge from the domain experts, and the canonical constraints under which these operations can take place. Assume that the resultant Problem Map is called PM_3, and the resultant conceptual Graph is named CG_3 then, this Problem Map has two definitions as follows:

pmindex(PM_3, CG_3, CG_3id, A, Aid, [_], [e -> d_1 -> c_1 -> b_1 -> a], [G],[ISRPM$_1$], PM_3 id$_1$).
pmindex(PM_3, CG_3, CG_3id, A, Aid, [_], [g -> f -> c_3 -> b_1 -> a], [G],[ISRPM$_2$], PM_3 id$_2$).

Physically, there is only one conceptual graph (i.e., CG_3) in the knowledge base, but we have two different Problem Map definitions to represent the two execution sequences. This enables the Problem Map Execution Algorithm to better select the execution sequence, since there is relevant strategic knowledge attached to the third node in the map.

4   Example of Knowledge Fusion

As described in Garner and Lukose (1990a), with the availability of the Goal Specification Graphs, Actor Graphs, and the Goal Interpretation Mechanism, the domain expert can now interact with the knowledge acquisition system which would elicit his/her expertise. Utilising this knowledge acquisition technique, we have been able to construct an expert advisor for career planning. The novelty of our approach to representing the domain expertise lies in the use of Problem Maps. A complete review of the entire Problem Map associated with this expert advisor is beyond the scope of this paper. Here we will review a portion of this Problem Map. This portion implements part of a career planning technique. Figure 7 and 8 depicts two of the Problem Maps that were elicited from the domain expert. The first phase of the career planning technique involves identifying the most suitable occupational category. The domain expert has identified two techniques to identify occupational category:

• identifying the frequency of occupations selected within a certain occupational category; or
• by querying the user for occupational category.

Both these Problem Maps achieve the same goal (ie, obtain occupational category of the user). There are similar objects in both the Problem Maps, and the role of all these objects are identical. Thus, these two Problem Maps are prime candidates for the Knowledge Fusion process. The resultant Problem Map is depicted in Figure 9.

5   Conclusion

This research demonstrates a new approach to knowledge modelling designed to reflect difference in views of experts on possible solution paths in problem solving. Goal space modelling identifies the total solution space known to experts in the field and permits the elaboration of strategic knowledge [4] [7] to disambiguate alternative steps (actions) in the solution process.

Knowledge fusion techniques are, consequently, believed to have wide application in goal space modelling!

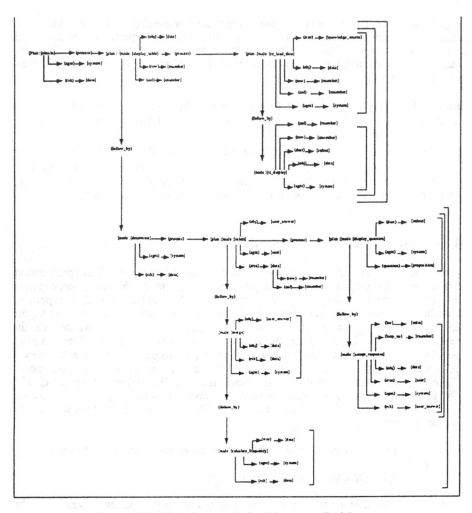

Figure 7. First Technique to obtain "job category" of the user

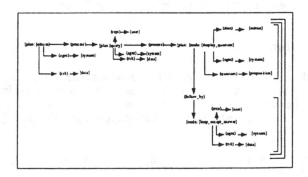

Figure 8. Second technique to obtain "job category" of the user

166

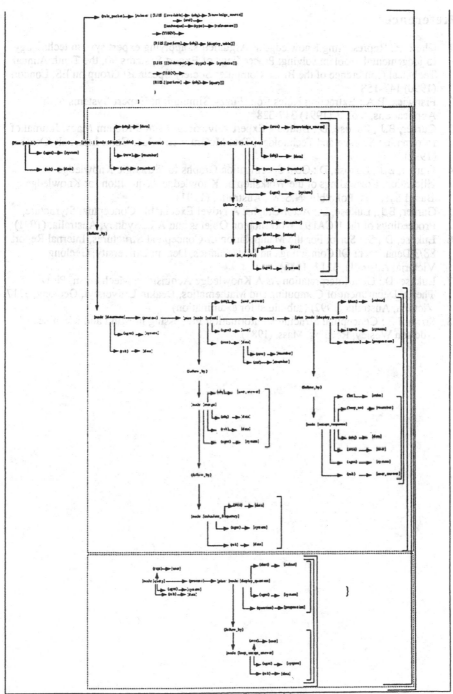

Figure 9. Resultant Problem Map

References

1. Clark, P.: Representing Knowledge as Arguments: applying expert system technology to judgemental problem solving; Proceedings of Expert Systems 90, the Tenth Annual Technical Conference of the Britsh Computer Society Specialist Group on ES, London (1990) 147-158
2. Fishwick, P.A.: Extracting Rules from Fuzzy Simulation; Expert Systems with Applications, Vol. 3 (1991) 317-328
3. Garner, B.J., Lukose, D.: Building Expert Advisor Utilising Problem Maps, Journal of Information Science and Technology, Tata McGraw-Hill Pub. Com. Ltd., July/October (1992) (to appear)
4. Garner, B.J., Lukose, D.: Goal Specification Graphs for Strategic Knowledge Elicitation, Proceedings of the Workshop on Knowledge Acquisition for Knowledge Based Systems, Pokolbin, N.S.W., Australia, (1991)
5. Garner, B.J., Lukose, D.: Actor Graphs: A Novel Executable Conceptual Styructure, Proceedings of the IJCAI'91 Workshop on Objects and A.I., Sydney, Australia, (1991)
6. Lukose, D.: Set Script For the Manipulation of Conceptual Structures, Internal Report 88/3,Department Of Computing and Mathematics, Deakin University, Geelong, Victoria, Australia, 3217 (1988)
7. Lukose, D.: Goal Interpretation As A Knowledge Acquisition Mechanism, Ph.D. Thesis, Department of Computing and Mathematics, Deakin Universiyt, Geelong, 3217, Victoria, Australia (1992) (submitted for examination)
8. Sowa, J.F.: Conceptual Structures: Information Processing in Mind and Machine, Addison Wesley, Reading, Mass. (1984)

IV. Databases and Modeling

Using Conceptual Structures to Translate Data Models: Concepts, Context and Cognitive Processes

Gary Berg-Cross
Advanced Decision Systems/BA&H
2111 Wilson BLVD
Suite 800
Rosslyn, VA 22201

and John Hanna
Vitro Corp.
194 Howard St.
New London, CT 06320

e-mail: garybc@chesapeake.ads.com
jhanna@mcimail.com

June, 1992

Abstract. Sowa's conceptual structures work is broadly framed in terms of logic, linguistics, knowledge representation, conceptual analysis and mental models. Recently the CG representation has been proposed as a normative language for metalevel descriptions of traditional data models by the ANSI Information Resource Directory System (IRDS) working group (X3H4). This paper explores some issues that arise from attempts to use ontologies and "crystallized" representations to perform such things as data model translation and integration. Issues and alternate views that grow out of Sowa's knowledge soup metaphor are described and a functional architecture is proposed that addresses some of these issues. The notions of model knowledge as a dynamic and cognitive construction based on contexts is developed and combined with the idea of global workspaces.

1   Introduction

Sowa's conceptual graphs, as originally described in Conceptual Structures, draws heavily on five interrelated sources: logic, language sciences, data/knowledge engineering, AI technology, and cognitive psychology. While there has been considerable direct work on representational issues, logic, and AI inspired implementations for a variety of applications, the cognitive basis for conceptual structures has received less attention, although the recent discussion of abductive reasoning in a microworlds framework using a "knowledge soup" metaphor (Sowa, 1990) provides a more complex view of classic logic's place and value within other forms of reasoning. It is the position of this paper that a deeper understanding of cognition and cognitive agents is fundamentally important to the continued evolution of conceptual structures. In particular, the artificial separation of knowledge and knowledge processing within conceptual models and expert system implementations disguises hard problems. Reliance on artificially "neat" approaches to knowledge have obscured difficult problems (Berg-Cross,1991) and has produced too narrow a definition of the knowledge level (Clancey, 1992).

In the past year there has been a substantial surge of interest in large KBs (e.g., Cyc) and in the interchange of data/knowledge and knowledge sharing (Mark, 1991). It is generally recognized that there are difficult issues which need to be addressed. These include the technical challenge of sustaining a major effort by multiple people over several years in order to accumulate the knowledge and methods of indexing knowledge without having to manually and exhaustively search for it. Some ideas on cooperative building of a CG-oriented KB was discussed by Berg-Cross (1991) using Cyc as a model of an evolving knowledge base building approach that helps anticipate or repair inconsistencies and awkward constructions. More recently the ANSI X3H4.6 committee, part of the ANSI Information Resource Directory System (IRDS), has developed a draft document discussing a conceptual schema for an Information Resource Directory System (X3H4.6/92-001). The thrust of this work concerns modeling facilities (including CASE tools) which are used to record and analyze enterprise knowledge in the form of data and process models; reference management (library science) facilities like dictionary, encyclopedia, thesaurus, glossary, and concordance; directory management facilities which are used to record data addresses and attributes and provide naming services; systems administration facilities which are used to install and manage computerized information repositories; and life cycle management facilities which are used to support systems engineering and software engineering planning, analysis, and implementation functions. Obviously the work is enormously broad and ambitious.

A growing consensus of this work is that conceptual schemas, represented by a "formal logic base," are necessary to structure and manage the semantics of increasingly complex and large information models. This and related knowledge sharing efforts seek to overcome practical difficulties in reusing knowledge-based systems by use of logic-based representations and complex, fixed ontologies. A central tenet is that ontologies can serve as pre-fabricated skeletons (conceptual structures) on which to hang details for specific applications. Ted Shortliffe, for example, noted that there might be the following relation between the ANSI and DARPA efforts assuming a logic-based knowledge representation standard serving as a neutral interface between systems:

"If we were lucky, we could just use a subset of the languages being developed by the AI community for the exchange of "knowledge" between programs, and by the database community for the representation of "conceptual schemas."

However, as evidenced by a series of "on-line" discussions there are substantial differences of opinion on how to do this, and whether we have languages with enough expressive power to do this. There is some consensus that a formalism can cover data expression and data communication. However, there is dispute about whether linguistic distinctions, made for human communication, are sufficient, or even necessary concepts for data model translation or knowledge reuse. There is even less consensus that the fit of data and knowledge to reasoning processes, the so called inference engines of AI systems, can be addressed by a formalism. In this paper we build on these issues to explore the notion of preserving a richer context as a further basis for data meaning.

While we recognize that this excursion might be considered a further complication of an already challenging approach, we believe that it highlights some tough issues that lurk in the wings and will need to be addressed in the long run. In the sections that follow (within a tight page limit) we touch on the recent conceptual schema work of the X3H4.6 committee to identify some issues involved in static knowledge approaches that support integrating representations between two different data models. Some model entities presented are inspired by the ANSI X3H4.6 committee to help understand this work. Following this, we illustrate underlying cognitive ideas within the debate, and discuss functional architectures to support the cognitive view. Finally, two ideas of global workspaces and situated contexts for agent action are outlined for incorporating the semantics of cognitive agents and process and are within a broader Conceptual Structures framework. These taken together, it is suggested, are part of an evolutionary approach to what will probably be a long road to effective knowledge sharing. Given a flexible framework, particular data structures can be seen as usable and relevant to particular processes and vice versa. The overall approach, which might be called a cognitive systems approach, is discussed within a broad view of conceptual structures and mental models.

2 Illustrations from the X3H4 Work on the Information Resource Directory System (IRDS) Conceptual Schema

Conceptualization of data models is used here to illustrate the role of metadata elements that define alternative data model elements in the X3H4 work. This is a direct attempt to describe a range of data model concepts using static elements. In order to create a conceptual schema for data models, a rudimentary ontology of data model concepts is being structured by the committee. As a starting point, simple descriptive summaries were used to capture the core meaning of a data model and its function. Fact statements on data modeling were developed by the X3H4.6 committee members and standard sources. A sample of starting descriptions include:

1. A data model is a representational tool created by a data modeler using data modeling methods.

2. A data model can be translated into other forms of data models using a conceptual schema.

3. Data models are abstract devices used to describe a domain of discourse, and obtain a reasonable interpretation of data (Tsichritzis and Lochovsky, 1984).

4. Representation is any use of some thing in place of some other thing. In humans, it is a result of fundamental cognitive processes (coding/assimilation) involving things in some real or imagined world and their mental reference by cognitive agents (adapted from X3H4.6 definition).

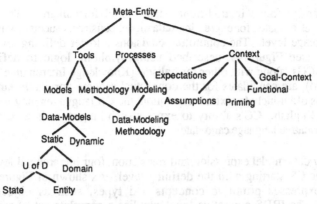

Figure 1. Ontology of Data Modeling Concepts

We note the presence of a variety of ambiguous, cognitive-style and macro terms even within standard definitions. There is considerable surplus meaning to the idea of tool and a structural definition of this term takes us into the ambiguity of the knowledge soup as rapidly as a discussion with Wittgenstein.

Recent work on data and knowledge models by Diederich and Milton (1991) provides an example of the rich knowledge texture of data modeling. They note the role of a data model as providing a "ready-made framework" for a designer to "use in specifying the structure, constraints and possible operations on the proposed database." It is observed, however, that popular modeling formalisms such as Entity-Relationship models often fail to represent all semantic aspects needed. Under these circumstances "the designer is forced to represent it (a data aspect) in an ad hoc manner." Taken together this captures the selective, and at times arbitrary, nature of data modeling. Their discussion also illustrates the need to understand the varied purposes of a data model including as a tool for "communication." A model must also be verifiable to humans. This is a non-trivial interaction and provides one of the natural appeals of CGs for data model description and interpretation.

While a variety of alternative structures can be used to describe the fact descriptions provided above, the following tri-part ontology is our own construction developed for exposition purposes. The ontology, shown below in Figure 1., has the following parts: tool/model, process/modeling, and a context portion that is loosely defined, but is used in more interpretive, cognitive aspects of model understanding that floats in a knowledge soup.

The large preponderance of the work of the ANSI IRDS committee has been concerned with modeling static entities that appear under the models portion of the CS ontology with some additional work on the modeling of process.

One way to view the ANSI work is building metadata model that can be used to translate other data model formalisms. This view is depicted in Figure 2. As shown there, a schema in a formalism such as NIAM, might be "translated" into an entity relation (ER) form using conceptual graphs and mapping grammars as intermediaries. A philosophical base for this work is a form of the "triangle of meaning" (Sowa, 1984). As shown in Figure 3, objects in the world can be described (indirect relation) using a variety of symbols which are directly connected to semantics provided by a conceptual schema. The ANSI work proposes that one can use a variety of logically equivalent conceptual schema (CS) languages to express the syntax and semantics found in traditional data model formalisms. The set of equivalent languages modeled for expressing data model constructs succinctly is called a normative language level. The committee envisions a lower defining level with basic constructs (see Figure 4), described using symbolic logic to define the normative level. CGs, along with other formalisms (Knowledge Interchange Format (KIF) and SUMM), are candidates for the conceptual schema normative language. Since data models often include textual annotation, the CS languages may include a form of stylized English. CG's ability to express such languages is a substantial advantage as a normative language candidate.

To allow data model expression and translation, four hierarchical levels are envisioned for the CS starting with the defining level. As shown in Figure 5, the defining level expresses primitive concepts and types, such as object type, proposition, etc. The IRDS normative level supplies a complete set of modeling constructs. These are expressed in terms of the structures of the defining schema. The modeling schema level, in turn, establishes framework for capturing the conceptual content of a specific model such as ER, NIAM, etc. This is based on the more general ontology established in the normative level. At the highest level is an application schema which defines the object in an application domain.

A small, preliminary portion of the CS ontology is shown below in the CG formalism, using some reasonably standard conceptual relations. These serve to provide points of connection between concepts used in the larger conceptual structure implied by the ontology figure above. We can think of these examples as a small crystallization of "noodles" out of our knowledge soup (Sowa, 1990).

thing > tool > abstract-tool> model> data model>conceptual-model

[data model] -
        (attr)->(context)-
                        (data-model-agent-view)
        (cont) -> [content] -
                        (parts) -> {[data-objects, relation, special-
elements]}
                        (name) -> [data-elements]
        (srce) -> [data modeling]
        (part) -> [static-description]
        (part) -> [dynamic-description]
        (purp) -> [communication]->(obj) ->[understanding]->(loc)-
>[enterprise]
        (char) ->[reduced-abstraction].
process > act> mental-act> modeling-activity> data-modeling-activity>data-
analysis

[data-modeling]-
        (rslt) -> [data-model]
        (actor) ->[data modeler]
        (inst) ->[data-modeling-method]

being>actor>data-modeler

Note: Prototype definitions for many concepts are useful. For example, a prototype for special-element (x) in the above data model definition, might include the attribute concept which is produced as a result of the experience of an actor/agent. Prototypes are useful concepts for people and for models. But as part of conceptual modeling, one often uncovers very preliminary examples of modeling entities. Thus, as part of modeling, the agent may try a particular formalism's attribute concept "fit" against a domain's reality. This heuristic and opportunistic aspect of human data modeling is not easily represented, but an automated tool will need to at least cooperate with a human on such a task. Automating the process will likely require elements of machine learning and hypothesis based reasoning that is not prescribed in fine detail. We believe that the ability to evolve formal models from scruffy and rudimentary ones is an important part of data modeling, but this dynamics is hard to capture. For AI systems it is analogous to providing an explanation trace. For knowledge acquisition tools it is like an illustrated record of "knowledge" revision. The dynamic evolution of model entities should probably be built in to the initial definitions of schemes so that the above definition of data model evolves for the changing views of data elements over time. We recognize that such dynamic views connect to hard problems like the frame problem.

One essential point of this admittedly surface model, is the ease with which the CG expands into a rich schema net that includes cognitive processes and cognitive agents. Is this a useless expansion of an otherwise useful and implementable model? While this is a broad research issue, we would suggest that the answer is no. We believe that any large, useful KB system, metamodel, or knowledge translating facility will require some explicit cognitive structure to be of use. It is in the spirit of data modeling and conceptual structures to tackle such functions. What might be added to the original conceptual structures approach to allow for this? We can think of two things in a preliminary sense. Both relate to the idea of the context - a global workspace approach that allows modular understanding of a complex cognitive space and a constructivist approach to concepts that recognizes the situated and evolutionary nature of some data and associated reasoning.

3   Language Models and Conceptual Modeling

One issue, for at least a vocal minority of the AI community, is the utility of language phenomena as a source for and basis of the "meaning" of " objects" used in data and knowledge bases. Does a natural language oriented view provide a path to expressive power for such KBs? If so, then the mapping from natural language to KBs is a central concern for knowledge engineering methods rather than an independent matter for linguistics. This issue has been discussed by Sowa and others in relation to data and knowledge standards. We noted before the possibility of using restricted natural language to express some concepts that are difficult for traditional data models. A sample of conflicting points concerns linguistic categories and language phrasing. For example, nouns, verbs and adjectives represent linguistic categories. A standing issue is whether these external categories are useful in cognitive models and in building KBs. Informally, such structures as noun phrases are widely used in conceptual analysis. It is important to note that this discussion often breaks into two parts --- the passive use of NL concepts for ontologies and the active use as part of methodology. In any extended discussion, the two parts often fuse. This often confuses the discussion, but if explicated may offer a more robust basis for understanding. Appeal to language also pulls in the idea of context within which linguistic constituents are processed. Language context exists at many levels --- phrase, sentence, paragraph, etc. Within the field of discourse, contexts are seen as enormously broad and complex environments that provide a range of different semantics to linguistic constituents. Based on recent discussion an operating assumption for CGs follows John McCarthy's proposal that context be a first-class object in representation. A CG-context can be simply defined as a concept whose referent is some collection of propositions. This collection is itself describable in a CG. As a CG-object, a context can therefore be talked about and operated on by actors. The contents of a context, as used in Cyc, may be a complete microtheory and thus be very broad. It may be usable in some form to represent hypothetical situations and versions of objects at different times. This idea has recently been discussed by Sowa in relation to object-oriented models.

One view of language is as a model of the world, but a very special model. Language has a favored position with humans since it serves a daily, social communication function and has "evolved" through selective adaptation of the form. An auxiliary hypothesis is that many, if not all, linguistic distinctions make a difference even if we, as language comprehenders, are not conscious of the reason for them. We can superficially explain these as due to the varying schematic contexts that they either evoke or exist in. Without trying to explore this particular argument here, we note the growing work in cognitive anthropology, for example, Keller and Lehman (1991). Their conceptual analysis finds many more distinctions than those of the English language when trying to capture the meaning of other cultures and *translate* them into broad, objective, scientifically based language models. Such efforts have found a need to modify and elaborate cognitive science concepts like category theory, prototype definitions, and basic-level categories. It is our speculation that the current state of data models can be thought of in cognitive, adaptive, or evolutionary terms itself. This idea is portrayed in Figure 6. Both data models and natural language are implicit models and are used to communicate between agents. But distinctions in this communication involve some cognitive system to keep the focus of attention, to respond to environmental changes and to actively perceive the external strings and relate them to internal states. The data model realm is, of course a small subset of what full natural language is about and thus data model languages can exist in rudimentary forms that are not as fully expressive as natural language. However, because data is ultimately about the real world, an amazing amount of complexity creeps in when converting from one scheme to another. At its extreme, it is like a knowledge acquisition transfer from an expert (one scheme) to a knowledge formalism (the other scheme). Indeed, knowledge discourse rather than use of mapping rules may be a more useful way of thinking about data model translation, integration and knowledge fusion.

Therefore, there are a number of reasons why a cognitive agent might want to distinguish between alternate NL expressions and why a seemingly philosophical difference that makes no difference in a truth-functional way, can make a difference in mapping to and from principal categories of a commonsense ontology which underlies understanding. Indeed the crux of this argument might be expressed as a central question for knowledge engineering. Is there a knowledge base and psychological reality to 'principal categories'? Is there a practical significance that matches the intuitive sense we have about these? The work of the ANSI X3H4.6 committee on data model concepts represents one testbed for this idea. Indeed, as an

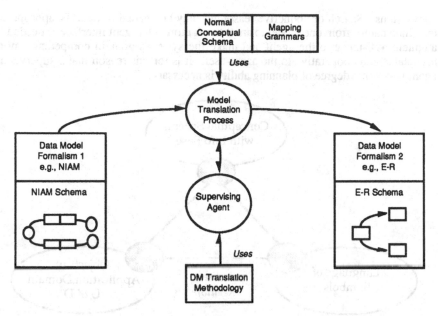

Figure 2. Model of Supervised Data Model Translation Supported by Housekeeping Type Aid

effort to foster data model conversions, it parallels the work of translation from one language to another. A stylized view of this process was shown in Figure 2. This simplified conceptualization shows an intervening process of human agent and machine processing that converts information from one data model formalism into another. In this version the agent has overall and detailed control and uses some data modeling methodology at least to guide the conversion, start it with identified areas, handle failures, etc. The automated process, under human supervision, accesses a normative schema and mapping grammars to aid the human in translation. A problem with this view is the assumption that data model translation can be entirely pre-analyzed and its semantics and procedural essence captured in a combination of static metadata structures and forward chaining type algorithms.

As previously noted, data models represent data that is about the real world. Thus, a certain amount of additional semantics arises from the meaning of data within organizational settings, specialists, etc. Data models have been moving towards more expressiveness, but still typically don't express the schematic contexts within which data modeling focus meanings are checked for broader fit . For example, nowhere in the typical data model formalism is the context of alternative choice made by a data modeler represented. A further complication is that data modeling, as an intellectual exercise, involves extensive analysis and human conceptualization. Efforts to handle such complexity in an automated fashion usually are described in terms of knowledge-based efforts. For example, we can view model translation support through an associate technology (Berg-Cross, 1991) which is made up of an intelligent interface and KB system for reasoning about data model conversions. Search or abductive reasoning may be needed to identify appropriate matching names from one model to another. A more intelligent interface is needed to augment assistance to the agent and make the system's domain competence more acceptable and cooperative to the agent-user. It is for this reason that a supervising agent with some degree of planning ability is necessary.

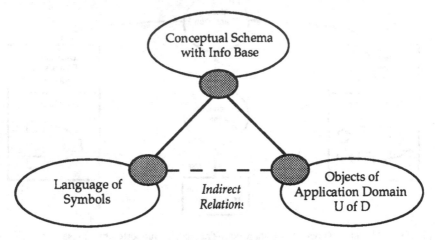

Figure 3. Conceptual Schema Triangle of Meaning

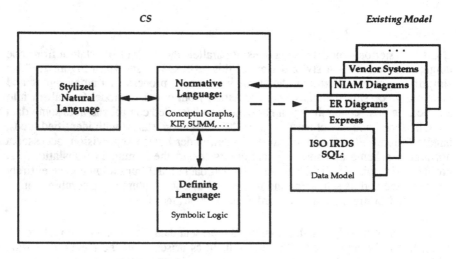

Figure 4. Conceptual Schema (CS) Languages

Such a cooperation needs to be based on some common meaning – for both humans and systems. One very simple view of the elements involved in such a communication is loosely based on the triangle of meaning as shown in Figure 4. This shows three types of elements - concepts, symbols and reality of which the first two are used as a basis of the discussion about the third. We view both humans and artificial cognitive systems (ACSs) as having a "concept" level that employs a large cognitive soup as a basis for its knowledge. We, of course, recognize that current AI implementations are deficient in this area. We envision the concepts as floating in an unimplemented soup developed as part of knowledge acquisition (see Figure 7). In the case of human agents (right side of Figure 7), these experiences --- schematic-based musing and background knowledge --- are further structured into a mental model. For ACSs (left side of Figure 7), they are developed into a knowledge level model we recognize as a KB. In our diagram, we distinguish a further structuring made popular by Cyc, in terms of a scruffy, heuristic level and a neat, logic level. The meeting point between agents and systems is through the symbols each uses to express their model. Each model describes a portion of reality, in this case, the enterprise. The appeal of CGs as a data model formalism is evident in this diagram since it serves both as a description of data and of some portion of language.

Implicit in this view of data models is the idea they do not exist by themselves but represent the translated product of sustained cognitive activity that is guided by models but grows out of inherently scruffy knowledge. While a data model seems static, it is dynamic like a sentence. Data model facts are always part of a richer context, including inference processes such as abductive reasoning. The structure versus process dichotomy can be more broadly viewed in terms of the traditional distinctions between declarative and procedural approaches (fixed and fluid) to knowledge and the corresponding structure versus processing debate from cognitive science. It also corresponds to the KB vs inference engine components of expert systems implementation. But the experience of early AI systems reveals deep problems lying under these simple dichotomies. In the next section we describe a functional model of data translation service that is driven by a context of interaction between a cognitive agent and a data modeling method.

4 Data Model Translations Viewed from a Three Level Functional Architecture

In this section, we discuss a functional architecture to understand and handle two types of things that arise from a broad view of data modeling - environmental interactions and conceptual models. Both views grow out of attempts to provide cognitive bases for complex, orchestrated behavior such as modeling analysis. The first, interactions, allows a translation system to be responsive and proceed without having complete knowledge. It represents a functionality separate from monolithic models and can be thought of as reactively driven by bottom up processes. The second, conceptual models, is more typically associated with CG work, but in this context represents a tighter, top-down, type of control of more traditional functions via a process that might be thought of as relatively abstract planning.

The environmental or reactions position of data (Agre, 1988) is that descriptions (knowledge) of entities in the real world, and any representations of them, necessarily involve a combination of a cognitive agent's view point (what it is doing) and the place these entities have in relation to it. Unlike a model-driven function, the representation of the problem state of translation need not be either objective or global. In this view of data modeling, the representations for a data model are treated not as static or object facts, but as objects indexed to a context that is a combination of other agent's activity (i.e., the data modelers actions through the interface) and the system's own dynamics. This view of data, which we believe is compatible with CG approaches to concept labeling, describes all data representations as indexed and functional. A very simple schema for this was evident in our definition of data model. Data models, and the formalized knowledge that makes them up, are acquired in interaction with the environment, as shown in Figure 6. Their context is dynamic and shifts, but provides a situated basis for the subsequent formalized models that are externally readable. One example of this shifting context is the priming effect, widely studied in human cognition. Cognitive processing, especially encoding, is highly focused by prior events. The processing and interpretation of succeeding events can be primed by earlier ones. Alternatively, active anticipation provides a context for modeling decisions made as a model is developed. Again, we find ourselves thinking of the modeling process as a *learning one in which cognitive products result from the processing of older knowledge and newer experiences.* It is our guess that the recent discussion of contexts with the AI and CG communities can be combined with the situated understanding movement of Cognitive Science to forge a useful ontology of contexts. While these provide a much more complex view of what goes on in forming a data model, the resulting understanding represents a significant crystallization of knowledge out of a dynamic soup.

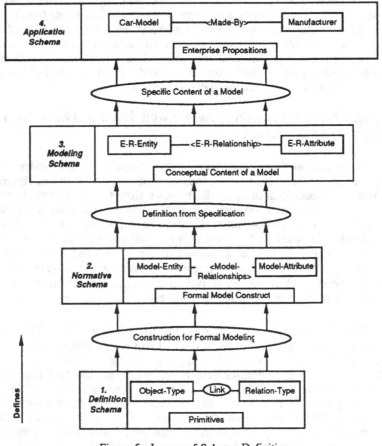

Figure 5. Layers of Schema Definition

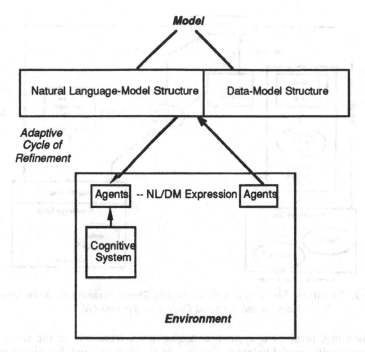

Figure 6. Adaptation of Model Structure To Agents Working In Environment

      While we can not yet offer a satisfactory approach to representing context, we can envision some functional architectures to support Data Model translation. This is shown in Figure 8. The model, an extension of our Figure 2. view, is based on Clancey's (1992) conceptualization of an alternate view of AI architectures. It is inherently cognitive in nature and describes a central level, that is model-based, responding to environmental demands above and using highly specific models below. At the highest (in the sense of being able to exert control over lower ones) level are the interaction oriented systems just discussed. This function provides a basis for perceptual-type cognitions that derive from interactions with the environment. In this case, interactions involve a very complex interface with the data modeler/agent, any dynamics with the data models themselves, and with its working translation. Representation of these interactions is probably very different from lower level representations and looks more like dynamic state sequence structures.   For interactions with an agent, it may involve high-level concepts of agent-intentions which are passed up from the mid-level. The operations at this level of the architecture are describable in terms of interaction such as monitoring external states. These can affect lower processes by providing a context for them. For example, a change in focus by the agent may require a substantial change in the model level process below. It may be useful to note that simple monitoring interactions don't require deep understanding, but interactions with an agent may.

      At the mid or model level, we can adequately deal with user intention issues and with the consequence of meaningful environmental changes. This level deals with the high cognitive processes that form and test conceptual descriptions of identified things and processes occurring in the environment. Clancey (1992)

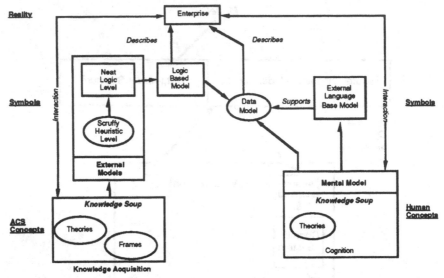

Figure 7. Identifying Elements and Views Within Shared Meaning Between Interacting
Humans and Artificial Cognitive Systems (ACS)

describes this level as a system-modeling level from which the most detailed
situation specific model below can work. At this lowest level, functionality is quite
powerful, but responds to what might be loosely called situated understanding given
the response of the general "model" to the interaction level's perception from the
environment. However, in the three-level architecture the overall work is now
situated and responsive. This lowest level functionality may also include very
powerful AI techniques such as machine learning as in SOAR and PRODIGY.
However, all such efforts are within a context generated by the higher levels.

In fact, base-level cognitive material such as goals and purposes of cognitive
agents, likely dominates the use of data. Since the extent of such cognitive agent's
mental contexts are large and varied, any data or knowledge standard must allow for
data incompleteness or varied mental contexts (mental models) that may be
inconsistent with one another. How those inconsistencies can be resolved is an
important research problem, since this is often a point of major *negotiation* between
data modelers. It is for this reason that it may be hazardous to elevate one prescribed,
static standard method, such as implied in Figure 6., over others. The current
approach under consideration by the X3H4.6 committee is that a data standard must
have a way of packaging knowledge in contexts and allowing somebody to apply
various reasoning systems to any part of it. That is, the method should have
qualities we loosely identify with cognition --- dynamic responses that follow
agendas but are context sensitive. We faced this problem in our original IMACS
work where we employed a methodology called MASER and partially structured
entities in MODAM (Berg-Cross & Hanna, 1986). The view outlined in Figure 8.
moves towards an explicit model of how such dynamic responses might be
achievable. Past work has shown that the conceptual structures approach provides a
good basis for representing the mid and lowest level, but what about the context of
situations? We believe that context can be modeled for this. That is, we believe that
there is *no contradiction in modeling this interactive and constructivist level.*

The functional architecture has grown out of a knowledge soup metaphor (see Figure 9), and we find it useful to represent the interactive portion of this as a "foam" that is the dynamic interface between the soup and the air (external world). In Table 1. below, we provide a summary of some of the distinctions that apply to the three components of such a foamed soup. The dynamic portion of the soup base provides a great range of the problem solving ability, but its use through computation is inefficient, may include inconsistent structures, and is highly dependent on context. This makes it difficult to generalize. The solid part of the soup, the noodles if you will, are theories and knowledge structures that we have crystallized out of this apperceptive mass, making clear distinctions. Such formalizations are efficient for isolated problems, and are relatively context insensitive and should be consistent. The foam portion we think of as working memory or even consciousness in a cognitive theory. It's contents are focused by a mixture of actions in the real world, and the noodles within the context of the soup base activities. In this metaphor, a noodle crystallizes out of the foam, but the record of the constructive activity is lost as the foam moves on to other activities. It is possible to reconstruct this formative context, but this itself is a foamy activity and not usually crystallized into structured understanding.

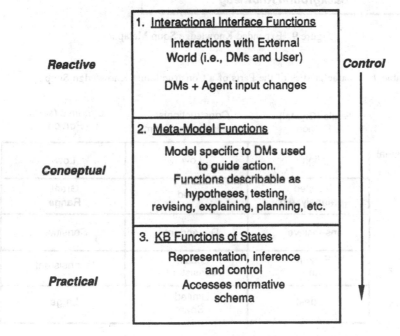

Figure 8. Functional View of Architecture Layers Used to Support DM Translation

Reality

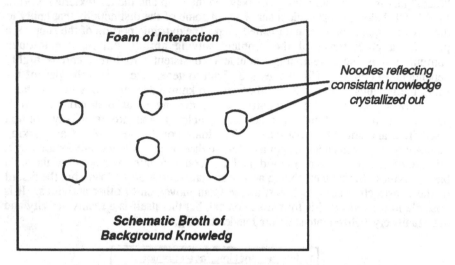

Figure 9. Expanded Knowledge Soup Metaphor

Table 1. Characteristics of the Parts of a Constructionist Knowledge Soup

	Static (Crystal) Portion	Constructionist Working area (Foam)	Dynamic (soup) Portion
Computational Efficiency	High	Low	Low
Problem Solving	Isolated (preanalyzed)	Mixed	Great Range
Context	Insensitive	Dependent	Sensitive
Consistency	Frozen in	Locally consistent	Inconsistent
Span	Modest	Limited Span	Large

5 Final Thoughts

In the future, we expect a more thorough exploration of some other implications of this revised metaphor, but we want to briefly note two directions. The first concerns an idea and an implementation model – global workspaces. Global workspaces have been employed in advanced AI implementations when a unitary expertise is not adequate to solve a complex problem. We agree with the position that such distributed, cooperative systems provide a dynamic way to bring diverse knowledge, such as found in separate data models, together. In this view, relevant data is not totally organized for overall function, but is brought into new relationships through locally orchestrated, dynamic processes. An ontology is usually thought of as a static way to organize knowledge, but ultimately local contexts are needed to add or modify semantics to these knowledge structures. Given the state of knowledge engineering (KE) practices, much captured, externalized knowledge is inherently ambiguous and cannot be unified with related knowledge without *active cognitive (learning) processes on the part of agents*. Perhaps language phenomena is such a useful tool for KE because it, like visual perception, is under determined – its meaning only emerges from numerous routinized linkings. Global workspace implementation allows such linkings through knowledge broadcasts. We believe that this process should be explicitly modeled and incorporated in the conceptual structures approach. The ACTOR linking with CONCEPTS provides a very high level mechanism to start this, but the idea needs to be extended to allow links to contexts at a variety of levels. We recognize that this may make models quite complex and difficult to manage. Further implementations from the models will be difficult, but we believe useful.

The second idea concerns adding a constructionist philosophy as a further addition to conceptual structures. This moves away from the older notion of cognitive science that cognition involves manipulating fixed representations which contain meaning within them. A more constructionist view is that meaning emerges from the interaction of cognitive processes (ACTOR processes) working on structures *in the context of actual or possible experience*. In this view, the older conceptual structures are just part of a cognitive structure. Cognitive processes working interactively are the other part. This leads to a new view of ontologies as dynamic structures. One view of such a structure is provided in Table 1. and our discussion of foam. We hope that operation of a conceptual schema in an environment of external models, linked to communication with data modelers, provides a domain for testing a tractable subset of this theory.

References

Agre, P.E. (1988). *The dynamics structure of everyday life*. Unpublished doctoral dissertation, Massachusetts Institute of Technology, Cambridge, MA.

Berg-Cross G. and Hanna J. "Database Design with Conceptual Graphs," First International Workshop on Conceptual Graphs, sponsored by IBM Systems Research Institute, Thornwood, NY, 1986.

Berg-Cross & Price "Acquiring and Managing Knowledge using a Conceptual Structures Approach" IEEE Transactions on Systems, Man and Cybernetics, vol 19:3 Pg 513-527, 1989.

Berg-Cross G. Multi-Agent Associates for SOAS Planning," paper for the Symposium on Associate Technology: Opportunities and Challenges, George Mason University, June, 1991.

Berg-Cross G. "Can a Large Knowledge Base be Built by Importing and Unifying Diverse Knowledge?: Lessons from Scruffy Work., presented at the 6th Annual Workshop on Conceptual Graphs, July 10-13, 1991, Binghampton, NY.

Clancey B. Model construction operators, *Artificial Intelligence*, 1992.

Diederich J. and Milton J. "Creating Domain Specific Metadata for Scientific Data and Knowledge Bases," *IEEE Trans on Knowledge and Data engineering* , Vol. 3, #4, Dec. 1991.

Mark, William Panel on Achieving Large Scale Knowledge Sharing, in Proceedings of KR'91 2nd Int. Conference , Principles of Knowledge Representation and Reasoning, 1991.

Sowa, John Crystallizing Theories out of knowledge soup, in Intelligent Systems: State of the Art and Future Directions (Eds.) Rasa and Semankova, Ellis Horwood, 1990, pp 456-487, 1990.

Tsichritzis D. & Lochovsky F., Data Models, Prentice -Hall Inc. 1984.

X3H4.6/92-001 Model Unification for Data Repositories, March 1992

Towards Deductive Object-Oriented Databases Based on Conceptual Graphs

Vilas Wuwongse and Bikash C. Ghosh

Division of Computer Science, Asian Institute of Technology
G.P.O. Box 2754, Bangkok 10501, Thailand
Fax: (66-2) 5245721, e-mail: vw@ait.th

Abstract. Deductive object-oriented databases (DOODBs) are an integration of deductive databases (DDBs) and object-oriented databases (OODBs). DOODBs could be considered to be a database system which is based on logic and object-oriented paradigm. Application areas of DOODBs include advanced information systems, natural language processing and knowledge bases. Conceptual graphs (CGs), a system of ordered-sort logic, have useful constructs which are suitable for the requirements of DOODBs. This paper employs conceptual graphs to develop a foundation for DOODBs. The DOODBs are characterized by data abstraction through objects, object identifiers, object types, type hierarchy, property inheritance, methods and message passing and a logical formalism with a sound inference system. Some restrictions and extensions are proposed for the general CGs so that they can be used to represent the DOODB concepts. These extended CGs are called deductive object-oriented conceptual graphs (DOOCGs). The object types, individual objects and object identifiers of DOODBs map into concept types, individual CGs and individual referents, respectively. Methods are defined using conceptual schema graphs with bound actors and interpreted in a success/failure paradigm. A set of extended derived rules of inference is formulated for DOOCGs which are proved to be sound. The semantics of DOOCGs is also briefly outlined. A DOODB is then defined to be a set of DOOCGs together with a set of axioms and a set of inference rules.

1 Introduction

In spite of its widespread use in business and industry since its introduction in 1970, the relational data model has certain limitations. It can only deal with data represented in terms of records which consist of a fixed number of fixed-length fields. Abilities to represent complex data are required in some advanced application areas like computer-aided design and manufacturing, integrated office information systems, computer-aided software engineering, and artificial intelligence. A detailed description of the major limitations of record-oriented data models can be found in [10]. The two of the major areas of extensions over the conventional databases are: i) capturing more semantics in the represention of data [3], and ii) storing intensional rules and performing reasoning over the database states [11]. The first extension leads to the research on *object-oriented database* (OODBs) systems [16], while the second one is handled by the *deductive databases* (DDBs) [12, 16].

Deductive databases integrate ideas from logic programming and relational databases [11] while object-oriented databases emerge from the integration of database concepts with object-oriented programming paradigm. Since each of the DDBs and ODDBs offers some advantages over conventional databases, there have been efforts to merge the two approaches to obtain the advantages of both the systems, and these efforts represent a challenge [1]. The notion of *deductive object-oriented database* (DOODBs) comes from some of these efforts. DOODBs can also serve as a starting point for a future *knowledge based management system* (KBMS) that integrates the ideas of artificial intelligence and databases [2].

The formalism of *conceptual graphs* (CGs), introduced by SOWA [13], offers some useful constructs which make it a likely platform for an integration of ideas from DDBs and OODBs. It is a powerful knowledge representation language in AI with a well-defined theoretical basis and a close mapping to both natural language and first-order logic [9, 13, 15]. The objective of this work is to develop a notion of a deductive object-oriented database system based on the formalism of CGs. Some restrictions and extensions to CG formalism are introduced and a theoretical basis for a DOODB system based on the extended CG formalism is developed.

The remainder of this paper is organized into four sections. Section 2 summarizes the characteristics and requirements of DOODBs. Section 3 shows how to construct DOODBs from CGs, while section 4 provides their semantics. Section 5 draws conclusions.

2 Deductive Object-Oriented Database Systems

Two general approaches towards research in DOODBs have been mentioned in [17], namely DOODBs as an extension of DDBs and DOODBs as an extension of ODDBs. We propose a third approach which is to integrate the concepts of DDBs and OODBs and map them into a third framework. In this work, the formalism of CGs is used to represent the features of DOODB systems whose characteristics are outlined as follows.

(a)   *Objects and object properties*: A DOODB should allow all conceptual entities to be uniformly modeled as objects. Objects should be the unit of access and control. An object has properties which could be static or dynamic. Static properties represent states (or structures) of the object. Dynamic properties (called behaviors) are represented by a set of methods. A DOODB system should provide mechanisms for generating and maintaining unique object identifiers for individual objects.

(b)   *Types (or classes), type hierarchy and property inheritance*: Objects are instances of types. A type describes the structure and behavior of all of its instances. Types should form a type hierarchy based on the subtype/supertype relationships. Objects of one type inherit properties from the type and its supertypes. A DOODB system should have mechanisms for:

(i)   resolving the problem of multiple inheritance, and
(ii)   handling exceptions to property inheritance.

In general, the definition of an object type in a DOODB should include:

- Specification of the structure of the objects that are instances of this type,
- Specification of the position of the type in the type hierarchy,
- Specification of the set of methods that are applicable to the instances or states of the instances of that type.

(c) A DOODB system should have a logical formalism to represent objects, methods, intensional rules or (deductive) laws, integrity constraints and queries, with a well-defined semantics of all of these components as well as answers to queries. The logical system should have a sound (deductive) inference mechanism.

(d) Some of the other desirable properties are : declarative expressions for databases, queries and answers to queries, and single language for expressing databases, queries and integrity constraints.

3 Deductive Object-Oriented Databases Using Conceptual Graphs

The features of DOODBs are to be represented by the *deductive object-oriented conceptual graphs* (DOOCGs) [7]. The major features of DOOCGs are explained in this section. More details about DOOCGs can be found in [7].

3.1 Primitives of Deductive Object-Oriented Conceptual Graphs - Syntax

Object types in DOODBs are represented by concept types in CGs, *individual objects* of a type are represented by individual CGs of the corresponding concept type. For example, an object type STUDENT can be defined as a subtype of PERSON as shown in Fig. 3.1, where all other concept types and conceptual relations are assumed to be defined already. An individual object of type STUDENT is shown in Fig. 3.2. The object type hierarchy is represented by the concept type hierarchy that is a complete lattice based on the subsumption relation $\leq_T$. The type definition specifies the position of the newly defined type in the hierarchy. Individual referents for the concepts that are parameters of the type definitions are used as *object identifiers*. For example, the individual referent #112 in Fig. 3.2 is an object identifier. Note that the individual referents include the individual markers as well as other externally communicable objects like strings and numbers.

Formulation of *methods* and *message passing* is one of the important problems in merging OODB and DDB concepts. For our representation, the major requirements are as follows:

- a mechanism for defining methods and associating the method definitions for a type with the corresponding type definition,

```
STUDENT = (λ x)
      [PERSON:*x] -
            (CHRC)-->[ID]
            (OBJ)<--[ENROLLMENT] -
                        (PTIM)-->[DATE]
                        (CHRC)-->[DEG]
                        (LOC)-->[DIVISION],
            (ADVISOR)-->[FACULTY].
```

Fig. 3.1 Definition of STUDENT type as a subtype of PERSON

```
STUDENT(#112) =
      [PERSON:#112] -
            (CHRC)-->[ID:3421]
            (OBJ)<--[ENROLLMENT:#321] -
                        (PTIM)-->[DATE:01/01/1990]
                        (CHRC)-->[DEGREE:"M Engg."]
                        (LOC)-->[DIVISION:"CS"],
            (ADVISOR)-->[FACULTY:#311].
```

Fig. 3.2 An individual object of type STUDENT

- a format for method calling or message passing,
- a backtrackable method evaluation mechanism (so that all possible values are returned) with a success/failure interpretation [6] of method evaluation,
- a mechanism that allows individual objects of one type to share the methods defined for its supertypes.

A method is defined using a *conceptual schema graph* [13] with bound actors. A method call or message is represented by a *message graph*, that is a simple CG with an appropriate actor bound to it. An example method definition for the age of a person is shown in Fig. 3.3, where <TODAY> and <DIFF_DT> are two actors. The actor <TODAY> is a system defined actor that asserts the current date in the referent field of its output concept. The actor <DIFF_DT> is a primitive actor (or can be defined by a functional dataflow graph) that asserts the difference of two of its input dates into the referent field of its output concept. An example message graph is shown in Fig. 3.4, where the type label in the actor box before the name of the actor specifies the receiving object type. The graph in Fig. 3.5 is the same as that in Fig. 3.4 except that the output concept of the actor <PERSON:DIFF_DT> contain a request mark ("?") which signifies the triggering of the execution of the corresponding method. In general, if m is a message graph then ?m is used to denote the corresponding method call. Note that the projection of a message graph is defined as the projection of the CG obtained from the message graph by erasing the actor bound to it.

A new CG operation is necessary in formulating the rules for method execution. The

SCHEMA for PERSON = (λ x)

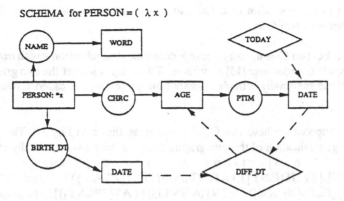

Figure 3.3: An example of method definition for the type PERSON

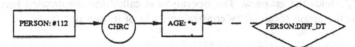

Figure 3.4: An example message graph

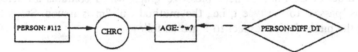

Figure 3.5: The message of Fig. 4.4 is being "passed"

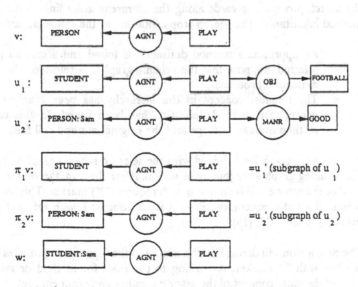

Figure 3.6: An example of constrained join - the graph w is
obtained by constrained join of u₁ and u₂

main purpose of this operation is to find out the *maximal common overlap (modulo restrictions)* between two CGs.

Definition 3.1 Let two CGs u_1 and u_2 have a common generalization v with *maximally extended compatible projections* [13] $\pi_1:v \to u_1$ and $\pi_2:v \to u_2$. A join of the two graphs $\pi_1 v$ and $\pi_2 v$ on these maximally extended compatible projections is called a *constrained join*.

Example 3.1 Suppose we have two CGs u_1 and u_2 as shown in Fig. 3.6. The graph v is a common generalization of the two graphs u_1 and u_2 with two maximally extended compatible projections π_1 and π_2, where π_1 = {([PERSON],[STUDENT]),(AGNT,AGNT),([PLAY])} and π_2 = {([PERSON],[PERSON:Sam]),(AGNT,AGNT),([PLAY],[PLAY])}. The graph $\pi_1 v$ is a subgraph of u_1 and $\pi_2 v$ is a subgraph of u_2. A join of $\pi_1 v$ and $\pi_2 v$ (on the projections π_1 and π_2) yields the graph w. The operation is called the *constrained join* and the graph w is the graph obtained by constrained join of u_1 and u_2.

Rules for method execution

Let S be a set of CGs and m be a message graph which contains an actor e. the message is "passed" to the type t, i.e., the method call ?m is issued. Steps taken in executing the method with respect to S are as follows:

1)   Search for a definition of method corresponding to the message starts at the schematic cluster [13] of type t. If a schema is not found for the type t, then the search proceeds upwards along the hierarchy according to the rules of method inheritance. The search stops with one of the following conditions:

-       An appropriate method definition is found and a schema graph is selected. Suppose that the schema graph u is selected. The method evaluation proceeds.
-       The topmost concept in the hierarchy has been reached but no appropriate method definition has been found. In this case, the method evaluation stops and the original method call fails.

2)   The message graph ?m is joined with the selected schema graph u. Let u' be the resulting graph after joining m with u. After the join, the output concept in u' of the actor e will be marked with request ("?") marker. This will trigger a goal directed computation. For all input concept i, with referent (i) = *, mark (i) is set to "?" [13].

3)   The next action will depend on the nature of the input concept nodes that are marked with "?" marker. According to the rules for method definition [7], each of the input concept of the actor e is either an output concept of another actor or a message to another type, or an "independent property" that is a part of the type defining graph.

- For each input concept that is an output concept of a message to another type, the message is passed to the corresponding type. If the evaluation of any of such message fails, then the actor e is blocked [13] and the original method call (?m) fails.
- For each input concept that is an output concept of another actor, the "?" maker is propagated in the backward direction.

The system would search for an individual graph to evaluate those concepts that represent independent property and are marked with request marker. This search can be performed in the following way:

- Suppose the corresponding type has been defined with an abstraction of the form λxv. The graph v is joined with u' using the constrained join. Let v' be the resulting graph. If v' does not contain all those concepts that are marked with "?" marker but are not output concepts of any actor or message, then the search stops and the original method call fails. Otherwise, the graph v' is to be projected on the set of individual CGs in S. Two possible cases are:

   (i)   The graph v' has no projection on the set of individual CGs in S. The original message call fails.
   (ii) The graph v' has a projection π' on the set of individual CGs in S.

- If the previous step succeeds, then the graph π'v' is joined (using maximal join) with u' yielding w for each projection π' of v' on the individual graphs in S. As a result of this, individual referents would be asserted into some or all the concept nodes that were marked with "?" marker.

4)   If all previous steps succeed, then the actor e in w must be enabled and fired. This firing would assert a referent in the output concept of the actor. Let w' be the final working graph after the actor has fired. The original method call succeeds with the answer <true,π> where π:m→w'. In general, there may be more than one return value depending on the number of distinct projections of v' on the set of individual graphs in S. Suppose m' is the CG obtained from m by erasing actor(s) bound to it. Then πm' is also true, i.e., the assertion made by the message graph m is true.

Success and failure of a method call

The value of the method execution ?m will be <false,{}> only if the method evaluation fails. The method evaluation fails if any of the following is true:

- The search for an appropriate schema fails in the first step, i.e., their no method definition corresponding to the message.
- The firing of the actor is blocked in step 4 for any reason whatsoever.

If the method does not fail, then it succeeds with the answer $<true,\pi>$ where $\pi{:}m{\rightarrow}w'$, where w' is the final working graph. Suppose m' is the conceptual graph obtained from m by erasing actor bound to it. Then $\pi m'$ also makes a true assertion.

Property inheritance

In CG notations, a branch of the type defining graph consisting of a conceptual relation and a concept type label corresponds to an attribute in the DOODB concepts and an individual referent in the concept node of that branch is the value of that attribute. The inheritance of properties and the exception rules are "naturally" handled by the *type expansion* [13] operation of CGs that use the maximal join and replaces a single concept node with the graph that defines that concept type. Methods defined for a particular type are also inherited by its subtypes. The search for a method definition starts at the type that is specified in the actor box with the actor name. If a method is called with the type label t, then the search starts at the schematic cluster of type t. If the method definition is not found, then it is searched in the set of method definitions for each of its direct supertypes. The search proceeds upward along the chains of the hierarchy until an appropriate method definition is not found, then it is searched in the set of method definitions for each of its direct supertypes. The search proceeds upward along the chains of the hierarchy until an appropriate method is found or the topmost type UNIV is reached. The first method encountered in the search is the one that is executed.

3.2 Conceptual Basis

The meanings of conceptual graphs in CG formalism are implicitly related to some basic CG constructs. These are collectively referred to as a conceptual basis, which is defined as follows.

Definition 3.2 A conceptual basis is defined as

$CB = <(R,\leq_R), (T,\leq_T), D, SC, P, A>$, where

CB	:	Conceptual basis,
R	:	A set of conceptual relation labels,
$\leq_R$	:	Subsumption relation between conceptual relation labels,
$(R,\leq_R)$	:	Conceptual relation hierarchy,
T	:	A set of concept type labels,
$\leq_T$	:	Subsumption relation between concept type labels,
$(T,\leq_T)$	:	Concept type hierarchy - the type lattice,
D	:	A set of concept type definitions for types in T,
SC	:	A set of conceptual schema graphs for types in T,

P   :   A set of prototype graphs for types in T,
A   :   A set of actor definitions.

3.3 Deductive Object-Oriented Conceptual Graphs (DOOCGs)

Before defining DOOCGs, we define atomic conceptual graphs and compound graphs as follows.

Definition 3.3 An atomic conceptual graph (ACG) [4] is a conceptual graph that contains no logical connective and no quantifier other than the implicit existential quantifiers. An atomic conceptual graph containing only individual referents in all the concept nodes is termed as a ground atomic conceptual graph.

Definition 3.4 An conceptual graph is called a compound graph if any or all of the following conditions hold: (a) to contain contexts of depth [13] higher than 0, (b) it contains ACGs connected by coreference links.

Example 3.2 An example of atomic conceptual graphs and ground atomic conceptual graphs is shown in Fig. 3.7(a) and Fig. 3.7(b) respectively. A compound conceptual graph represented by nested negative contexts is shown in Fig. 3.7(c).

Now we define DOOCGs.

Definition 3.5 The deductive object-oriented conceptual graphs (DOOCGs) include the following graphs:

[PERSON] <--(AGNT) <--[SIT]--> (LOC)--> [PLACE]
(a) An atomic conceptual graph

[PERSON:Sam] <--(AGNT) <--[SIT:#543]--> (LOC)--> [PLACE:Cafeteria]
(b) A ground atomic conceptual graph -
all concept nodes contain individual referents

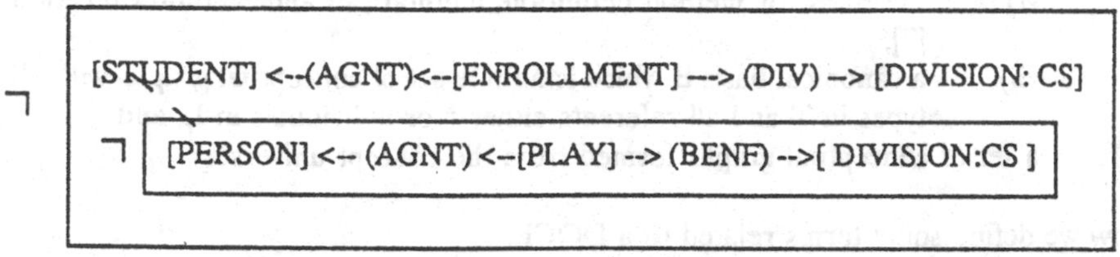

¬ [STUDENT] <--(AGNT)<--[ENROLLMENT] ---> (DIV) --> [DIVISION: CS]

¬ [PERSON] <-- (AGNT) <--[PLAY] --> (BENF) -->[DIVISION:CS]

(c) A compound graph that says "If a person is a student of CS division
then he/she can play for CS division"

Fig. 3.7 Three conceptual graph examples

(a)   Atomic conceptual graphs.
(b)   Message graphs.
(c)   Compound graphs; represented in nested negative contexts with the following constraints:
-       The graphs consist of only two levels of nesting.
-       The context at depth 1 contains one or more atomic conceptual graphs and/or message graphs in addition to one negative context.
-       The context at depth 2 contains exactly one atomic conceptual graph.
-       There may be coreference links among concepts of the conceptual graphs at various contexts.

These graphs have the general form: $\neg[v_1 \ldots v_n \; \neg[w]]$ where, v_i ($i = 1,\ldots,n$) is an atomic conceptual graph or message graph, w is an atomic conceptual graph. If $n = 0$, the above graph becomes $\neg[\neg[w]]$ or simply w that represents a single atomic conceptual graph w in the outermost context.

3.4 Deductive Object-oriented Conceptual Graph Language (DOCL)

Definition 3.6 A *deductive object-oriented conceptual graph language* (DOCL) consists of the following:

(a)   a *concept universe* consisting of,
   i)   a conceptual basis CB,
   ii)   a non-empty set of individual referents I_r that consists of the individual markers $\{\#1, 2, \#3, \ldots\}$ as well as other externally communicable objects like strings, numbers and times,
   iii)   the conformity relation :: that relates types in T to individual markers in I_r, the function ltype that relates concepts and conceptual relations to type labels in T and R respectively and the function referent that relates concepts to the elements of $I_r \cup \{*\}$, where * denote the generic marker,
   iv)   the set of canonical formation rules together with the maximal join operation and the constrained join operation,
   v)   the rules for actors and dataflow graph definitions [7, 13],
   vi)   the rules for method definition, method call and method execution [7],
   vii)   a canonical basis B which is a finite set of conceptual graphs with all types in T and all referents either * or individuals in I_r, and
(b)   a set of conceptual graphs formed from the concept universe.

Now we define some terms related to a DOCL.

Definition 3.7 A *conceptual program* CP is a finite set of deductive object-oriented conceptual graphs.

Definition 3.8 The *degree* of DOOCGs is a mapping from a set of DOOCGs into the set of integers. For a DOOCG C, the degree of C, denoted by DEG(C) is the number of conceptual graphs in the context at depth 1. For a DOOCG of the general form C = ¬[v_1 v_2 ... v_n ¬[u]], DEC(C) = n. For an atomic CG or message graph u in the outermost context, DEC(u) = 0.

Definition 3.9 A *query graph* is either an atomic conceptual graph or a message graph.

Definition 3.10 A *query* consists of a single query graph or a number of query graphs in conjunction.

Thee may be coreference links among concepts of the query graphs in a query. For example Q = q_1...q_n is a query where each of q_i is either an ACG or a message graph.

Definition 3.11 A *goal graph* is either an atomic conceptual graph or a message graph.

Definition 3.12 A *goal* consists of a set of goal graphs in conjunction. There may be coreference links among concepts of the goal graphs in a goal.

Definition 3.13 The goal G corresponding to a query Q = q_1...q_n is represented as {q_1,...,q_n}, which is the assertion made by the query graph q_1,...,q_n in conjunction (together with the coreference links of Q).

Definition 3.14 Let u and v be two conceptual graphs. The graph u is a referent-specialization of v if u is a specialization of v with the following property: u is the same as v except that zero or more generic concepts in v are specialized to individual concepts in u. Let π be a projection of v. The referent-specialization operator (rso) ρ for the projection π consists of a set of pairs of the form (c,c"), which is obtained from π in the following way:

a)   initially ρ is set equal to { },
b)   for each pair (c,c') such that πc = c' where referent(c) = * and referent(c') ≠ *, a pair (c,c") is added to ρ, where type(c") = type(c) and referent(c") = referent(c').

Note that, ρ = {} represents the identity referent-specialization.

Theorem 3.1 [7] Let v be a conceptual graph, π be a projection of v and ρ be the referent-specialization operator for the projection π. Then, $\pi v \leq_G \rho v \leq_G v$.

Definition 3.15 Let CP be a conceptual program and G = {g_1,...,g_n} be a goal. An answer for G from CP is G' = {g'_1,...,g'_n}, where each g'_i, called an answer graph for the goal graph g_i is a referent-specialization of g_i. It is to be noted that if G contains only individual concepts then the only possible answer is obtained by the identity referent-specialization.

3.5 Extended Rules of Inference for DOOCGs

The propositional rules of inference or alpha rules, the first-order rules of inference or beta rules and a set of derived rules or delta rules are presented in [13]. Two fundamental properties of DOOCGs are stated in the form of rules (α0) and δ0). Let S be a set of DOOCGs.

(α0)   Let $u \in S$ and v be any atomic conceptual graph. If u is a specialization of v, (i.e., v is a generalization of u), then the assertion made by v is true if the assertion made by u is true. That means, if $u \leq_G v$, then $\emptyset u \supset \emptyset v$ where $\emptyset$ is the formula operator defined in [13].

(δ0)   Suppose m is message graph and the method call ?m succeeds with respect to S with a projection π. Let m' be the graph obtained by erasing the actors from the graph πm. The graph m' makes a true assertion in the outermost context.

Now a special case of the rule (α0) is stated here as (Δ0) as follows.

(Δ0)   Suppose, u and v are two atomic conceptual graphs, and the graph u makes a true assertion. Then the assertion made by v must be true if any of the following statements holds:

   (1)   Subgraph: v is a subgraph of u.
   (2)   Subtypes: u is identical to v except that one or more type labels of v are more general than to subtypes of the corresponding concept types in u.
   (3)   Individuals: u is identical to v except that one or more generic concepts of u are restricted to individual concepts of the same type in v.

The extended derived rules of inference are formulated as follows.

Theorem 3.2 Let S be a set of DOOCGs and u and v be any conceptual graphs where $u \leq_G v$ with a projection $\pi : v \to u$. Let m be a message graph, and the method call ?m succeeds with respect to S with a projection π_m. Any graph derived from S by the following *extended derived rules of inference* is said to be provable from S.

(Δ1)   In an oddly enclosed context, v may be replaced with πv where each coreference link to a concept c of v is transferred to the corresponding concept πc in πv. In an oddly enclosed context, m may be replaced with π_mm and actors bound to π_mm may be erased, where each coreference link to a concept c of m is transferred to the corresponding concept π_mc in π_mm.

(Δ2)   Let P be a graph $\neg[u \; \neg[v]]$. Then P itself is a theorem, and P with coreference links $<\pi c, c>$ for any c in v is also a theorem. The graph $\neg[\pi v$ $\neg[v]]$, and $\neg[\pi v \ \neg[v]]$ with coreference links $<\pi c, c>$ for any c in v are also theorems.

(Δ3)   In an evenly enclosed context, u may be replaced with v where each coreference link to a concept πc of u is transferred to the corresponding concept c of v and the other coreference links attached to u are erased.

(Δ4)   Generalized modus ponents: If the outer context contains the graph u as well as a graph of the form $\neg[v \ \neg[w]]$, possibly with some coreference links from v to w, then the graph w may be derived with each coreference link attached to a concept c in v reattached to the corresponding concept πc in u. If the outer context contains the graph $\neg[m \ \neg[w]]$, possibly with some coreference links from m to w, then the graph w may be derived with each coreference link attached to a concept c in m reattached to the corresponding concept $\pi_m c$ in $\pi_m m$.

The following theorem holds [7].

Theorem 3.3 The extended derived rules of inference are sound.

4. Semantics of DOOCGs

4.1 Interpretations and Models

To define interpretations of DOCL, we need to have interpretations of actors. An actor can be abstractly considered as a function, which is called *coneptual function*.

Definition 4.1 A conceptual function is obtained for each actor in a DOCL L in the following way:

- The function name is the same as the corresponding actor name. The function name is followed by a pair of balanced parentheses that encloses the arguments separated by comma.
- For each input concept of the actor, an argument is added which is the concept itself.
- The function maps the input concepts into a set of referents that conforms to the type of the output concept of the actor.

Suppose we have a method definition that is shown in Fig. 3.3 with two actors <TODAY> and <DIFF_DT> which are assumed to be present in the system as primitive actors. Now, we will have the following two conceptual functions:

$$f_1 = \text{TODAY}(), \text{ and}$$
$$f_2 = \text{DIFF\_DT}([\text{DATE}],[\text{DATE}])$$

where f_1 conforms to DATE type and f_2 conforms to AGE type. The evaluation of a conceptual function corresponds to the execution of the corresponding actor.

Although, an interpretation is defined, in general, with respect to any domain, we consider the domain called *universe of interpretation*, as the domain for the interpretations of DOOCGs.

Definition 4.2 Let L be a DOCL. The universe of interpretations for L, UI_L is defined as $<CB,f_c,I_r::>$, where CB is the conceptual basis of L, f_c is the set of all conceptual functions for actors in L, I_r is the set of individual referents that includes individual referents explicitly present in the graphs in L as well as those obtained by evaluating the conceptual functions in f_c, and :: is the conformity relation of L. I_r must have at least one referent that conforms to some concept type in CB.

Definition 4.3 Let L be a DOCL. The interpretation base for L, IB_L is defined as $<A_b,I_b>$, where

A_b : the set of ground atomic conceptual graphs that can be obtained from the graphs in L and individual referents and :: relations in UI_L.

I_b : the set of individual referents of A_b.

Definition 4.4 An interpretation *In* of a DOCL L consists of the following:

a) A domain, $UI_L = <CB,f_c,I_r::>$.
b) For each generic concept, assignment of an element in I_r wrt to the relation ::.
c) For each individual concept, assignment of an element in I_r wrt to the relation ::.
d) Different graphs having concepts with the same individual referents are joined on the common individual concept so that one individual referent in I_r occurs at most in one concept in *In*.
e) For each atomic conceptual graph and message graph in L, assignment of a truth value from {true,false}, that is, a mapping from the set of atomic conceptual graphs and message graphs into {true,false}.

Note that an interpretation is represented by a pair $<A,I>$ where A is a set of ground atomic conceptual graphs and I is the set of individual referents in A.

Definition 4.5 For any conceptual graph u of a DOCL L, a *ground instance* of u is a graph obtained from u by replacing the referent field of all generic concepts in u by some individual referents in UI_L.

Suppose in forming a ground instance of graph u, a generic concept c is being specialized with an individual referent i, then the following conditions hold,

1)   -   the referent i is copied to the referent field of all other concepts dominated by c,

        -     any newly added referent that is identical to a referent previously present is erased,

        -     for concepts that have more than one distinct individual referents, '=' sign is inserted between the referents,

        -     all coreference links dominanted by c are erased, and

2)   if u' is the ground instance of u and if there is no concept in u' with more than one distinct individual referents, then $u' \leq_G u$.

Definition 4.6 Let *In* be an interpretation of a DOCL L and C be any DOOCG or a set of DOOCGs of L. Then *In* is a model of C if C is true wrt to *In*.

Now we define the notion of logical consequence and correct answer for DOOCGs.

Definition 4.7 Let S be a set of DOOCGs of a DOCL L and v be any DOOCG of L. Then v is a logical consequence of S if v is true in every model *In* of S.

Definition 4.8 Let CP be a conceptual program, G be a goal and G' be an answer for G from CP. The answer G' is a correct answer if G' is a logical consequence of CP.

The game theoretic semantics for general CGs with respect to a closed world [13] has been extended for the DOOCGs. A model-theoretic semantics and a fixpoint semantics of DOOCGs have been defined and they are proved to be equivalent for a conceptual program. A proof procedure, called direct derivation proof has been formulated for DOOCGs and it is proved to be sound and complete. Details about the semantics of DOOCGs and the proof procedure can be found in [7]. A notion of a DOOCG theory has been developed, and a DOODB is formally represented as a special DOOCG theory.

Definition 4.9 A DOOCG theory T_G consists of a DOCL L, a set of axioms and a set of inference rules,

- Logical axiom : {}, the empty set.
- Proper axioms : A set of DOOCGs.
- Set of inference rules : the set of first-order rules and the set of extended derived rule.
- Proof procedure : The direct derivation proof procedure [7].

Definition 4.10 A *deductive object-oriented database*, DOODB is a DOOCG theory whose proper axioms are:

- A conceptual program CP that consists of:
  - a set O of elementary facts or individual objects which are ground atomic conceptual graphs.

> - a set S that consists of deductive laws, which are compound DOOCGs and message graphs.
- A set of integrity constraints IC, which are DOOCGs.

5 Conclusions

A subset of general conceptual graphs has been extended to form deductive object-oriented conceptual graphs that includes simple graphs, message graphs and compound graphs with only two levels of nesting and exactly one simple graph in the context at depth 2.

The approach taken for this work towards representing DOODBs is fundamental in the sense that most of the other known approaches are either to extend DDBs to incorporate some of the constructs of OODBs or to extend OODBs with some of the features of DDBs [17]. Being a relatively new area of research, there is no generally accepted notion of DOODBs yet. A more concrete notion of DOODBs has been specified in this work in terms of characteristics of DOODBs. The use of conceptual graphs allows a large amount of "knowledge" to be statistically captured in the representation itself through the use of type definition, type hierarchy and property inheritance.

One of the major problems of integrating ideas of DDBs and OODBs is the representation and interpretation of methods [6]. The solution to this problem has been formulated with restrictive use of conceptual schema graphs, actors and dataflow graphs of conceptual graph formalism. This offers several advantages:

- Methods can be easily shared among different objects along a chain of type hierarchy without need for any additional cost, because conceptual schema graphs are associated to the type labels in the type hierarchy in the CG formalism by their definitions.

- The use of actors and dataflow graphs offers a natural means to merge the "procedural" notion of methods in OODBs with the declarative paradigm of DDBs, because actor and dataflow graphs in Cg formalism can be used declaratively when treated as a descriptions and procedurally when they are executed [13].

- The mechanism of method evaluation is formulated in a success/failure paradigm which is considered as one of the requirements for merging object-oriented concepts with deductive database concepts [6].

This work created a basis for a database system that would be useful for knowledge representation as well as many advanced database applications. However, much work has to be done over the basis founded by the results of this work to make it evolve into a complete operational database system.

References

1. S. Abiteboul, S. Grumbach: A Rule-Based Language with Functions and Sets, ACM Transactions on Database Systms, Vol. 16, No. 1, pp. 1-30, March 1991.
2. M.L. Brodie, J. Mylopoulos, eds,: On Knowledge Base Management Systems-Integrating Artificial Intelligence and Database Technologies, Springer-Verlag, 1986.
3. K.R. Dittrich: Object-Oriented Database Systems: The Notion and the Issues, Proceeding of International Workshop on Object-Oriented Database Systems, Dittrich, K.R. & Dayal, U. (eds.), IEEE Company Society Press, pp. 2-4, 1986.
4. G. Ellis: Compiled Hierarchical Retrieval, Proceeding of the Sixth Annual Workshop on Comceptual Structures, pp. 187-207, July 1991.
5. J. Fargues, Marie-Claude Landau, A. Dugourd, L. Catach: Conceptual Graphs for Semantic and Knowledge Processing, IBM Journal of Research and Development, Vol. 30, No. 1, pp. 70-79, January 1986.
6. H. Gallaire: Merging Objects and Logic Programming: Relational Semantics, Proceeding of Fifth National Conference on Artificial Intelligence, AAAI-85, Vol. 2, pp. 754-758, 1986.
7. B.C. Ghosh: Towards Deductive Object-Oriented Databases Based on Conceptual GRaphs, Masters Thesis, Division of Computer Science, Asian Institute of Technology, 1991.
8. T.R. Hines, J.C. Oh, M.L.A. Hines: Object-Oriented Conceptual Graphs, Proceeding of the Fifth Annual Workshop on Conceptual Structures, pp. 81-89, July 1990.
9. M. Jackman, C. Pavelin: Conceptual Graphs, Approaches to Knowledge Representation: An Introduction, Ringland, G.A. & Duce, d.A. (eds.), John-Wiley & Sons Inc., pp. 161-174, 1988.
10. W. Kent: Limitations of Record-Based Information Models, Readings in Artificial Intelligence and Databases, Mylopoulos, J. & Brodie, M.L., (eds.), Morgan Kaufmann Publishers, Inc., pp. 85-97, 1989.
11. J.W. Lloyd: An Introduction to Deductive Database Systems, The Australian Computer Journal, Vol. 15, No. 2, pp. 52-57, May 1983.
12. J.W. Lloyd: Foundations of Logic Programming, Second Extended Edition, Springer-Verlag, 1987.
13. J.F. Sowa: Conceptual Structures: Information Processing in Mind and Machine, Addison-Wesley Publishing Company, Inc., 1984.
14. J.F. Sowa, E. Way: Implementing a Semantic Interpreter using Conceptual Graphs, IBM Journal of Research and Development, Vol. 30, No. 1, pp. 57-69, January 1986.
15. J.F. Sowa: Knowledge Representation in Databases, Expert Systems and Natural Language, Artificial Intelligence in Databases and Expert System, Meersman, R.A., Shi, Zh. & Kung, C.H. (eds.), North-Holland Publishing Co., Amsterdam, pp. 17-50, 1990.

16.   J.D. Ullman: Principles of Database and Knowledgebase Systems, Vol. 1, Computer Science Press, 1988.

17.   K. Yokota: Outline of a Deductive and Object-Oriented Language: Juan. SIGDB No. 78, July 1990.

AERIE:
Database Inference Modeling and Detection Using Conceptual Graphs

Harry S. Delugach and Thomas H. Hinke

Computer Science Department,
University of Alabama in Huntsville,
Huntsville, AL 35899, USA
Electronic mail: delugach@cs.uah.edu

Abstract. Secure databases are ones in which classified information is protected from access by unauthorized persons. Although the information itself may be secure from direct access, data within the database may be combined along with external data to permit classified data to be inferred. This problem, called the *database inference problem*, can be addressed by analyzing a database and its design. An inference detection model based on conceptual graphs is introduced, and explained in terms of different kinds of inference that may be performed. An automated inference analysis tool (IAT) is introduced and its overall architecture described.

1 Introduction

This paper discusses an application of conceptual graphs to a current problem in database security; namely, the *database inference problem*. Since it is a description of an application, we will first introduce the problem, outline our strategy of solving that problem, and then briefly show how conceptual graphs are used to implement that strategy. This work represents an on-going research effort, and the current results, while they appear promising, are still preliminary.

Data that in isolation may be unclassified can be combined with other unclassified data to infer data of a higher classification. This problem cannot be solved with multilevel secure database management systems since the individual data items taken taken in isolation may be correctly classified and protected. However, this data combined with other data may reveal data that is classified at a higher level than any of the individual data items themselves. The problem lies in the inability to determine if the assigned classification of the data is sufficient to prevent the inference of more highly classified data. This is called the *database inference problem* and is the focus of this paper.

Database inference research has been classified by [4] into the following categories:

1. efforts to discover fundamental laws that determine whether the potential for undesirable inferences exists within a given database,
2. efforts to discover automatically inference rules from fundamental relationships among data that pertain to a domain,
3. efforts to automate (via expert systems) the process of inferring sensitive data within a specific domain.

The first category of research is characterized by that of [8] which considered the inference potential of the various dependencies that exist in relational database systems, [6] which proposed a theoretical model of the inference problem. Research by [9] describes the use of semantic nets and conceptual structures for describing multilevel data and then develops rules for reasoning about inferences in such structures and [5] observes that some problems which appear to be inference problems are in reality problems of misclassification.

The second category of research is characterized by that of [3] and [4] which views inference detection as the discovery of paths through conceptual structures that describe the data associated with an enterprise.

The third category of research, characterized by [1], is concerned with the construction of expert systems to detect inferences based on inference detection rules that are built into the system.

Inference research at the University of Alabama in Huntsville has addressed the second and third categories of inference research. It has resulted in the development of a new inference model called AERIE (Activities, Entities, and Relationships Inference Effects) Model which is the subject of this paper. The AERIE model builds on the work of [3], [4] and [9]. Research on the AERIE Model represents work in both the second and third category of inference research since it is used to discover inference rules from conceptual graph descriptions, but will also contains a significant knowledge-based and associated low-level inference rules that provides an expert-system-flavor to the system that will implement the AERIE model.

The conceptual graphs approach provides a promising basis for examining this problem, because we are not required to restrict our attention just to information that is available in other databases in order to be able to characterize the information brought to bear on an inference. In a practical sense, someone trying to crack into a data base will use all the knowledge at his/her disposal. This approach therefore assumes no particular limitations on the knowledge that might possibly be used to analyze a database's inference problems. The approach also uses conceptual graphs as the basis to find inference paths that can be used to build a classified association using unclassified data.

The AERIE model is intended to provide the basis for the implementation of an Inference Analysis Tool (IAT). The intent is that the IAT would have the capability to evaluate a database and advise the database designer of potential inference problems. There are three issues involved in this approach. First, since an adversary is not limited to the data within a single database or even a particular group of databases in making his inference attack, any solution to the problem must assume the use of knowledge that goes well beyond that available in the database. The second issue is that multiple techniques can be used in performing the inference; hence, any solution must recognize this possibility. The third issue is that the problem must be bounded, since the generation of all possible inferences would overwhelm any computational system.

The remainder of this paper elaborates on the AERIE model and the use of conceptual graphs as the basis for an IAT which is being designed. In line with these objectives, the next section presents the basic concepts of the AERIE model while the following section describes the use of conceptual graphs as the basis for the design of an inference analysis tool. The final section presents our conclusions to date.

2 AERIE Model

AERIE Model views inference detection as a two-phase process. During the first phase entities and activities are materialized from the conceptual graphs which describe the data. The entities are analogous to the English nouns which represent physical objects. Activities are analogous to verbs in that they describe action of a state of being. The word "materialization" is used because, in effect, performing an inference is making visible something that is not clearly articulated in the database. If it had been clearly articulated and it represented sensitive data, then presumably it would have been properly protected. However, since it is not clearly articulated, it may not be properly protected, and hence is available for an adversary to materialize this information and thus gain access to classified information.

The second phase of inference detection is the determination of relationships between various materialized entities and/or activities.

The AERIE model will be presented in the next two sections. The next section presents the classes of inference that AERIE is to address. These are called the inference target classes. The following section presents some of the methods that AERIE will use to perform these inferences.

2.1 Inference Target Classes

The research on the AERIE Model has identified seven inference target classes. In addition to assigning each target class a number, we will use to refer to it, each class can be characterized in terms of the entities, indicated with an "E," and/or activities, indicated with an "A", that are involved in the inference. Using this notation, the seven inference target classes are as follows:

1. E: The materialization of an entity,
2. A: The materialization of an activity,
3. (E,E): The materialization of a sensitive relationship between two or more materialized entities,
4. (A,A): The materialization of a sensitive relationship p between two or more materialized activities,
5. (E,A) or (A,E) (we will assume that the two are equivalent): The materialization of a sensitive relationship between one or more materialized entities and one or more materialized activities,
6. ((W,X),(Y,Z)): The materialization of a sensitive relationship between sensitive relationships (W,X,Y,Z refer to unknown entities or activities), and
7. $[C_1, C_2, ..., C_j] \rightarrow C$: The materialization of a sensitive rule from existing classes.

The Class 1 and 2 inferences are the simplest, since they deal with the inference of only a single entity or activity. The Class 3, 4 and 5 inferences deal with relationships between entities and activities. They are of a more complex nature than the first and second classes since they may require that Class 1 or 2 inferences be performed to materialize required entities or activities. It is believed that Classes 3, 4 and 5 are similar in complexity. The Class 6 inference deals with relationships between relationships and is more complex than the other five classes since it includes these more primitive classes as well as the additional requirement that the earlier relationships be related to each other. Finally, the Class 7 inference is somewhat distinct from the others, since it deals with the ability to infer rules from previously known classes or class instances. More will be said about the Class 7 inferences later.

Examples of each of these inference types will be presented in what follows:

Class 1: An example of a Class 1 inference target can be found within a logistics database. If a site orders a part that is unique to a particular type of equipment, such as a certain radar unit, then an adversary with access to this database could infer that the site has this particular type of radar unit.

Class 2: An example of a Class 2 inference target is the ability to infer that a construction project is occurring based on the class of equipment that is being ordered.

Class 3: A Class 3 inference target is the determination of a sensitive association between two entities, such as the fact that a particular company is supporting a classified project.

Class 4: A class 4 inference represents a sensitive association between two or more activities. For example, a new type of tank could be under development. Characteristics of this tank could be revealed if the designer of the tank ordered both blades and a bucket, indicating that this tank could participate in the associated activities of pushing and digging.

Class 5: A Class 5 inference represents a sensitive association between one or more entities and one or more activities. For example, if an intelligence activity required that a certain fixture be placed in the space shuttle's payload bay to support a particular type of sensor, then the association of this fixture with the intelligence gathering activity would represent a sensitive relationship that should be protected.

Class 6: A Class 6 inference represents a sensitive relationship between sensitive relationships. An example of this class is a student grade inference. Assume that grades are posted by student numbers, to preserve the confidentiality of the grade that a particular student received. This represents a class 1 relationship between the entity student and the entity grade (e.g., (E,E)). However, if these posted grades are sorted by the last name of the student, then this represents a sensitive relationship, called "Sorted-by-name" between the (Student_number, Grade) relationship which is public knowledge and the (Student_name, Grade) relationship which is sensitive. However, if this Sorted_grade relationship can be inferred, then the very sensitive (Student_name, Grade) relationship can also be inferred.

Class 7: A Class 7 inference represents the inference of a sensitive rule. An example of a sensitive rule might be one used by a credit card company that says to approve all charges for food, using the reasoning that since the food is already consumed, there is no reason not to approve it. In general, this class of inference incorporates any inference that results in a rule, rather than an entity, attribute or relationship. This class of inference target represents a considerably different target than the previous ones that are shown, but it is included since it represents another type of information about which one could launch an inference attack.

Having presented the various inference target classes, the next section will present some methods by which these classes can be inferred.

2.2 Inference Method Classes

For each inference target class, one or more methods will be applicable to performing the inference. These various methods can be clustered into classes, such that each member of the class has the same basic characteristics. It is also the case that method classes applicable to one inference target class may also be applicable to other inference target classes.

For the Class 1, the entity target class, the following methods are applicable to inferring an entity:

1. Statistical inference
2. E, (E,E)
3. A, (A,E)

For example to infer a particular radar unit from a part, we can use the second method : E1 AND (E1, E2) → {E2}, where E1 is some unique part and the (E1,E2) relationship is the part-of relationship that breaks down the parts that are contained in each piece of equipment. The set {E2} represents all of those end-products on which the part is used. If the part is used only a single end-product, then the cardinality of the set {E2} is one and we have an inference that results in the unique identification of a piece of equipment.

For the Class 2, the entity target class, the following methods are applicable to inferring an entity:

1. Statistical inference
2. E, (E,A)
3. Traffic flow analysis (analogous to the traffic flow analysis performed for networks, but here concentrating on a high volume of activity on, for example, data about Iraq)

As an example, we will consider how the activity of pushing can be inferred from an entity using the second method. Assume that a blade is used for pushing. Thus, if a requisition is received for a blade, then the inference can be made that the pushing activity is to be performed.

Method classes for the remaining target classes are omitted here for brevity. These examples illustrate one of the areas of investigation that is being pursued by this research, with the goal of of identifying a set of methods for each target class. The next section will describe the overall design of an Inference Analysis Tool and then show how it will use conceptual graphs to perform the more complex inferences of the type just described.

3   Inference Analysis Tool

The tool makes use of information obtained from several sources. We presume that this information is in the form of conceptual graphs. These *knowledge sources* are:

- Description of the database under analysis. There are two parts to this:
  - Database specification graphs (DBSG): describe the classes of information that are immediately available (i.e., directly inferrable), and
  - Database instance graphs (DBIG): capture actual instances in the database (if known) which are facts in the inference analysis.

- Domain-independent knowledge (DIK): General knowledge (i.e., information that is considered publicly available), either so-called "common sense" knowledge, or encyclopedic knowledge. (We discuss this further below.)
- Domain specific knowledge (DSK): Information that a domain expert would know.

In addition, the tool will have access to the following:

- Sensitive targets (ST), instances of some inference target class. The database designer supplies facts and/or relationships that he wants to be protected against being inferred from the data base.
- The AERIE Model, described above, that guides the IAT's steps in seeking and identifying inferences between the source information and the database designer's sensitive targets.

The overall goal of the inference analysis process is to determine whether or not sensitive targets can be inferred from the original database, and if so, what information enables the inference to be performed. Since the analysis tool will be based upon the conceptual graph representation, each sensitive target will be kept as a graph. The sequence of inference steps that derives a sensitive target graph from the database and other graphs will be called an *inference path*. An inference path therefore contains the originating information as well as any intermediate inferences (and their enabling graphs).

The tool maintains a set of inference paths that it has currently derived. An *inference path* is a sequence of one or more inference steps. An *inference step* is composed of the following:

- *Enabling graphs*. A set of graphs (identified as to their origin) that enables the inference to be performed;
- *Inferred graphs*. A set of graphs representing the facts that were derived;
- A list of the materialization methods used to perform the inference;
- A characterization of the materialization method class(es) and
- A characterization of the target class(es).

Within an inference path, an inferred graph for one step may also serve as an enabling graph for a subsequent step. We define a *path-enabling graph* to be an enabling graph that is not also an inferred graph of some other step in the path (i.e., independent information that allows the inference to be made). The inference path can therefore be envisioned as a function whose input consists of path-enabling graphs, and whose output is the inferred graphs of the last step in the sequence.

The tool's basic operation is to construct inference paths. For each sensitive target, the tool seeks an inference path whose path-enabling graphs appear in one of the knowledge sources. If no such path-enabling graphs are found, then the tool postulates some intermediate graph(s), and seeks an inference path that can infer the intermediate graph(s), which can then be connected to a previous path.

Once an inference path is established, interpretations can be drawn and advice to the database designer/administrator will result. For example, if all path enabling graphs for a piece of sensitive information appear in the general knowledge base (DIK), then the sensitive information is based entirely upon common knowledge, making it nearly impossible to keep secret. Likewise, if all path-enabling graphs are in either the general

(DIK) or domain-specific knowledge base (DSK), then the information can be inferred by a knowledgeable specialist, and will be difficult to keep secure. If at least one path-enabling graph appears in the database specification (DBSG), it would show the designer that mere knowledge of the database's structure allows the sensitive information to be inferred.

The AERIE Model gives a taxonomy of target classes, where each class consists of one or more method classes, and each method class consists of one or more methods that lead to inference of the target. The tool uses the AERIE Model to organize its search and storage of information. The tool starts by classifying each sensitive target into one or more target classes. For each of these, the model has one or more materialization method classes that are used to infer it. Each method class has one or more actual methods, that can be applied one by one. The order of search corresponds to the order in the model: The tool seeks graphs for which the materialization method will produce either the sensitive target or some intermediate target.

The following discussion describes an example to give the flavor of the IAT. The graphs themselves are omitted; they are only paraphrased here. This example shows how we would materialize a senstive entity, making use of a sample database that consists of an equipment manufacturer's inventory and shipping information. The hypothetical database contains records with information such as part number, description, quantity on hand, and destination.

The domain-independent knowledge base (DIK) of general knowledge consists of the following:

- There is type hierarchy representing a set of classes, where each type is a subtype (sub-class) to the class of its parent(s). The type T at the top represents the universal type. The existence of a type hierarchy implies the inference rule that if some instance X is of type A and type A is subtype to type B, then instance X is of type B also.
- If some instance Y is a part of some instance X, and some instance Z is a part of the instance Y, then instance Z is a part of instance X.
- A piece of equipment exists with the following typical associated information:
  - It costs some money;
  - It is operated by a person;
  - It is located in some place;
  - It is owned by either an organization or a person, and
  - It is not part of something else.
- "Pushing" (an act) is typically caused by some animate instance Y and acts upon some instance X. The size of instance X is less than the size of instance Y.
- If some instance X exists, and some Y is a part of X, then an instance of Y exists.

The domain-specific knowledge base (DSK) consists of the following:

- A construction project (an activity) exists with the following typical associated information called a schema:
  - It costs some financing;
  - It is located in some place;
  - It employs one or more persons, and
  - It has as its parts digging, carrying or pushing.

- A blade is used for pushing, a bucket is used for digging, a hitch is used for pulling, and a loader is used for carrying or pushing.
- A tractor has as its parts a hitch, a blade and a tire.
- A backhoe has as its parts a loader, a bucket and a tire. A tooth is part of a bucket.
- A cotton-harvester has as its parts a hitch and a tire.

Suppose the database designer requires that a construction project is a sensitive activity, that is, the existence of any construction project is deemed sensitive. Given the manufacturer's database and knowledge sources above, we briefly show how the tool proceeds to materialize the construction project.

To analyze the secure information, the tool must identify what (if any) knowledge might be used to infer the existence and/or location of a construction project. First, the DIK is scanned for any appearance of CONSTR_PROJECT; there is none. Then the DSK is scanned; since it contains a schema for CONSTR_PROJECT, the existence of a construction project can be inferred if some of its associated concepts are present, (e.g., FINANCING, PLACE, PERSON, DIGGING, CARRYING or PUSHING). One of these concepts, PLACE, appears in both the specification graph and the DSK's schema. Suppose we were to decide that the existence of one concept in a schema is sufficient to infer the rest of the schema, then we would have just shown that the location of a construction project could be inferred from the database.

The reader can quickly see that such an inference is not particularly valid; as an old saying goes, "Everybody's got to be somewhere," so that mere sharing the concept of location (as opposed to sharing some particular location instance) is insufficient to infer an entire schema. For each schema, some heuristic must therefore be chosen that determines what components of the schema must be present before we can infer the central concept (in this case, CONSTR_PROJECT).

If we adopt some general heuristic that at least half of a schema's components must exist before we are willing to infer the entire schema, we immediately encounter problems. For this schema, it is clear that many activities involve financing, are located in some particular place and employ persons (e.g., a sports event or a newspaper publisher). What makes a construction project distinguishable is that it involves all those relations, plus it is made up of digging, carrying and pushing activities. So to infer a construction project's existence, we must establish the existence of one of those sub-activities.

We must therefore search the DIK for any graphs containing the concepts DIGGING, CARRYING, and PUSHING. We find a schema for PUSHING, so that PUSHING can be inferred if we find some ANIMATE entity operating upon something. We see that the database does not contain any concepts that are sub-types of animate, so PUSHING cannot be inferred using the schema.

As there are no other graphs in the DIK containing any of the sub-activities we seek, we look in the type hierarchy. Some of the activities are subtypes of ACT, which is of limited value since assertions about ACTs are not necessarily true for all of its subtypes. None of the activities has any subtypes of its own to spur further searches in the DIK, so there is no further DIK knowledge that can be used.

We now scan the DSK for any appearance of the three activities. We see that all three of them appear, associated with various concepts: a BLADE is used for PUSHING, a BUCKET for DIGGING and a LOADER for either CARRYING or PUSHING. None of these concepts appears in the specification graph.

Can any of the concepts BLADE, BUCKET or LOADER be inferred from the DBSG? We scan the DIK for any occurrence of these concepts, and there are none. We scan the DSK for them, and find each of them included in a part relation. BLADE is part of a TRACTOR, BUCKET is part of a BACKHOE, and LOADER is part of a BACKHOE. By the general inference rule in the DIK, we can therefore infer the existence of a part if we can infer the existence of its whole. Since these new concepts do appear directly in the specification graph as subtypes of PART, then for any instance of BACKHOE and TRACTOR in the database, we can infer, by a direct path, the existence of a construction project.

Figure 1 illustrates the conceptual graphs that support this inference example.

4 Conclusion

While the research is currently in the preliminary stages, we believe that is has already made a contribution to inference research by providing the classification of inference classes.

A second contribution is the use of coupling of conceptual graphs to support this application domain. At this preliminary stage, the use of conceptual graphs shows promise, but much work remains to be done to validate this promise. One of the major concerns in this research is whether the approach will scale to the large volume of relevant data that will be required to support real database applications. Our current approach is to first ensure that this conceptual graph approach will adequately handle a real database example. To accomplish this, we are continuing with the design of the inference analysis tool and plan on constructing a small prototype which will be applied to a logistics-type database.

References

1. Buczkowski, L. J.: Database Inference Controller In Database Security III, Status and Prospects, North-Holland, 1990.
2. Garvey, Thomas D. and Teresa F. Lunt, Cover Stories for Database Security, Proceedings 5th IFIP WG 11.3 Working Conference on Database Security, Shepherdstown, WV, USA, Nov 1991.
3. Hinke, Thomas H.: Inference Aggregation Detection in Database Management Systems. 1988 IEEE Symposium on Security and Privacy, Oakland, CA, USA, April 1988.
4. Hinke, Thomas H.: Response to Research Question 3 in "Research Questions List, Answers, and Revision," (e.d., Carl E. Landwehr) in Database Security, III: Status and Prospects, Results of the IFIP Working Group 11.3 Workshop on Database Security, Monterey, CA, September 1989, North-Holland, 1990.
5. Lunt, Teresa: Aggregation and Inference: Facts and Fallacies. Proceedings of the 1989 IEEE Symposium on Research in Security and Privacy, Oakland, CA USA, May 1989.
6. Morgenstern, Matthew: *Security and Inference in Multilevel Database and Knowledge-Base Systems*, Proceedings of SIGMOD 1987, Association of Computing Machinery, 1987.
7. Sowa, John F.: Conceptual Structures: Information Processing in Mind and Machine, Addison-Wesley, Reading, MA. 1984
8. Su, Tzong-An and Gultekin Ozsoyoglu: *Data Dependencies and Inference Control in Multilevel Relational Database Systems*, IEEE Symposium of Security and Privacy, Oakland, CA., April 1987.
9. Thuraisingham, Bhavani: The Use of Conceptual Structures for Handling the Inference Problem, Cover Stories for Database Security, Proceedings 5th IFIP WG 11.3 Working Conference on Database Security, Shepherdstown, WV, USA, Nov 1991.

215

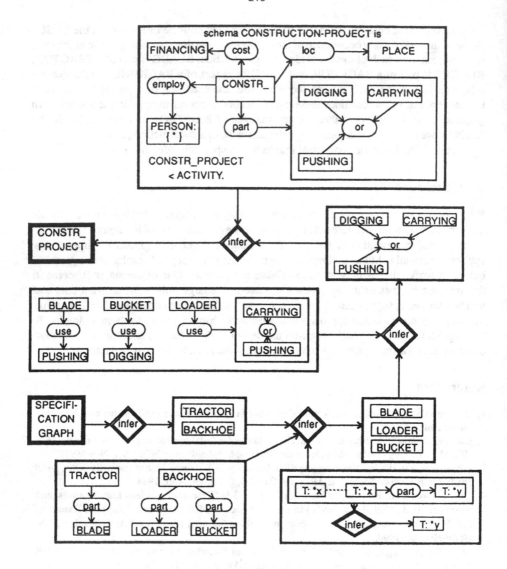

Fig. 1. Example Conceptual Graphs Used by IAT

V. Tools and Algorithms

The Birth of PEIRCE: A Conceptual Graphs Workbench

Gerard Ellis[1] and Robert Levinson[2]

[1] Comp. Sc. Dept., Univ. of Queensland, Brisbane, QLD 4072, Australia. ged@cs.uq.oz.au
[2] Dept. of Comp. and Inf. Sc., Univ. of California, Santa Cruz, 229 Appl. Sc. Building,
Santa Cruz, CA 95064, USA. levinson@cse.ucsc.edu

Abstract. PEIRCE is a project aiming to build a state-of-the-art, industrial strength conceptual graph workbench. PEIRCE is integrating the conceptual graph development efforts that are taking place around the world. There are already over 40 researchers from over 8 countries involved in the project. This paper describes the PEIRCE project and the structure for managing the distributed development of PEIRCE. PEIRCE will provide a robust, portable, freely available conceptual graph workbench that will fast track new techniques into the community; facilitate comparison of competing techniques; help researchers cooperate in development; and speed application development.

1 Introduction

The PEIRCE project is named after the American philosopher Charles Sanders Peirce who in 1896 developed the logic Existential Graphs [11]. Peirce described existential graphs as "the logic of the future". It is nearly a century since Peirce made that statement, and only now is his logic poised to become a reality with the widespread use of powerful graphics workstations that can run the PEIRCE workbench.

John Sowa introduced Peirce's Existential Graphs to the field of artificial intelligence, as the foundation of his Conceptual Graphs theory [12]. Conceptual graphs is an order-sorted logic with a standard mapping to natural language. A diverse, active research community has grown up around Sowa's work. Each year an international conceptual graph workshop is held, this being the seventh. Next year the first international conceptual graphs conference will be held. A survey of current research and practice in the conceptual graph community can be found in [9].

There are several very good implementations of subsets of conceptual graph theory, and there are many people doing innovative work on particular aspects. But there is not as yet a robust, widely available set of tools for developing applications based on conceptual graphs.

The PEIRCE project is an open, nonproprietary, international, collaborative effort to build an industrial strength, portable, freely available conceptual graph workbench (PEIRCE). The project is open to collaboration from all researchers in the conceptual graph community and others.

PEIRCE will allow developers to write/draw/parse/learn large conceptual graph programs/databases/ontologies. A module will be included to demonstrate PEIRCE's support for natural language processing - a major application for many conceptual graph researchers. An adaptive, pattern-oriented chess playing system [7], Morph, will be used to verify and show the high-level support for learning mechanisms and databases of conceptual graphs. Morph is currently implemented in C. Much of the technology in Morph is built on top of a generic retrieval, learning and search model known as APS [2]. APS and Morph will form a valuable part of the PEIRCE Workbench. PEIRCE will also be used for knowledge based systems, graphical programming, computer assisted theorem proving, specification and verification.

PEIRCE will fast track the introduction of new techniques into the community. New techniques will be directly evaluated against old by using the same data, platforms and many of the same support routines. The project will nurture synergism between competing mechanisms. The developers will be able to assemble hybrid tools from basic tools using well documented interfaces. Prototyping of methods will be done in PEIRCE, thus the prototypes will be available immediately for evaluation by the community.

There is a group dedicated to interfacing PEIRCE to large ontologies such as CYC. PEIRCE will accept ontologies in many formats, KIF (Knowledge Interchange Format), CYC, KL-ONE, SNePS, etc. Access to large collections of knowledge will prompt and validate new techniques.

The PEIRCE project at the time of writing involves over 40 researchers from over 8 countries each developing part of the workbench. The diversity of development environments is itself a test of portability. But, portability is not left to chance, programming standards are being set.

This paper describes the main high-level modules contained in PEIRCE, each of which is being developed by researchers spread around the world. The paper describes the organization and glue needed to successfully complete the distributed development of a large software project such as PEIRCE.

2 Parts of PEIRCE

The PEIRCE project is divided up into modules including

- Programming Standards
- Programming Interfaces for C, C++, Prolog, and Other Languages
- Database Storage and Retrieval
- Linear Notation Input and Output
- Graphical Editor and Display
- Conceptual Catalogs (Ontologies)
- Programming in Conceptual Graphs with Constraints
- Inference/Theorem-Proving Mechanisms
- Learning Mechanisms
- Natural Language Parsers and Generators
- Vision System

Each of these groups is described in terms of their dependency on other groups and the sorts of functions they provide. A working document on PEIRCE [5] further details the work of each group.

2.1 Programming Standards (STDS)

The core implementation language of PEIRCE is C++. Advanced services can be implemented in either C++, Prolog or other high-level languages on top of the core. The development environment of choice is X-Windows on UNIX workstations. There are plans for portability to MacIntosh and IBM PC compatibles. This group is establishing guidelines for language and environment versions, naming conventions, system calls, and software release procedures.

2.2 Programming Interfaces for C, C++, Prolog, and Other Languages (FACE)

The FACE group is charged with investigating, documenting, and standardizing calls across programming languages used in PEIRCE. The languages supported initially are C++ and Prolog. Section 3 describes some of the general abstract data types needed for PEIRCE. For example a graph ADT in Prolog may call C++ routines to do graph matching. A C++ conceptual graph ADT may call a Prolog routine for parsing conceptual graphs.

2.3 Database Storage and Retrieval (DB)

The DB group is involved in manipulating, storing and retrieving large numbers of conceptual graphs. This group is involved in implementing core operations such as graph matching (projection), unification (join), indexing, semantic distance, etc. Many of the PEIRCE modules are dependent on the DB module. The DB module is a core service of PEIRCE.

Testing the database requires large numbers of conceptual graphs, thus the database is dependent on the conceptual catalog group. In the short term the graph mechanisms are being tested on graphs (not necessarily conceptual) that are automatically generated. For example the chess graphs generated by Morph [7] are being used to test classification and compilation techniques in the generalization hierarchy [4].

Database operations are encapsulated in a partially ordered set of conceptual graphs ADT, CGPoset (see Section 3), so that implementations of each of these methods can be easily replaced. Current implementations of the generalization hierarchy (CG database) [6, 4] are data structures kept in memory. This is hidden from the users of CGPoset. In the future, large, external, distributed conceptual graph databases can be used as a replacement or alternative to current implementations of CG databases.

2.4 Linear Notation Input and Output (LINEAR)

The LINEAR group is charged with standardizing the linear notation for conceptual graphs and providing parsers and generators for the notation. [5] contains a grammar for the standard notation. The LINEAR group records any extensions to the syntax, such as including the programming languages: Prolog, and CGC (Conceptual Graphs with Constraints).

2.5 Graphical Editor and Display (CGED)

The CGED group is standardizing the graphical notation and developing graphical editors and displays. The editor will look something like the palette system for XFig, MacDraw etc. The CGEDitor will allow the user to draw assertions and queries (conceptual graphs) on the sheet of assertion. The palette will contain concepts, relations, contexts, etc. Answers to queries will give a view of the sheet assertion, that is, the query will act as a filter on the sheet of assertion. CGED will include 3-Dimensional rendering of knowledge. There will also be a standard package for printing conceptual graphs in documents. A linear notation parser will be written in PostScript. The brackets, parenthesis, and arrows in

```
[type:ref]->(type)->[type:ref]...
```

will be implemented by PostScript macros. Thus PostScript will parse the linear form of conceptual graphs and print the intended graphical form.

2.6 Conceptual Catalogs (CCAT)

The CCAT group needs a database system and a linear/graphical interface for encoding ontologies in conceptual graphs. As well as developing their own ontologies and some support tools, CCAT is interfacing PEIRCE to large collections of knowledge such as CYC. CCAT is involved in designing maintenance routines for merging ontologies; building translators for knowledge representations other than conceptual graphs: CYC, KIF (Knowledge Interchange Format), KL-ONE, SNePS, etc. These data sets will be valuable in testing the other modules, as well as forming the basis for many applications.

2.7 Programming in Conceptual Graphs with Constraints (CGC)

The language CGC will include conceptual graphs as a subset, but add notation for specifying focus and functional constraints. CGC will have a similar control structure to Prolog but will replace Prolog's Herbrand terms and predicates with conceptual graphs. Prolog can be viewed as a restricted set of conceptual graph theory. A compiler for CGC will also compile Prolog. CGC would eventually replace Prolog after it has bootstrapped itself from a Prolog implementation. People implementing software in conceptual graphs can be viewed as already programming in CGC. With the addition of the CGEDitor, people will be able to truly program in conceptual *graphs.*

The philosophy behind CGC is that the tools available to the programmer should not be restricted. That is, the compiler should be clever enough to determine when the more powerful features are not being used and compile the program as efficiently as Prolog or simpler structures in these cases. We believe that strategies designed to compile and index large sets of graphs will outperform the general search technique of Prolog when applied to graphs. This is being investigated in the ongoing work on compilation of conceptual graphs in [4].

CGC will have all the powerful data structures of Prolog such as lists, trees, records etc, compiled efficiently. As well as having access to general graphs.

If PEIRCE members style their Prolog programs so that their representation of graphs can be replaced with conceptual graphs, they will be able to take advantage of the graph indexing and compilation techniques developed in the future. For example the following CGC program

```
[CAT: Tom]<-(AGNT)<-[EAT]->(OBJ)->[MOUSE].
?- [CAT:*x]<-(AGNT)<-[EAT]->(OBJ)->[MOUSE:*y],
    write('The cat '),
    write_name(*x),
    write(' is eating a mouse '),
    write_name(*y).
```

can be translated to the following Prolog program by a preprocessor

```
cg(c('CAT':'Tom'/[1], 'EAT':*/[1,2], 'MOUSE':'Jerry'/[2]),
    r('AGNT'/r(2,1), 'OBJ'(2,3))).
?- call_cg(cg(c('CAT':X/[1], 'EAT':*/[1,2], 'MOUSE':Y/[2]),
    r('AGNT'/r(2,1), 'OBJ'(2,3)))),
    write('The cat '),
    write_name(X),
    write(' is eating a mouse '),
    write_name(Y).
```

In the interim, the program will call the predicate call_cg

```
call_cg(QueryCG) :-
        clause(cg(Cs, Rs)),
        match(QueryCG, cg(Cs, Rs)).
    match(QueryCG, DatabaseCG) :-
        ...
```

By building this level of encapsulation, calls to call_cg can be removed by the CGC compiler in the future and the graph will be directly compiled into C++code to do unification/matching within the CGC database when such a compiler is available.

PEIRCE will be a conceptual graphical programming system that will allow developers to draw programs (using conceptual graphs) or write programs (using natural language) that use large bodies of knowledge in many different formats.

2.8   Inference/Theorem-Proving Mechanisms (PROOF)

The PROOF group will develop proof procedures for conceptual graphs. The alpha and beta rules of Existential Graphs [11] will make excellent interactive proof tools. Eventually proof techniques developed by this group will be written in CGC.

2.9   Learning Mechanisms (LEARN)

LEARN is providing mechanisms for learning search control knowledge, structural concepts from noisy examples, conceptual clustering and classification. LEARN requires a database system, conceptual catalog, and programming language. The LEARN group is intimately involved with the database group DB. Classification of conceptual graphs in the generalization hierarchy is the foundation of structural learning. The LEARN group will provide graph generalization, construction and extraction routines.

Ultimately LEARN will use CGC to encode it's learning mechanisms. By using conceptual graphs to encode learning mechanisms, eventually PEIRCE will learn meta-rules, that is, learning strategies. For example the following is an example of a rule to forget conceptual graphs.

```
IF   [CG:*x]->(DEGREE-OF-USE)->[REAL]->(LESS-THAN)->[REAL: 0.1]
THEN [*x]-><FORGET>.
```

LEARN is vital to the long term research prospects of PEIRCE. LEARN will supply the large numbers of graphs that are necessary to test the KB module. Computer generated graphs are currently the only source of large sets of complex graphs needed to test the KB module.

2.10 Natural Language Parsers and Generators (NLP)

One of the goals of PEIRCE is to support natural language processing. PEIRCE supplies the NLP group with a database system; large conceptual catalogs; matching, unification, and other graph manipulation routines; a theorem-prover; a high level programming language, CGC; linear and graphical editors; and learning mechanisms. The NLP group will specify its requirements implementing much of its own high level requirements.

2.11 Vision System (VISION)

The PEIRCE project hopes to get people involved in a third application, vision, to demonstrate the processing power of the system as well as to provide a useful set of library routines. Applications developers of all kinds are encouraged to get involved now, so PEIRCE can supply the tools they need.

3   The Foundation of PEIRCE

At the heart of PEIRCE is a collection of abstract data types (ADTs). One such ADT is the conceptual graph ADT. The conceptual graph ADT include methods for

1. subsumption - an operation for comparing two conceptual graphs.
2. read - a conceptual graph parser (either linear (supplied by LINEAR) or graphical (supplied by CGED).
3. write - a conceptual graph display system (linear or graphical).
4. decompose - decompose a conceptual graph into a set of canonical formation rules.
5. compose - compose a set of canonical formation rules into a conceptual graph.
6. join - the canonical formation rule for joining two conceptual graphs.
7. copy - copy a conceptual graph.
8. restrict-type - restrict a concept's type.
9. restrict-referent - restrict a concept's referent.

10. simplify - remove a duplicate relation.
11. generalize - generalize a graph by removing structure or generalize concepts.
12. specialize - specialize a graph by adding structure or specializing concepts.
13. semantic-distance - calculate the semantic distance given another conceptual graph.

A possible CG ADT will be built on a general graph ADT. General graph algorithms that are written for this graph class are inherited by the CG class. The graph class will be relatively stable, any changes to conceptual graph theory will generally be independent of the basic graph formalism. Conceptual graphs can be seen as general graphs with labels of the following format

```
Label        :: C.
C            :: Concept | Relation.
Concept      :: '[' Abstraction ':' Referent ']'.
Relation     :: '(' Abstraction ')'.
Abstraction  :: Type | Lambda.
Referent     :: Coref | Individual | CGList.
Coref        :: '*' Var.
Individual   :: '#' Int.
Lambda       :: 'lambda' CorefList CGList.
CorefList    :: '[' Corefs ']'.
Corefs       :: Coref | Coref ',' Corefs.
CGList       :: '[' CGs ']'.
CGs          :: CG | CG CGs.
```

One problem is that many of the routines work on different subsets of conceptual graph theory. For instance many database methods do not consider nested graphs, quantifiers, type definitions. That is, they only support flat graphs with (undefined) types and individual markers. It is important that methods work for general graphs, as these can be extended to the CG subclass level by working on routines for the labels.

There is a partially ordered set ADT (poset) that is used to implement the hierarchies used in conceptual graphs such as the type hierarchy, and the generalization hierarchy. The set of methods for the Poset ADT will include

1. insert - insert an element into the poset.
2. delete - delete an element from the poset.
3. immediate-successors - return the set of immediate successors of an element.
4. immediate-predecessors - return the set of immediate predecessors of an element.
5. merge - merge two posets.
6. statistics - print statistics on cost of queries, structure of poset, etc.
7. display - display the poset in a hierarchical form.
8. encode - encode the hierarchy using the bit encoding method for posets described in [1].
9. code - find the binary code for a element in the poset.
10. decode - find the element(s) of the poset given a binary code.
11. glb - compute the greatest lower bound (code) of two given elements of the poset.
12. lub - compute the least upper bound (code) of two given elements of the poset.

226

The Poset ADT handles the type hierarchy, the generalization hierarchy over conceptual graphs, and hence the generalization hierarchy over Prolog terms, and feature terms. Using a graph's decomposition rules, the hierarchy and the graph elements can be compiled directly into C++ code rather than a data structure that requires interpretation. Figure 1 illustrates the hierarchy over ADTs in PEIRCE.

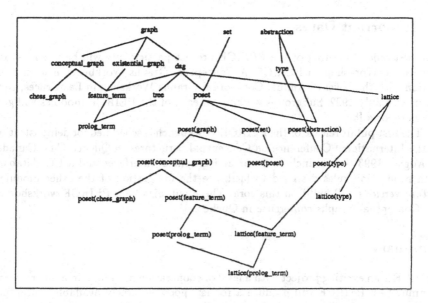

Fig. 1. A Hierarchy of ADTs in PEIRCE

The advantage of this object-oriented approach is that much of the useful work can be done in the graph, poset, poset(graph), and lattice ADTs that will remain independent of any changes to conceptual graph theory, and can be reused for other graphical languages.

The Poset ADT can be used to implement feature logics and Prolog-Like languages as well as conceptual graphs. The Poset ADT will form the basis of the database provided by the DB group. It will also form the core of the CGC language. CGC will be compiled down to C++ code using the poset(conceptual_graph) ADT.

This work will make use of the retrieval methods described in [3, 8]. Some of the ideas in these papers have already been implemented and are in use in Morph [7]. There is also an implementation in C++ that will form the basis of PEIRCE. [3] Through restructuring and compilation of graph data these retrieval methods are expected to satisfy the efficiency requirements that will allow PEIRCE to be a viable tool for both research and industrial-strength application development.

---

[3] The linear notation parser described in [10] is being integrated with the database to demonstrate the database at the conceptual graph workshop and is the first test of PEIRCE's distributed development.

4   Distribution and Control

The PEIRCE workbench will be stored at several FTP sites around the world: `crl.nmsu.edu` (United States), `cluster.cs.su.oz.au` (Australia). Subgroups will send their code to a team of people for vetting before release on the FTP sites.

5   Important Dates

The first official meeting of the PEIRCE project participants will be the First International Workshop on PEIRCE: A Conceptual Graphs Workbench held in association with the Seventh Annual Conceptual Graphs Workshop in Las Cruces, New Mexico, in July 1992. Subgroups working on each of the PEIRCE modules will give status reports [5].

The first public release of the PEIRCE Workbench is scheduled for demonstration at the International Conference on Conceptual Structures in Quebec City, Canada, in August 1993. It will include linear and graphical interfaces and a CG database with production rules designed to facilitate the integration of the other modules. Future versions will build on this core. There will also be a PEIRCE workshop at the Conceptual Graphs conference in Quebec.

Summary

PEIRCE is an exciting project with a level of cooperation not found in many research communities. PEIRCE will produce a robust, portable, freely available conceptual graph workbench. This shared effort will allow many of the researchers in the conceptual graph community to get on with the research they are really interested in: natural language processing, knowledge based systems, graphical programming, data modeling, vision, learning, planning, etc.

The international PEIRCE TEAM so far includes

Alex Bejan	IBM Kingston, New York	United States
Anthony Cheng	Univ Queensland	Australia
Key-Sun Choi	Korea Adv Inst Sc Tech	Korea
Tae-Sang Choi	Univ Missouri-Kansas City	United States
Young Bae Choi	Univ Missouri-Kansas City	United States
Peter Creasy	Univ Queensland	Australia
Keith Campbell	Stanford Univ	United States
Harry Delugach	Univ Alabama-Huntsville	United States
Peter Eklund	Univ Adelaide	Australia
Gerard Ellis	Univ Queensland	Australia
John Esch	Paramax Systems	United States
Norman Foo	Univ Sydney	Australia
Brian Gaines	Univ Calgary	Canada
Steve Graham	Univ Missouri-Kansas City	United States
Young-Suk Han	Korea Adv Inst Sc Tech	Korea
Tim Hines	Univ Missouri-Kansas City	United States
Kees Hoede	Univ Twente	The Netherlands
Wen-Jung Hsin	Univ Missouri-Kansas City	United States
Myung-Cheol Kim	Korea Adv Inst Sc Tech	Korea
Pavel Kocura	Loughborough Univ Tech	United Kingdom
Fritz Lehmann	GRandAI Software	United States
Bob Levinson	Univ California at Santa Cruz	United States

Illias Mavreas	Univ Missouri-Kansas City	United States
Guy Mineau	Laval Univ	Canada
Robert Muehlbacher	A C Nielsen	Austria
Sung Hyon Myaeng	Syracuse Univ	United States
Jonathan Oh	Univ Missouri-Kansas City	United States
Maurice Pagnucco	Univ Sydney	Australia
Heather Pfeiffer	New Mexico State Univ	United States
Simon Polovina	Loughborough Univ Tech	United Kingdom
Kwang-Jun Seo	Korea Adv Inst Sc Tech	Korea
John Sowa	IBM Thornwood	United States
Bill Tepfenhart	Bell Labs	United States
Bosco Tjan	Univ Minnesota	United States
Harmen van den Berg	Univ Twente	The Netherlands
Wei Wei	Univ Missouri-Kansas City	United States
Yin Min Wei	Univ Ohio	United States
Michel Wermelinger	Univ Nova de Lisboa	Portugal
Mark Willems	Univ Twente	The Netherlands
Gi-Chul Yang	Univ Missouri-Kansas City	United States

References

1. Aït-Kaci, H., Boyer, R., Lincoln, P., Nasr, R.: Efficient Implementation of Lattice Operations. ACM Trans. Prog. Lang. Sys. 11(1) (1989) 115–146.
2. Gould, J. and Levinson, R. A.: Experience-Based Adaptive Search to appear in Machine Learning IV: A Multi-Strategy Approach, R. Michalski and G. Tecuci editors, 1992.
3. Ellis, G.: Compiled hierarchical retrieval. In *Current Directions in Conceptual Structures Research* Nagle, T., Nagle, J., Gerholz, L., Eklund, P. (Eds.), Ellis Horwood (1992)
4. Ellis, G.: Compiling Conceptual Graphs. IEEE Trans. Know. Data Eng. (to appear)
5. Ellis, G., Levinson, R. A.: Proceedings of The First International Workshop on PEIRCE: A Conceptual Graphs Workbench. Held in association with the Seventh Annual Conceptual Graphs Workshop, Las Cruces, New Mexico, July 8-10, (1992)
6. Levinson, R. A.: Pattern Associativity and the Retrieval of Semantic Networks. J. Comp. Math. with Appl. 23 (part 2, pp. 573-600) (1992)
7. Levinson, R. A., Snyder, R.: Adaptive pattern oriented chess. Proc. AAAI-91. Morgan-Kaufman (1991) 601-605
8. Levinson, R. A., Ellis, G.: Multi-Level Hierarchical Retrieval. Know. Based Sys. J. (1992)
9. Nagle, T., Nagle, J., Gerholtz, L., and Eklund, P.: Conceptual Structures: Current Research and Practice. Ellis Horwood (1992)
10. Pagnucco, M.: CoGNO - A Graphical User Interface to Conceptual Graphs. Honours project, Sydney University, Nov (1990)
11. Roberts, D. D.: The Existential Graphs of Charles S. Peirce. Mouton, The Hague (1973)
12. Sowa, J. F.: Conceptual Structures: Information in Mind and Machine. Addison-Wesley, Reading, MA (1984)

Specialization: where do the difficulties occur?

M. Chein, M.L. Mugnier
LIRMM (Université Montpellier and CNRS)
860 rue de Saint-Priest
34090- Montpellier (France)
chein@crim.fr, mugnier@crim.fr

Abstract. An analytical study of the Sowa's elementary specialization operations is presented in this paper. Each operation is studied independently, then combined with the others. Some complexity results are proven. For instance, the elementary decision problems, related to these operations, are shown to be isomorphic-complete or NP-complete. The main objective of this paper is a better understanding of the reasons of the difficulties occurring in the algorithmic study of these operations.

1 Introduction

Elementary specialization operations on conceptual graphs are fundamental to Sowa's proposed theory. In this paper an analytical study of these operations is presented. Each operation is studied independently, then combined, in order to specify where the difficulties occur.

The study is carried out in four directions:

is it possible to find a maximal specialization of a conceptual graph - a normal form - related to an operation?

is the transitive closure of the relation - related to an operation - a partial order?

is it possible to find global characterizations of specific specialization operation sequences? (i.e. is it possible to specialize the Sowa's correspondance between projection and specialization?)

what are the algorithmic complexities of the elementary decision problems related to these operations?

A simple conceptual graph (Scg) is a connected ordered bipartite graph with vertex labels, with the labelling fulfilling some constraints. One does not take into account the structures of the label sets because difficulties occur even if the restriction operation is not taken into consideration.

We call c-vertices and r-vertices the concept and relation vertices respectively. The isomorphism between two Scg G and H is denoted by $G \approx H$.

2 Specialization relations

2.1 Définitions

The elementary join operation ([S] p.94) is decomposed into two operations.

The *internal join* is the unary operation defined by the identification of two identically labelled c-vertices c and c' of a Scg G. The result of this operation is denoted by $j_I(G,c,c')$.

The *external join* is the binary operation defined by the identification of two identically labelled c-vertices c_1 and c_2 belonging to two different Scg G_1 and G_2. The result of this operation is denoted by $j_2(G_1,c_1,G_2,c_2)$.

Two r-vertices are *twin vertices*, if they have the same labels and the same neighborhoods. The *simplification* operation is the unary operation defined by the deletion of a twin r-vertex. $s(G,r,r')$ denotes the Scg obtained from G by the deletion of r', where r and r' are twin r-vertices.

Using j_1, j_2 and s, and Scg isomorphism, one can define specific sequences, which correspond to the following binary relations between Scg:

$j_1*(G,H)$ iff H can be obtained from G by a finite sequence of internal join, more precisely if there is a sequence $G=G_0G_1...G_iG_{i+1}...G_n \approx H$ with $G_{i+1} = j_1(G_i,c,c')$ for all i.

$j_2*(G,H)$ iff H can be obtained from G by a finite sequence of external join, more precisely if there is a sequence $G=G_0G_1...G_iG_{i+1}...G_n \approx H$ with $G_{i+1} = j_2(G_p,c,G_q,c')$, p and q <i, for all i.

$s*(G,H)$ iff H can be obtained from G by a finite sequence of simplification, more precisely if there is a sequence $G=G_0G_1...G_iG_{i+1}...G_n \approx H$ with $G_{i+1} = s(G_i,r,r')$ for all i.

Combining these relations one can define:
>from j_1 and s the relation $(j_1s)*$
>from j_2 and s the relation $(j_2s)*$
>from j_1 and j_2 the relation $j*$
>from j_1 and j_2 and s the relation $(js)*$

2.2 Normal forms and orders

A *confluent relation* (Church-Rosser relation, cf. [H] or [CC]) is a binary relation R on a set E such that: each pair of elements of E having a common ascendant have a common descendant (i.e. if $R*$ is the reflexo-transitive closure of R then $\forall x \forall y \exists v$ $(vR*x) \wedge (vR*y) \rightarrow \exists w (xR*w) \wedge (yR*w)$).

An element x of E is a *normal form* (for R) if no element y of E exists such that xRy. If x and y are elements of E, y is a normal form of x if y is a normal form and if $xR*y$.

One has: "If R is confluent and if an element has a normal form then it is unique" (cf.[H]).

It is straightforward to prove the following properties.

(1.1) j_1 *is a confluent relation and every Scg has a unique normal form* - which is the maximal specialization related to j_1* and where the c-vertex labelling is injective - and this Scg can be computed with a naïve algorithm in $O(n^2)$, where n is the cardinality of the c-vertex set. Furthermore j_1* *is a partial order* (an internal join decreases the cardinality of the c-vertex set).

(1.2) s *is a confluent relation and every Scg has a unique normal form* - which is the maximal specialization related to $s*$ and where there are no twin r-vertices - and this Scg can be computed with a naïve algorithm in $O(km^2)$, where k is the maximal arity of an r-vertex and m is the cardinality of the r-vertex set. Furthermore $s*$ *is a partial order* (a simplification decreases the cardinality of the r-vertex set).

(1.3) j_2 *is not a confluent relation*. Nevertheless j_2* *is a partial order* (an external join increases the cardinality of the c-vertex set).

(1.4) *j is not a confluent relation*. Nevertheless *j* is a partial order* (the cardinality of the r-vertex set strictly increases if there is at least one external join).

(1.5) j_1s *is a confluent relation and every Scg has a unique normal form* - the maximal specialization related to $(j_1s)*$ having injective c-vertex labelling and no twin r-vertices - and this Scg can be naïvely computed in $O(\max(n^2, km^2))$. Furthermore $(j_1s)*$ *is a partial order*.

(1.6) j_2s *is not a confluent relation*. Nevertheless $(j_2s)*$ is a partial order.

(1.7) *js is not a confluent relation and (js)* is not a partial order*. For instance, in Figure 1: G is obtained from H by a j_1 operation and an s operation, and H is obtained from G and G', a disjoint copy of G, by a j_2 operation.

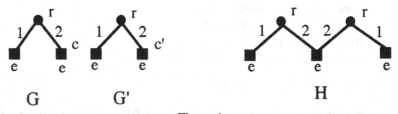

G   G'   H

Figure 1.

The result implies that the specialization relation is not a partial order. This fact has been noted several times (see for instance [E], [W], wherein [J] is quoted, or [C]). The specialization being transitive, one can define an equivalence relation (G is equivalent to H if G is a specialization of H and H a specialization of G) and the quotient specialization relation modulo this equivalence is a partial order. But if one obtained ([CM]) a characterization of equivalent Scg, one also proved that this characterization is not efficient since determining whether two Scg are equivalent is a NP-complete problem.

3 Morphisms

In this paragraph one only considers binary Scg (all the relations are binary relations). This restriction leads to characterizations of the preceding specific specializations in terms of specific multigraph morphisms. Multigraph morphisms will also be used in the last paragraph in order to obtain complexity results.

One can consider different classes of binary Scg in one-to-one correspondance onto classes of multigraphs. For instance:

the class of binary Scg without twin r-vertices is in one-to-one correspondance onto the class of multigraphs with colored arcs and labelled vertices,

the class of binary Scg with only one relation type is in one-to-one correspondance onto the class of vertex labelled multigraphs,

the class of binary Scg with only one relation type and without twin r-vertices is in one-to-one correspondance onto the class of vertex labelled graphs, etc....

The following representation of binary Scg allows simple definitions of these different classes by adding constraints to the labelling function.

A *conceptual multigraph* is a triple $G = (X,U,l)$, where (X,U) is a connected ordinary graph (directed and with loops) and where l, the labelling function, associates:

to each vertex x of X a couple (concept type, referent),

to each arc u of U a set of couples (relation type, arity), $\{(r_1,m_1),...(r_k,m_k)\}$ with, for all i and j, $r_i \neq r_j$ and m_i being an integer >0.

For an arc u of U, if $l(u) = \{(r_1,m_1),...(r_k,m_k)\}$,

$R(u)$ denotes the set $\{r_1,...,r_k\}$, and $m_r(u) = m$ if (r,m) is into $l(u)$.

The notion of Scg morphism, which is Sowa's projection with no restriction of the c-vertex labels ([S] p.99, [CM]), is translated in the following conceptual multigraph morphism.

A *conceptual multigraph morphism* from $G = (X,U,l)$ to $G' = (X',U',l')$ is an application f from X to X' with:

(i) for all arc u=xy of U, for all r of R(u): f(x)f(y) is into U' and r is into $R(f(x)f(y))$.

Thus one can extend f to U.

One considers the classical surjectivity (onto), injectivity (one-to-one), and bijectivity (one-to-one and onto) properties for a morphism f from G to G', and this is used for the vertex set or the arc set.

One also considers the following additional properties.

A morphism f from G to G' is said to be:

covering iff it satisfies

(ii) for all u' into U', and for all r into R(u'): the sum of the $m_r(u)$ for all the u into U with f(u) = u' is $\geq$ to $m_r(u')$;

exact iff it satisfies

(iii) which is obtained from (ii) by restricting the inequality to an equality; (a covering or an exact morphism is such that f(U) = U' i.e. surjective for the arc sets - and the vertex sets since the graphs are supposed to be connected)

strongly covering iff it satisfies

(iv) for all u into U, and for all r into R(u): $m_r(u) \geq m_r(f(u))$.

One has the following characterizations.

(2.1) *$j_1*(G,G')$ iff a morphism f from G to G' exists which is exact.*

If such a morphism exists, by identifying all the vertices of G having the same image by f, one obtains $j_1*(G,G')$. The converse is trivial too.

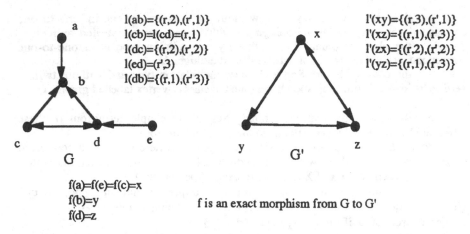

l(ab)={(r,2),(r',1)}
l(cb)=l(cd)=(r,1)
l(dc)={(r,2),(r',2)}
l(ed)=(r',3)
l(db)={(r,1),(r',3)}

l'(xy)={(r,3),(r',1)}
l'(xz)={(r,1),(r',3)}
l'(zx)={(r,2),(r',2)}
l'(yz)={(r,1),(r',3)}

G

G'

f(a)=f(e)=f(c)=x
f(b)=y
f(d)=z

f is an exact morphism from G to G'

Figure 2.

(2.2) $j_2*(G,G')$ *iff a partition of U' exists in subgraphs of G' isomorphic to G, the quotient graph G' modulo this partition being a tree* (connected and without undirected cycles).

The proof is trivial, one only illustrates the property by the following example:

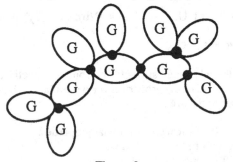

Figure 3.

(2.3) $s*(G,G')$ *iff a morphism from G to G' exists which is bijective for the vertex and the arc sets, and strongly covering.*

The deletion of a twin r-vertex, having x and y as neighbors, corresponds to the decrement of the m related to the relation type in l(xy).

(2.4) $(j_1s)*(G,G')$ *iff a morphism f from G to G' exists which is covering.*

(2.5) $(j_2s)*(G,G')$ *iff one has the property (2.2) where the isomorphisms are replaced by the same morphisms as in (2.3).*

(2.6) $j*(G,G')$ *iff* $k \geq 1$ *morphisms* f_i *from G to* G_i *exist, with:*
                    each G_i *is a subgraph of G',*
                    each f_i *is an exact morphism,*
                    and the sum (one adds the arity of the arcs) of G_i
is equal to G'.

The sufficiency condition can be proven by induction on k. If k = 1 then an exact morphism exists from G to G' and one concludes by 2.1. Let us suppose that the property is true for all k<n and let us consider G and G' with n morphisms f_i satisfying the conditions of the property. If H denotes the union of $G_1,...,G_{n-1}$ then, by the induction hypothesis, one has j*(G,H). Let us consider the vertex sets $\{x_1,...,x_p\}$ and $\{y_1,...,y_p\}$ of H and G_n, which correspond to vertices of G' belonging to H and to G_n. By identifying, for instance, x_1 and y_1 by a j_2, then the others x_i and y_i by j_1, one gets G' from H and G_n. Note that this operation, which combines a j_2 followed by j_1, is frequently used and would be interesting to investigate.

(2.7) $(js)*(G,G')$ iff one has the property (2.6) where exact morphism is replaced by covering morphism.

This last result can be extended to canonical Scg - one can consider that, in this paper, the canon is always restricted to a unique graph - allowing to add details to the Sowa's correspondance between projection and specialization ([S] thm 3.5.4).

4 Complexity results

4.1 Simple relations

Let us consider the elementary decision problems related to the different relations.

Problem 1.
Instance: two Scg G and G'.
Question: G ≈ G'?

With the correspondance between Scg and multigraphs introduced in the preceding paragraph, it is straightforward to prove that P_1 is isomorphic-complete. Indeed G ≈ G' iff the conceptual multigraphs corresponding to G and G' are isomorphic. Furthermore the latter problem is clearly polynomially equivalent to the isomorphism graph problem.

For the following problems one implicitly uses the correspondance between Scg and conceptual multigraphs when necessary.

Problem 2.
Instance: two Scg G and G'.
Question: $j_1(G,G')$?

Let A and B be two conceptual multigraphs. Figure 4 illustrates the transformation of A into A' and of B into B'.

235

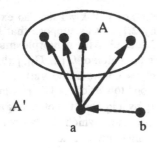
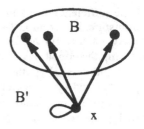

Figure 4.

It is trivial to verify that A ≈ B iff $j_1(A',B')$.

Problem 3.
Instance: two Scg G and G'.
Question: $j_2(G,G')$?

Let A and B be two conceptual multigraphs. Figure 5 illustrates the transformation of A into A' and of B into B' and B".

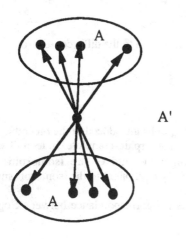
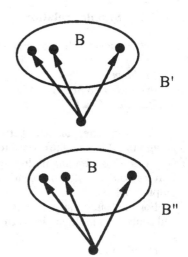

Figure 5.

It is straightforward to verify that A ≈ B iff $j_2(B',B'',A')$.

Problem 4.
Instance: two Scg G and G'.
Question: $s(G,G')$?

Let A and B be two conceptual multigraphs. Figure 6 illustrates the transformation of A into A' and of B into B'.

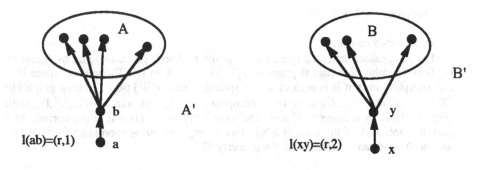

Figure 6.

It is straightforward to verify that A ≈ B iff s(A',B').

the elementary decision problems P_1, P_2, P_3, P_4 are all isomorphic-complete. Thus the complexity of P_i is polynomially equivalent to the isomorphism problem between graphs.

4.2 Other relations

Problem 5.
Instance: two Scg G and G'.
Question: $j_1*(G,G')$?

P_5 is NP-complete.
Actually the property (2.1) $j_1*(G,G')$ holds iff an exact morphism exists from G to G'. Let us consider the particular case where G is a multigraph with l(u)=(r,1) for every arc u of G, G' being a symmetric triangle without any loops, with vertex set {1,2,3}, with l(12)=(r,m_1), l(21)=(r,m_2), l(13)=(r,m_3), l(31)=(r,m_4), l(23)=(r,m_5), l(32)=(r,m_6), and the sum of the m_i being equal to the cardinality of the arc set of G. Then an exact morphism exists from G to G' iff G has a 3-coloring of its vertex set, and this last problem is NP-complete.

Problem 6.
Instance: two Scg G and G'.
Question: $j_2*(G,G')$?

P_6 is isomorphic-complete.
One can use the property (2.2). Let us consider the particular case where G is a 2-connected multigraph, then $j_2*(G,G')$ holds iff each bloc of G' is isomorphic to G. Furthermore computing the blocs of a graph is a polynomial problem, therefore one can conclude.

Problem 7.
Instance: two Scg G and G'.

Question: s*(G,G')?

P₇ is NP-complete.

Let us consider the isomorphic partial graph problem: given two graphs A and B, is B isomorphic to a partial graph of A? This problem is NP-complete since it is NP-complete even if B is restricted to a spanning tree ([GR] pb.). Given a graph H= (X,U) one can transform it to a complete multigraph cm(H)=(X,X²,l), with l(xy)=(r,2), if xy belongs to U and otherwise l(xy)=(r,1). Then B is isomorphic to a partial graph of A iff there exists a bijective strongly covering morphism from cm(A) to cm(B). One can conclude with the property (2.3).

Problem 8.
Instance: two Scg G and G'.
Question: (j₁s)*(G,G')?

P₈ is NP-complete.

Let us consider the particular case where G is a graph (every arc label is equal to (r,1)). Then the problem P₈ is the graph homomorphism problem ([GJ] pb.) which is NP-complete.

Problem 9.
Instance: two Scg G and G'.
Question: (j₂s)*(G,G')?

P₉ is NP-complete.

Let us consider the particular case where the number of vertices of G is equal to the number of vertices of G'. In this case the problem P₉ is equivalent to P₈.

Problem 10.
Instance: two Scg G and G'.
Question: j*(G,G')?

P₁₀ is NP-complete.

Let us consider the particular case where the number of arcs of G is equal to the number of arcs of G'. In this case the problem P₁₀ is equivalent to P₅.

Problem 11.
Instance: two Scg G and G'.
Question: (js)*(G,G')?

5 Conclusion

In this paper no structure on the label set is used. The elementary restriction operation requires at least a partial ordering of the c-vertex label set. Willems [W] proposed avoiding use of the projection notion by linking to a c-vertex all labels greater than its own label. Thus the usual notion of graph morphism can be used, but this transformation may be expensive if the type concept lattice is very high. The restriction relation is not confluent even if partial ordering of the c-vertex label is a lattice because the null label is forbidden.

We proved [CM] that almost all interesting problems concerning Scg are NP-complete. Here we prove that almost all the elementary problems related to the elementary specializations are NP-complete too. Even if this kind of analysis has well-known limitations - the worst cases, or the order of magnitude required to give sense to the asymptotic analysis, can be irrelevant to the graphs involved in specific applications - it leads to a better understanding of the notions, and tractable particular cases may give useful heuristics [MC]. We obtained these complexity results without using labelling properties. Thus, as noted by Sowa (forum on CG 19 May), one of the interesting algorithmic directions for conceptual graphs involves construction of efficient algorithms using both the combinatorial structure of the graphs and the structure of the label set.

Pattern matching operations on conceptual graphs are often defined in terms of projection. Since there is a projection from G to H iff H is a specialization of G, pattern matching operations might be considered as sequences of elemntary specialization operations. The generalization operations - dual of the specialization operations - constitute a complete set of inference rules on the set of logical formulas associated with conceptual graphs [CM], this point of view may be useful if one is interested in logical interpretation of pattern matching operations. Otherwise, we believe that it is worthwhile to study pattern matching operations as kinds of labelled bipartite graph morphisms.

6 References

[CC]   G. Chaty, M. Chein, Réductions de graphes et systèmes de Church-Rosser. R.A.I.R.O. Oper. Res., vol.15, n°2, 1981, 109-117.

[C]   M. Chein, S-graphes(1): Notions de base. Res. Rep. CRIM, 87, oct. 1989.

[CM]   M. Chein, M.L. Mugnier, Conceptual Graphs: Fundamental Notions. Res. Report LIRMM, n°91-188, Nov 1991, 35p. (submitted to the R.I.A.).

[E]   G. Ellis, Compiled Hierarchical Retrieval. 6th Annual Workshop on Conceptual Graphs, 1991, 187-207.

[FLDC]   J. Fargues, M.C. Landau, A. Dugourd, L. Catach, Conceptual Graphs for Semantic Information Processing. IBM Journal of Res. and Dev., 30, 1, 1986, 70-79.

[GJ]   Garey, Johnson

[J]   M.K. Jackman, Inference and the Conceptual Graph Knowledge Representation Language. In S. Moralee, edt, Research and Development in Expert Systems IV, Cambridge University Press, 1988.

[H]   G. Huet, Confluent Reductions: Abstract Properties and Applications to Term Rewriting Systems. Journal of A.C.M., vol.27, 1980, 797-821.

[MC]   M.L. Mugnier, M. Chein, Polynomial Algorithms for Projection and Matching Problems. 7th Annual Workshop on Conceptual Graphs, New Mexico University, July 1992.

[S]   J. Sowa, Conceptual Structures. Information Processing in Mind and Machine. Addison-Wesley, 1984.

[W]   M. Willems, Generalization of Conceptual Graphs. 6th Annual Workshop on Conceptual Graphs, 1991.

Polynomial algorithms
for projection and matching

M.L. Mugnier and M. Chein

LIRMM (CNRS et Université Montpellier II)
860 rue St-Priest, 34090 Montpellier, France

Abstract. The main purpose of this paper is to develop polynomial algorithms for the projection and matching operations on conceptual graphs. Since all interesting problems related to these operations are at least NP-complete —we will consider here the exhibition of a solution and counting the solutions—, we propose to explore polynomial cases by restricting the form of the graphs or relaxing constraints on the operations. We examine the particular conceptual graphs whose underlying structure is a tree. Besides general or injective projections, we define intermediary kinds of projections. We then show how these notions can be extended to matchings.

1 Presentation

This paper deals with efficient algorithms implementing basic operations on conceptual graphs, namely *projection* and *matching*.

Since the associated problems are NP-complete, several approaches may be used to come to efficient algorithms: describe algorithms which are exponential on classical graphs hoping that they are good on conceptual graphs [MLL91], use backtracking techniques [McG82] or exploit efficient preprocessing [LE91]. We propose another perspective which consists of restricting constraints on the operations or the form of the graphs, and which provides polynomial algorithms.

The paper is organized as follows. The next part describes our framework. Part 3 is devoted to polynomial algorithms for the projection operator. These algorithms are based on the tree notion. We describe algorithms for computing a projection from a conceptual graph whose underlying graph is a tree, to another graph, for counting or for enumerating these projections. We then focus on injective projections. The decisional problem associated with an injective projection is NP-complete even if the first graph is a tree. But we define intermediary notions of projections, said to be locally injective, which are polynomially computable. We end with some extensions of preceding algorithms to matching problems.

2 Our Framework

2.1 Conceptual Graphs

For this study we restrict our framework to simple conceptual graphs (3.1.2 of [Sow84]). For the sake of brevity, we call c-vertices and r-vertices the concept and relation vertices respectively. Given any c-vertex c, its label —*label(c)*— is a pair

(t, m), where t is a concept type and m is a marker. Let L be the set of all possible labels for c-vertices. L is a lattice (in the classical order theory sense) whose order is the product of the concept type lattice and the marker set with an added absurd marker. Let $e = (t, m)$ and $e' = (t', m')$ be two labels of L, then $e < e'$ if $t < t'$ and $m < m'$ [CM91].

Given a conceptual graph G, C_G and R_G denote its c-vertex set and its r-vertex set respectively.

For the *general case*, we consider that a conceptual graph admits

- *n-ary relations, $n \geq 1$* —The edges between an r-vertex r and its neighbours are totally ordered; we label these edges $i = 1, \dots, degree(r)$.
- *multi-edges* —*i.e.* several edges between an r-vertex and one of its neighbours. Thus the number of neighbours of an r-vertex may be strictly smaller than its degree.
- *twin r-vertices* —*i.e.* r-vertices with the same type and exactly the same i-ith neighbour, for all $i \in \{1, \dots, degree(r)\}$ ("duplicate vertices" of [Sow84], 3.4.3).

Besides this general case, we will also consider a *restricted case*, where we only have binary r-vertices, no multi-edges —so each r-vertex has exactly two distinct neighbours— and no twin r-vertices. There is a trivial bijection between this sort of conceptual graphs and directed graphs without loops, labelled on vertices and edges, and such that there is at most one edge with a given label.

We have chosen to present algorithms in the restricted case before the general case. This distinction aims at not hiding the simplicity of the basic principles. Indeed, from an algorithmic point of view, the general case mainly causes formulation difficulties.

2.2 Different kinds of Projections and Matchings

Projections. A projection is a graph morphism (in a classical graph theory sense) provided with additional rules on labels.

Given two conceptual graphs G and G', a *projection* Π from G to G' is an ordered pair of mappings from (R_G, C_G) to $(R_{G'}, C_{G'})$, such that:
(i) For all edge rc of G with label i, $\Pi(r)\Pi(c)$ is an edge of G' with label i.
(ii) $\forall r \in R_G$, $type(\Pi(r)) = type(r)$; $\forall c \in C_G$, $type(\Pi(c)) \leq type(c)$.

There is a projection from G to G' if and only if G' can be derived from G by the elementary specialization rule [Sow84] [CM91]. Nevertheless, the projection operator is often used in a restricted form which is an *injective projection*: the image of G in G' is then a subgraph of G' isomorphic to G. The decisional problem associated with the projection operator is NP-complete under the two forms.

Matchings. Matching operations represent a second class of operations whose one variant is the extended join of [Sow84]. Generally, these operations are definable as follows: given two graphs G_1 and G_2, find a *correspondence of a certain type* between two *maximal* subgraphs of G_1 and G_2, say G'_1 and G'_2. Several parameters specify the above definition.

1. Compatibility criteria between the c-vertices labels: typically, if we are interested in a common specialization of G_1 and G_2 then the condition may be: two labels are compatible if their lower bound is not 0. If we want a common generalization of G_1 and G_2 then *a priori* no condition is imposed. If the aim is to map G_1 onto G_2 then the label of any c-vertex of G'_1 may have to be greater or equal to its image label.

2. The type of correspondence between the vertex sets of G'_1 and G'_2: may be a bijection, a surjective mapping from G'_1 to G'_2, or a general correspondence between G'_1 and G'_2.

3. The meaning of maximality: there are at least three; either the cardinal of G'_1 and/or G'_2 is maximal (*cardinal maximality*); or: there are no subgraphs of G_1 and/or G_2 which are solutions and extend G'_1 and G'_2 respectively (*inclusion maximality*); or: there is no extension of the correspondence which keeps the connectivity of G'_1 and G'_2 (*correspondence inclusion maximality*).

For extended join as defined by J. Sowa, and the variants we study, the associated decisional problems are NP-complete, whether we consider cardinal or inclusion maximality. Only the third sense provides polynomial algorithms.

2.3 Trees

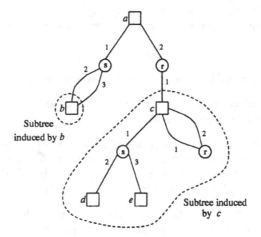

Fig. 1. Tree with root a

An interesting case concerning the algorithmic complexity of the projection operator —and matching operations as seen later— is the use of the tree notion. Let us define a *conceptual tree* as a conceptual graph without cycles, except for cycles created by multi-edges between an r-vertex and one of its neighbours (Figure 1). A specific c-vertex —the root— is usually a starting point for the operations. Consequently, we can speak in terms of father/son relationships between c-vertices. Let us point out that this direction from the root to the leaves is independent of the order put by the r-vertices on their neighbours. Given a rooted tree, and one of its c-vertices, say *c*, the *subtree induced by c* is the rooted subtree with root *c*.

Canonical graphs and type definitions are often rooted trees. Furthermore, the linear form of a conceptual graph can be considered as a tree with additional constraints.

3   Algorithms for the Projection Operator

Computing a projection from a graph to another graph is a tractable problem when the first graph is a tree. We develop the following points: given a graph G and a tree T, decide whether a projection exists from T to G, exhibit one or all projections from T to G, count the projections from T to G. Firstly, we describe the algorithms in the restricted binary case, then we extend them to the general case.

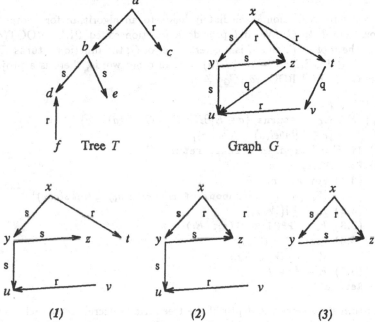

There are eight projections from T to G. *(1)* corresponds to the only injective projection,
$\Pi = \{(a, x), (b, y), (c, t), (d, u), (e, z), (f, v)\}$. Projection *(2)* is locally c-injective (see section 3.3.2) $\Pi = \{(a, x), (b, y), (c, z), (d, u), (e, z), (f, v)\}$. *(3)* is a projection with no particular properties, $\Pi = \{(a, x), (b, y), (c, z), (d, z), (e, z), (f, x)\}$.

Fig. 2. Projection from a tree to a graph (binary case)

3.1 Projection(s) from a tree to a graph (binary case)

As indicated before, binary conceptual graphs can then be seen as directed labelled graphs (Figure 2). Given two c-vertices, say a and b, we note $links(a, b)$ the set of edge labels between a and b, taking the edge direction into account —for instance r_+ (resp. r_-) if there is an edge ab (resp. ba) labelled r. "$links(a, b) \subseteq links(c, d)$" thus means that for all edge ab (resp. ba), there is an edge cd (resp. dc) with the same label.

First notice that there is a projection from T to G iff, given any c-vertex a of T chosen as the root, there is a c-vertex c of G and a projection from T to G which maps a onto c. Such a projection is easily defined recursively as follows.

Π is a PROJECTION from T to G, with $\Pi(a) = c$, if it satisfies:
- *(i)* $label(a) \geq label(c)$,
- *(ii)* for all a_i son of a, $v = \Pi(a_i)$ is a neighbour of c such that
  - *(ii-a)* $links(a, a_i) \subseteq links(c, v)$
  - *(ii-b)* Π_{T_i}, the restriction of Π to the subtree T_i induced by a_i, is a PROJECTION from T_i to G (and by construction, $\Pi_{T_i}(a_i) = v$).

This recursive definition immediately leads to an algorithm for computing a projection from T to G. We first provide a function called PROJ-ROOT(a, E), where a is the root of T, and E is a c-vertex set of G; the function returns $E' \subseteq E$ such that $\forall c \in E'$, $\exists \Pi : T \to G \mid \Pi(a) = c$. In other words, there is a projection from T to G iff PROJ-ROOT$(a, C_G) \neq \emptyset$.

```
PROJ-ROOT(a, E)
//a ∈C_T; E ⊆ C_G; returns {c ∈ E|∃Π : T → G, Π(a) = c}.
    (1) E ← {c ∈ E|label(a) ≥ label(c)}
    (2) If E = ∅ or a is a leaf, return E
    (3) For all a_i sons of a,
        (3.1) For all c ∈ E
              W_{c,a_i} ← {v neighbour of c|links(a, a_i) ⊆ links(c, v)}
        (3.2) E_i ← ∪{W_{c,a_i}}_{c∈E}
        (3.3) E_i ← PROJ-ROOT(a_i, E_i)
        (3.4) For all c ∈ E,
              V_{c,a_i} ← W_{c,a_i} ∩ E_i
        (3.5) E ← {c ∈ E|V_{c,a_i} ≠ ∅}
    (4) Return E
```

The algorithm stops since the depth of the tree strictly decreases at each recursive call.

Complexity of PROJ-ROOT. We suppose in the following that one can compare the labels of two c-vertices in a constant time. If not, let $comp$ be this complexity. Then, the supplementary cost for all algorithms described in this paper is at most $|C_T| \times |C_G| \times comp$.

Property 1 *The complexity of PROJ-ROOT(a, C_G) is $O(|C_T| \times |R_G|)$ —$|R_G|$ is here equivalent to the number of edges of G.*

Proof -By induction on $|C_T|$. For $|C_T| = 1$, only (1) and (2) are executed. (1) is in $O(|C_G|)$, (2) is in constant time, so the property is trivially true. Let T be a tree with root a, and $|C_T| = n$, $n > 1$. Let us analyse step (3) by summing at each substep the operations done for all sons of a. (3.1) is in $O(degree(a) \times \sum_{c \in C_G} degree(c))$. (3.2) is $O(degree(a) \times \sum_{a_i \; son \; of \; a} V_{c,a_i}) = O(degree(a) \times |R_G|$. For all a_i son of a, $|E_i| \leq |C_G|$. Using the induction hypothesis, (3.3) is in $O(|C_T| \times |R_G|)$: (3.4) can be done in $O(|sons \; of \; a| \times |R_G|)$, for instance: for all a_i, mark the vertices of E_i (one can use a vector with size $|C_G|$, that is initialized before the first recursive call, and used to mark the c-vertices of C_G, with different marks for different uses); then restrict all V_{c,a_i} to the marked elements. (3.5) is in $O(|C_G|)$. Since T has no multi-edges and its r-vertices are binary vertices, $degree(a) = |sons \; of \; a|$. Then, the complexity of PROJ-ROOT(a, C_G) is $O(|C_T| \times |R_G|)$.
□

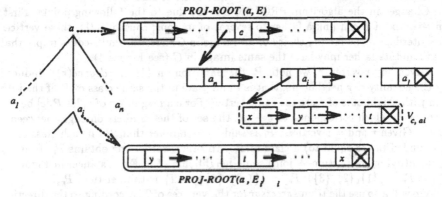

With each $a \in A$ is associated the set E' =PROJ-ROOT(a, E). With each $c \in E'$ is associated the set V_{c,a_i}, for all a_i son of a.

Fig. 3. The structure built by PROJ-ROOT(a, E)

Let us suppose that the V_{c,a_i} sets computed by PROJ-ROOT(a, E) are kept outside the function. Consider for instance, the structure built in Figure 3. This does not imply increasing complexity. In order to exhibit a projection Π, one only has to traverse this structure from the root of T to its leaves. One chooses $\Pi(a)$ in PROJ-ROOT(a, E), then recursively: if b_i is a son of b, and $\Pi(b) = c$, then choose $\Pi(b_i)$ in V_{c,b_i}. If one desires to enumerate all solutions, the structure is traversed in the same way, combining all possible choices for all vertices.

Rather than (or before) enumerating all projections from T to G, it may be interesting to count the number of such projections. The number of projections mapping the root a of T onto a vertex of $E \subseteq C_G$ is defined as follows.

$$NB(a, E) = \sum \{NB(a, c), \; c \in E\}$$
$$NB(a, c) = 0 \text{ if the condition "}label(a) \geq label(c)\text{" is not fulfilled}$$

1 if a is a leaf

Otherwise, $\prod\{NB(a_i, V_{c,a_i}), a_i$ son of $a\}$

—See (3.4) of PROJ-ROOT for definition of V_{c,a_i}—

$NB(a, E)$ is computable by simply adding counters to PROJ-ROOT. For instance, instead of returning a vertex set, the function returns a set of couples $(c, NB(a, c))$, where $NB(a, c) \neq 0$. This is done without increasing the complexity.

3.2   Projection(s) from a tree to a graph (general case)

Let us now consider general conceptual graphs (as defined in 2.1). r-vertices play a preponderent role. Indeed, assuming that Π is a projection from T to G, given r and $\Pi(r)$, there is a unique way for mapping each neighbour of r onto the neighbours of $\Pi(r)$. The converse is not true: when there are twin vertices, knowing the images of all c-vertices does not determine the image of the r-vertices (although it is usually no important whether one twin vertex or the other is chosen).

Changes in the algorithm PROJ-ROOT are due to the following points. First, an r-vertex of T can link a father to several of its sons; consequently, these vertices are interdependent for the choice of an image in G. Second, multi-edges imply that a son and its father may have the same image in G (see Figure 4).

Given any r-vertex r, we note P_r the partition on $\{1, \dots, degree(r)\}$, induced by the equality of r neighbours, that is i and j are in the same class of P_r if the i-ith and j-ith neighbours of r are the same vertex. For c a neighbour of r, let $P_r[c]$ be the class of P_r which corresponds to c —i.e. the set of the numbers on edges between r and c. Given r and r' r-vertices, P_r is said to be *thinner* than $P_{r'}$ if each class of P_r is included in (or equal to) a class of $P_{r'}$. In other words, one obtains $P_{r'}$ from P_r by identifying (restriction + join) some neighbours of r. For instance, in Figure 4, $P_{r_1} = P_r = \{\{1\}, \{2\}, \{3\}\}; P_{r_2} = \{\{1,3\}, \{2\}\};$ so P_r is thinner than P_{r_2}.

We will also use the term *successor* for the vertices of T, according to the direction from the root to the leaves. The successors of a c-vertex c are the r-vertices which link c and its sons, or admit c as a unique neighbour. The successors of an r-vertex are its neighbours, except for the one which is the father of the others.

A projection can now be defined as follows.

Π is a PROJECTION from T to G, with $\Pi(a) = c$, if it satisfies:

I   *(i)* $label(a) \geq label(c)$,

    (ii) for all r successor of a, $r' = \Pi(r)$ is a neighbour of c such that

        (ii-1) $type(r) = type(r')$

        (ii-2) $P_r[a] \subseteq P_{r'}[c]$, and,

II   *(ii-3)* for all a_i successor of r, $v = \Pi(a_i)$ is the neighbour of r' (unique if it exists) such that

        (ii-3-1) $P_r[a_i] \subseteq P_{r'}[v]$,

        (ii-3-2) Π_{T_i}, the restriction of Π to the subtree T_i induced by a_i, is a PROJECTION from T_i to G

        (and by construction, $\Pi_{T_i}(a_i) = v$).

(ii-2) and *(ii-3-1)* ensures that P_r is thinner than $P_{r'}$. Parts I and II correspond to decomposition of the algorithm in two crossed recursive functions. I keeps the name PROJ-ROOT and II is called PROJ-r.

246

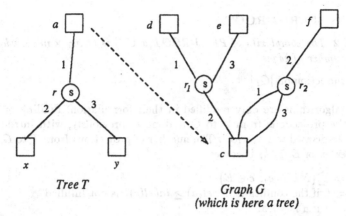

Tree T

Graph G
(which is here a tree)

All c-vertices have the same label. There is only one projection from T to G
mapping a on c. It is defined by $\Pi(r) = r_2$ (this implies that $\Pi(y) = \Pi(a) = c$
and $\Pi(x) = f$).

Fig. 4. Projection in the general case

```
PROJ-ROOT(a, E)
//a ∈C_T; E ⊆ C_G; returns {c ∈ E|∃Π : T → G, Π(a) = c}.
    (1) E ← {c ∈ E|label(a) ≥ label(c)}
    (2) If E = ∅ or a is a leaf, return E
    (3) For all r successors of a,
        (3.1) For all c ∈ E
              W_{c,r} ← {r' neighbour of c|type(r)=type(r') and P_r[a] ⊆ P_{r'}[c]}
        (3.2) E_r ← ∪{W_{c,r}}_{c∈E}
        (3.3) E_r ← PROJ-r(r, E_r)
        (3.4) For all c ∈ E,
              V_{c,r} ← W_{o,r} ∩ E_r
        (3.5) E ← {c ∈ E|V_{c,r} ≠ ∅}
    (4) Return E

PROJ-r(r, E)
//r ∈R_T; E ⊆ R_G and fulfills conditions defined by PROJ-ROOT;
//returns {r' ∈ E|∃Π : T → G, Π(r) = r'}.
    (1) E ← {r' ∈ E|P_r is thinner than P_{r'}}
    (2) If E = ∅ or |P_r| = 1, return E
    (3) For all a_i successors of r,
        (3.1) E_i ← ∪{c_{r'}|P_r[a_i] ⊆ P_{r'}[c_{r'}]}_{r'∈E}
        (3.2) E_i ← PROJ-ROOT(a_i, E_i)
        (3.3) E ← {r' ∈ E|c_{r'} ∈ E_i}
    (4) Return E
```

Complexity of PROJ-ROOT

Property 2 *The complexity of PROJ-ROOT(a, C_G) is $O(m_T \times m_G)$, where m denotes the number of edges.*

Proof -By induction on $|C_T|$.

The other algorithms are only modified in their formulation. Finding one or enumerating the projections from T to G is done as previously, with introduction of structures associated to r-vertices. The number of projections from T to G mapping a onto a vertex of $E \subseteq C_G$ becomes:

$NB(a, E) = \sum \{NB(a, c),\ c \in E\}$
$NB(a, c) = $ 0 if the condition "$label(a) \geq label(c)$" is not fulfilled
                1 if a is a leaf
                Otherwise, $\prod \{n_r | r$ successor of a, and $n_r = \sum \{NB(r, r')| r' \in V_{c,r}\}$
$NB(r, r') = $ 0 if the condition "P_r thinner than $P_{r'}$" is not fulfilled
                —For all a_i successor of r, let $c_{r'}$ be the neighbour of r' such that
                   $P_r[a_i] \subseteq P_{r'}[c_{r'}]$ —
                1 if there is an a_i such that $NB(a_i, c_{r'}) = 0$
                Otherwise, max $\{NB(a_i, c_{r'})\}$

3.3 Injective projection and variations

Instead of any projection, one often prefers to obtain an *injective* projection. More generally, given the two graphs G and G', one wants the image $\Pi(G)$ of G in G' to be as close as possible to G.

Deciding whether there is an injective projection from G to G' is an NP-Complete problem, even if G is a tree, and the graphs have only binary r-vertices without multi-edges (by reduction to "subgraph isomorphism" [GJ79]). But an injective projection from a tree to another tree is polynomially computable. And this provides a basis for defining polynomially computable kinds of projections from a tree to a graph.

Injective projection from a tree to a tree. Let us first give the skeleton of an algorithm from a tree T to a tree T', in the binary case (the algorithm is a trivial adaptation of the one analyzed in [Rey77]). It involves the notion of bipartite graph matching. A matching of a bipartite graph is a set of its edges, such that no two edges share a common endpoint.

Let a and b be the roots of T and T' respectively.
There is an INJECTIVE PROJECTION from T to T', mapping a onto b if
     (i) $label(a) \geq label(b)$
     (ii) there is a matching of cardinality $|A|$ in the bipartite graph on A and B
         where:
         $A = \{$sons of $a\}$, $B = \{$sons of $b\}$
         there is an edge $a_i b_j$, $a_i \in A$, $b_j \in B$, if
     (ii-1) $links(a, a_i) \subseteq links(b, b_j)$
     (ii-2) there is an INJECTIVE PROJECTION from the subtree
            induced by a_i to the subtree induced by b_j mapping a_i onto b_j.

Property 3 *"Injective Projection" can be implemented in $O(|C_T|^{1,5} \times |C_{T'}|)$ (or less depending on the complexity of the matching algorithm).*

Proof-Such an algorithm is easily implemented with complexity $O(\sum_{a \in T,\ b \in T'} match_{a,b})$, where $match_{a,b}$ is the cost of the matching between the sons of a and b. Indeed, we assume that step *(ii-1)* is in constant time since there is only one r-vertex linking a father and one of its sons. A good matching algorithm is, for instance, the algorithm of [HK73] (also analyzed in [PS82]) which has complexity $O(|A|^{1,5} \times |B|)$, with $|A| \le |B|$. Since $\sum_{a \in T,\ b \in T'} |sons(a)|^{1,5} \times |sons(b)| \le |C_T|^{1,5} \times |C_{T'}|$, we have the property.
□

Notes

 – Actually, we do not need to compute matching between A and B —see *(ii)*—, but only between subsets of c-vertices with the same links. But, for a complexity study, the worse case occurs when all vertices have the same links and the same label.
 – When coming to the general conceptual trees, the bipartite matching is computed on sets of r-vertices, and two corresponding r-vertices must have the same partition P_r. Using the same reasoning as previously described, we can roughly bound the complexity with $O(m_T^{1,5} \times m_{T'})$.
 – Let us note that counting the injective projections between two rooted trees is an NP-hard problem (by reduction to the problem of counting the perfect matchings in a bipartite graph ([GJ79] theorem 7.9))

Locally injective projections from a tree to a graph. Adapting the previous definition to the case where T' is any graph, leads to a projection which is *locally injective*. More precisely, it is injective on the successor set of each vertex of T.

In the binary case, sucessors are c-vertices (see Figure 2). PROJ-ROOT is modified as follows. Add a forth step (it acts as a filter) before returning E.

(4) For all c in E,
    Build the bipartite graph (A, B, U) such that:
        $A = \{$sons of $a\}$, $B = \{$neighbours of $c\}$
        (B can also be defined as $\bigcup\{V_{c,a_i},\ a_i \in A\}$
        $U = \{a_i v \mid v \in V_{c,a_i}\}$
    If this graph admits a matching with cardinality $|A|$,
        c is a solution
(5) Return all c-vertices which are solutions of (4)

In the general case, let us distinguish between two notions. A projection is said to be *locally c-injective*, if it is injective on the successor set of each r-vertex. It is said to be *locally r-injective*, if it is injective on the successor set of each c-vertex.

A transformation of PROJ-ROOT by adding a forth step as in the binary case provides a locally r-injective projection.

Obtaining a locally c-injective projection is an obvious change of PROJ-ROOT: let r be a successor of a, then $P_r \setminus \{P_r[a]\}$ has to be thinner than $P_{\Pi(r)}$. More generally, for obtaining a projection which is injective on the r-vertex neighbourhood, one only has to impose that for all r, $P_r = P_{\Pi(r)}$, instead of $P_r \subseteq P_{\Pi(r)}$.

But, computing, in the general case, a projection which is injective on the sons of each c-vertex of T is a difficult problem, as asserted below. Let *sons-injective projection* be the problem "given T a conceptual tree and G a conceptual graph, with n-ary r-vertices, is there a projection Π from T to G satisfying $\forall c \in C_T$, $\forall c_i$, c_j sons of c, $\Pi(c_i) \neq \Pi(c_j)$?"

Property 4 *"Sons-injective projection" is NP-complete.*

Proof -[Mug92]. Transformation from "partition into triangles" [GJ79]. It follows from the proof that the problem remains NP-complete even if the degree of the r-vertices is bounded by 4, and graphs have no multi-edges and no twin r-vertices.
□

4 Application to Matching operations

Two kinds of matching, say between G_1 and G_2, seem particularly relevant to us.

(a) The aim is to find a maximal common part to G_1 and G_2. We call this operation *isojoin* because the two corresponding subgraphs G_1' and G_2' are isomorphic. Either one merges each vertex of G_1' and its image in G_2', and the label of the obtained vertex is the lower bound of the two previous labels —one obtains a specialization of G_1 and G_2 [Sow84]; or, in G_1', one simply replaces each c-vertex label by the greater bound of its label and the label of its image in G_2'. The graph obtained from G_1' is a generalization of G_1 and G_2.

(b) The aim is to map G_1 onto G_2. G_1 and G_2 no longer have a symmetric role, and the corresponding subgraphs are not isomorphic. Indeed there is a surjective projection from G_1' to G_2'. The result of the join is built from G_1 and G_2 in the following manner: merge in G_1' all vertices having the same image; one obtains $G"_1$; then merge each vertex of $G"_1$ and its image in G_2'. Note that if $G_1' = G_1$, there is a projection from G_1 to G_2. Then the result of the join is simply G_2.

There are at least two ways of coming to polynomial algorithms for these matchings.

1. If we restrict ourself to a correspondence inclusion maximal join, then we obviously obtain polynomial algorithms [Sow84] [FLDC86]. A decisive parameter for the *form* of the result is the manner of searching the graph. For instance, using a breadth first principle tends to privilege G_1' and G_2' with minimal diameter. In other words, one focuses on vertices with minimal distance to the starting vertices.
2. From an algorithm viewpoint, matching is a generalization of the projection operation. We can thus use the previously studied algorithms. Note that the rules on labels play an unimportant role.
   - Building a join —version (*b*)— from a tree to a graph is obtained by a slight change of the projection algorithm from a tree to a graph.
     In the binary case for instance, the maximal number of vertices of T that can be joined with vertices of G is obtained by CardJoin(a, C_G) as follows.

$\text{CardJoin}(a, E) = 0$ if $E = \emptyset$

Otherwise, $\max\{\text{CardJoin}(a, c), c \in E\}$

$\text{CardJoin}(a, c) = 0$ if the condition $label(a) \geq label(c)$ is not fulfilled

1 if a is a leaf

Otherwise, $1 + \sum\{\text{CardJoin}(a_i, W_{c,a_i}), a_i \text{ son of } a\}$

—See (3.1) of PROJ-ROOT for definition of W_{c,a_i}—

- Computing a maximal isojoin between two trees, say T and T', is a polynomial operation, whatever the maximality definition may be.

In the binary case, the cardinal maximal of an isojoin from T to T' mapping the two roots, say a and b is defined as follows.

$\text{CardIJoin}(a, b) = 0$ if the condition $label(a) \wedge label(c) \neq \emptyset$ is not fulfilled

1 if a or b is a leaf

Otherwise, $1 +$ the weight of a maximal weight matching of the bipartite graph (S, T, U, p), where:

$S = \{ \text{sons of } a \}$, $T = \{ \text{sons of } b \}$,

$\forall s \in S, t \in T, st \in U$ if $\text{CardIJoin(s,t)} \neq 0$,

p is a mapping which associates with each edge $st \in U$,

the weight $p(st)=\text{CardIJoin(s,t)}$

Using the heuristic of local injectivity, one obtains a join algorithm between a graph and a tree.

More details can be found in [Mug92]. Extensions of these algorithms we are currently working on are: Enumerating all joins between two graphs with maximality for correspondence inclusion and privileging breadth-first order. Developing algorithms on graphs which are not trees, but almost trees.

References

[CM91] M. Chein and M.L. Mugnier. Conceptual graphs : fundamental notions. Research Report 188, LIRMM, Nov. 1991. 30 p.

[FLDC86] J. Fargues, M.C. Landau, A. Dugourd, and L. Catach. Conceptual graphs for semantics and information processing. *IBM Journal of Research and Development*, 30(1):70–79, 1986.

[GJ79] M.R. Garey and D.S. Johnson. *Computer and Intractibility - A Guide to the Theory of NP-Completeness*. W.H. Freeman and co, 1979.

[HK73] J.E. Hopcroft and R.M. Karp. A $n^{5/2}$ Algorithm for Maximum Matching in Bipartite Graphs. *J. SIAM Comp.*, 2:225–231, 1973.

[LE91] R. Levinson and G. Ellis. Multi-level hierarchical retrieval. In *Proceedings of the 6th Annual Workshop on Conceptual Graphs*, pages 67–81, 1991.

[McG82] J.J. McGregor. Backtrack search algorithms and the maximal common subgraph problem. *Software-Practice and Experience*, 12:23–34, 1982.

[MLL91] S.H. Myaeng and A. Lopez-Lopez. A flexible matching algorithm for matching conceptual graphs. In *Proceedings of the 6th Annual Workshop on Conceptual Graphs*, pages 135–151, 1991.

[Mug92] M.L. Mugnier. Quelques aspects algorithmiques en représentation des connaissances - Algorithmes incrémentaux d'héritage multiple, Opérations sur les graphes conceptuels. PhD. thesis, to appear in October, 1992.

[PS82]   C.H. Papadimitriou and K. Steiglitz. *Combinatorial Optimization: Algorithms and Complexity.* Prentice Hall, 1982.

[Rey77]   S.W. Reyner. An analysis of a good algorithm for the subtree problem. *SIAM J. Computer,* 6(4):130–132, 1977.

[Sow84]   J.F. Sowa. *Conceptual structures - Information Processing in Mind and Machine.* Addison-Wesley, 1984.

CGMA: A Novel Conceptual Graph Matching Algorithm

Gi-Chul Yang, Young Bae Choi and Jonathan C. Oh

Computer Science
University of Missouri-Kansas City
Kansas City, MO 64110-2499

Abstract: A novel Conceptual Graph Matching Algorithm (CGMA) is proposed along with a knowledge base organization scheme. It allows the user to select the Degree of Matching and Degree of Inheritance to increase the user's satisfaction. A conceptual graph is represented in a special linear form (called U-Form) for CGMA. U-Form eliminates the necessity of explicit arcs in the graph to achieve the simplicity.

1 Introduction

Information retrieval in a large knowledge base is a bottleneck in most of the current information retrieval systems. There is, also, a great demand to retrieve relevant information even when the exact information in a knowledge base is absent. To solve these problems we propose a novel algorithm for organization and retrieval of information based on conceptual graphs [Sowa84]. The algorithm is rooted in [Yang90] and called Conceptual Graph Matching Algorithm (CGMA).

CGMA allows the user to select the Degree of Matching (DoM) and Degree of Inheritance (DoI) for retrieval of relevant information where exact information is absent in a knowledge base. DoM and DoI shows the closeness between a query graph and an answer graph(s) from the different point of view. We calculate DoM = number of matched concept(s) / number of concept(s) in a query graph. DoM is crucial to retrieve the necessary information in some cases. For example, assume that the knowledge base contains 'Mary talked with Tom yesterday.' and a user want to know that 'Who talked with Tom yesterday at the park ?'. If DoM is not allowed or a user requests 100% of DoM then CGMA will not be able to return a correct answer, even though the knowledge base contains the relevant information. CGMA, however, will return 'Mary talked with Tom yesterday.' for the query with 80% of DoM.

DoI indicates the distance of inheritance between concepts in the type hierarchy. Distance of the immediate predecessor of a concept is +1, and the distance of immediate successor concept is -1. The default value of DoI is 0. Therefore, we can use integers for DoI.

Rau proposed a conceptual graph matching algorithm which is able to retrieve relevant information in the absence of an exact match by using semantic relations between concepts [Rau88]. Her algorithm utilizes the inheritance hierarchy in order to retrieve the relevant information in case of no exact matching is exist. The algorithm, however, did not considered DoM .

Levinson and Ellis [Levi92] remove the redundancy for efficiency of storage space and speed by using multi-level hierarchy. However, redundancies are still remained in the representation notation of a graph. For instance, the same concept appears many times in a node descriptor of a single graph because of the associativity of the node descriptors.

For CGMA, a conceptual graph is represented in a special linear form (named U-Form) which is described in section 2. A control structure for knowledge base is organized as Relation Index Set (RIS). The RIS and the knowledge base called C-Base are described in section 3 and 4. CGMA allows exact matching as well as partial matching. The algorithm is explained with an example in section 5. Section 6 describes the implementation of CGMA and section 7 concludes the paper.

2 Representation

Unlike the linear form of conceptual structure in [Sowa84], for CGMA, any conceptual graph will be stored in a knowledge base, which is called a C-Base, as a list structure called U-Form. U-Form is unique if we restrict the order of concepts and relations. The U-Form is compact in the sense that all the arcs are represented implicitly. All the implicit arcs are outgoing arcs. U-Form starts with a concept which has no incoming arc which is called the head concept (the head concepts will heve dummy relation 'h'). If there is more than one head concept then choose the concepts in alphabetical order for the first head concept.

After select the first head concept, choose the relation on outgoing arc(s) of the chosen concept in alphabetical order (the ordering restriction makes the U-Form unique). We keep writing relation and concept alternatively according to the syntax given below with a depth first manner. If we reach at the concept which has no more outgoing arcs, then consider the second relation of the first head concept. After visiting all the relations and concepts attached on the first head concept via chains of outgoing arcs, take the next head concept if there is any.

U-Form	::= (((h head-concept) Subgraph)+)
Subgraph	::= Slist I Nlist I Slist Subgraph I Nlist Subgraph
Slist	::= (relation concept) I Nil
Nlist	::= (relation concept (Slist))

Different drawing forms of a graph does not affect the structure of U-Form. For example, a sentence "Monkey eat a walnut with a spoon made out of the walnut's shell." can be represented as two different linear forms of conceptual graphs as shown below [Sowa84].

(1) [EAT] -
        (AGNT) -> [MONKEY]
        (OBJ)   -> [WALNUT:*x]
        (INST) -> [SPOON] -> (MATR) -> [SHELL] <- (PART) <- [WALNUT:*x].

(2) [SPOON] -
       (INST) <- [EAT] -
                    (OBJ) -> [WALNUT] -> (PART) -> [SHELL:*y]
                    (AGNT) -> [MONKEY],
       (MATR) -> [SHELL:*y].

There is, however, only one corresponding U-Form:

(3)   ((EAT   (AGNT MONKEY)
               (INST   SPOON (MATR SHELL:*X))
               (OBJ WALNUT (PART SHELL:*X)))).

The arcs are implicitly represented in U-Form and all of them are outgoing arcs (left to right). Only binary relationship is presented in (3), however, n-ary relationship can be represented by duplicating the relation. For instance, the relation 'Brother' in a sentence "Tom and Bob are the brothers of John." should be duplicated. The corresponding U-Form is

          ((John   (brother Bob)
                  (brother Tom))).

     It is possible to extract a corresponding conceptual graph from any U-Form. Take the above U-Form as an example. We know that there are 3 outgoing arcs (to AGNT, INST, OBJ) from the concept EAT in the conceptual graph represented as above U-form, since the concept EAT has 3 (children) sublists in the U-Form.
     From the first line of the above U-Form, part of the whole graph is extracted as shown below:

After considering the second line of the U-Form, the partial graph becomes:

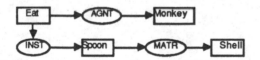

and finally the complete graph as given below is constructed by the third line of the U-Form.

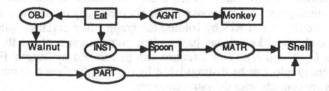

     If there is more than one concept which has no incoming arc(s), then U-Form will have more than one head concept.

3 C-Base

The C-Base is a part of a knowledge base which contains conceptual graphs in the order they arrive. Therefore, no special organization technique is needed for C-Base. Each conceptual graph in C-Base has a unique graph identifier (Gid) which can be used as a pointer (or actual surface word in case of natural language processing system).

Relation Index Set (RIS) and Gid will be used to retrieve the actual full fledged answer graph(s). For example, suppose that C-Base contains four U-Forms that have the following meaning.

(1) Jack is Charlie's philosophy teacher.
(2) Joe buys a necktie from Hal for $100.
(3) The cat is warm in the kitchen.
(4) Monkey eat a walnut with a spoon made out of the walnut's shell.

The corresponding U-Forms for these sentences are as follows:

```
G1 ((TEACH      (AGNT   TEACHER:Jack)
                (OBJ    PHILOSOPHY)
                (RCPT   PERSON:Charlie)))

G2 ((BUY        (AGNT   PERSON:Joe)
                (INST   MONEY:@$100)
                (OBJ    NECKTIE)
                (SRCE   PERSON:Hal)))

G3 ((WARM       (EXPR   CAT:#345)
                (INST   ~ANIMATE)
                (PLACE  KITCHEN:#54)))

G4 ((EAT        (AGNT   MONKEY)
                (INST   SPOON (MATR  SHELL))
                (OBJ    WALNUT (PART  SHELL))))
```

4 Relation Index Set

Relation Index Set (RIS) is the heart of CGMA. RIS provides tuples each of which contains a set of (Gid, concept, length) for each relation in a conceptual graph. The *length* represents the number of concepts in a graph. The relations will be stored only once in RIS regardless of how many times a certain relation is repeated in the knowledge base. Hence, the n-ary relationship does not create any redundancy in RIS. This fact will reduce the size if RIS by folding the graph. In a certain application such as an organic chemical knowledge base, often every graph contains the same concepts (such as C, O, and H). This situation can be handled by 'length' field in RIS. Two different graphs can be distinguished by the length of each graph, even though the graphs are consists with the same concepts.

We will illustrate the RIS organization with a simple example. The conceptual graphs (represented in U-Form) in section 3 will be used as an example. When the first conceptual graph is added to the C-Base, the Gid G1 will be assigned to it and stored into the C-Base as the following by using U-Form.

C-Base :   G1 ((TEACH   (AGNT   TEACHER:Jack)
                          (OBJ   PHILOSOPHY)
                          (RCPT   PERSON:Charlie)))

In the mean time, the RIS contains only three tuples as shown below. The number, in front of each concept in RIS, indicates Gid and the number appended after each concept is the length of the graph. Also, the length field will be used for speed up of exact match procedure.

RIS :   AGNT   = { (1 TEACHER:Jack 4) }   ---> Tuple 1
        OBJ   = { (1 PHILOSOPHY 4) }   ---> Tuple 2
        RCPT   = { (1 PERSON:Charlie 4) }   ---> Tuple 3

When the second conceptual graph:

            ((BUY   (AGNT PERSON:Joe)
                    (INST   MONEY:@100)
                    (OBJ   NECKTIE)
                    (SRCE   PERSON:Hal))),

is added, the Gid G2 will be assigned to that graph and two additional tuples for 'INST' and 'SRCE' are introduced into the RIS since the second graph has the relation 'INST' and 'SRCE' which are not in the first graph G1. The C-Base and the RIS will be updated as follows:

C-Base :   G1 ((TEACH   (AGNT TEACHER:Jack)
                              (OBJ PHILOSOPHY)
                              (RCPT PERSON : Charlie))).

           G2 ((BUY   (AGNT PERSON:Joe)
                              (INST MONEY:@$100)
                              (OBJ NECKTIE)
                              (SRCE PERSON:Hal))).

RIS :   AGNT   = { (1 TEACHER:Jack 4) (2 PERSON:Joe 5) }
        INST   = { (2 MONEY:@$100 5) }
        OBJ   = { (1 PHILOSOPHY 4) (2 NECKTIE 5) }
        RCPT   = { (1 PERSON:Charlie 4) }
        SRCE   = { (2 PERSON:Hal 5) }

The concept 'PERSON:Joe' is the agent of second graph, therefore (2 PERSON:Joe 5) is appended to AGNT tuple in RIS. The final C-Base will be constructed as shown in section 3 and the final RIS is given below after all four graphs are added:

RIS :   AGNT   = { (1 TEACHER:Jack 4) (2 PERSON:Joe 5) (4 MONKEY 6) }
        EXPR   = { (3 CAT:#345 4) }
        INST   = { (2 MONKEY:@$100 5) (3 ~ANIMATE 4) (4 SPOON 6) }
        MATR   = { (4 SHELL 6) }

```
OBJ    = { (1  PHILOSOPHY 4) (2  NECKTIE 5) (4  WALNUT 6) }
PART   = { (4  SHELL  4) }
LOC    = { (3  KITCHEN:#54   4) }
RCPT   = { (1  PERSON:Charlie  4) }
SRCE   = { (2  PERSON:Hal  5) }
```

List has been used as a data structure for RIS since it is simple and easy to explain the algorithm, however, more efficient data structure such as B+ trees should be used for a very large knowledge base.

5 The Algorithm

CGMA can perform exact matching as well as partial matching efficiently. CGMA is flexible enough to select the DoM by a user. For instance, if there is a query graph contains 10 concepts and a user wants to retrieve any answer graph with 80 % of DoM, then CGMA will return all the graphs with more than 8 equivalent concepts for each corresponding relations in the query graph. Matching can be accomplished by the following steps:

(Step 0) Assign the input query graph in U-Form to a variable IQ and
        Initialize CURNT,TEMP, and ANS as empty lists.
(Step 1) CURNT = Last (relation concept) pair in IQ.
        IQ = IQ - CURNT.
(Step 2) Pick the tuple which has same relation name with the relation
        of CURNT from RIS.
(Step 3) If there exists the element(s) in the tuple which contains the equivalent
        concept with the concept of CURNT
                Then TEMP = found Gid(s)
                Else check the type hierarchy according to DoI and
                TEMP = found Gid(s).
(Step 4) If TEMP = NIL or TEMPANS = NIL
                Then ANS = ANS U TEMP
                Else ANS = ANS ∧ TEMP
        If DoM = 100% (i.e., exact match) and only one Gid is in ANS
                Then get the graph from C-Base and determine whether it is the
                expected graph or not.
                        If the retrieved graph satisfies the condition of DoI
                                Then return the graph
                                Else stop
                Else ANS = ANS U TEMP.
(Step 5) If IQ = NIL
                Then calculate the acceptability of the retrieved graph(s) in ANS.
                (accept if DoM is <= number of matched concepts / total number of
                concepts in IQ). Delete the unacceptable Gid(s) from ANS and
                return ANS.
                Else go to step 1.

Here is a simple example. Assume that we have a query "Who is Charlie's philosophy teacher ?" and C-Base and RIS which are constructed in section 4 are used as a knowledge base.

```
(TEACH (AGNT  TEACHER:?)
       (OBJ  PHILOSOPHY)
       (RCPT  PERSON:Charlie)).
```

Above U-Form is the representation of the query. If a user does not specify the value of DoM and DoI then, DoM = 100% and DoI = 0 will be used as their default value. After accepting the query graph, take the last (relation, concept) pair from the query which is (RCPT PERSON:Charlie). Now we have one concept 'PERSON:Charlie' from the query to search the answer graph in the knowledge base. The concept 'PERSON:Charlie' has 'RCPT' as a corresponding relation, hence, look at the RCPT tuple in RIS. There is only one graph (graph number 1) in the final RIS in section 4. Therefore, TEMP will hold that graph's Gid which is 1. Append this to ANS, initially ANS is NIL. At this point, there is only one candidate graph, so if DoM=100% (i.e., exact match) then get the graph from C-Base and determine whether it is the expected graph or not, without considering the rest of the concepts to save the execution time of the procedure described in the rest of this section.

If DoM < 100%, then the rest of the concepts should be considered. So, take the next pair which is (OBJ PHILOSOPHY) from the query, and look at the 'OBJ' tuple in RIS. In this case there are three graphs in that tuple which are graph (1, 2, and 4) but only one has the matching concept which is graph 1. These will be assigned to TEMP. Now we will perform the intersection with the content of TEMP and the content of ANS and ANS will hold the result which is 1. Do the same thing for the next pair (AGNT TEACHER:?). In this case the concept is the question concept so can be matched with any concept in the corresponding tuple. The graph number 1, 2, and 4 will be retrieved and assigned to TEMP for this question concept. Get the final answer graph 1 after perform the intersection between TEMP and ANS. We retrieve the graph 1 and find the answer 'Jack'.

6 Implementation

This section will describe the implementation of the algorithm CGMA. The CGMA consists of several components such as knowledge base builder, graph retriever, type hierarchy handler, and query interface. It was implemented on VAX/VMS by using Quintus Prolog Release 2.4.

6.1 Architecture of CGMA System

The basic function of Knowledge Base Builder is performed by a set of submodules to construct a C-Base for the input conceptual graph given by a user.

U-Form User Interface: Get the conceptual graph written in U-Form.
Lexical Analyzer: Check the lexical correctness.
Conceptual Graph Parser: Check the syntax of a given conceptual graph based on the conceptual graph grammar.
Relation Index Set Constructor: Construct the RIS.

C-Base Constructor: Construct the C-Base for the all syntactically correct conceptual graphs given by a user

Query Interface: This module accepts the input query graph and necessary information (e.g., DoI, DoM, etc.) from a user and return the results.

CGMA: This module contains 2 types of conceptual graph matchers, i.e., *Exact Conceptual Graph Matcher* and *Partial Conceptual Graph Matcher*. The parameter DoM can have a value which is in the range of $0 < DoM <= 100$ %) and the parameter DoI can have a value which is in the range of $-7 <= DoI <= 7$.

DoM Processor: This submodule calculates the DoM for each candidate graph and compares it with the DoM value given by a user. If the value calculated by DoM Processor is bigger than or equal to the user's DoM value, then the corresponding candidate graph is accepted.

Type Hierarchy Handler: This module contains the *Type Hierarchy Base* for the concepts. The concepts are stored in the form of Prolog facts like

concepts(concept_a, concept_b).

where there is a subtype relationship between concept_a and concept_b. For example, the relationship between the concept 'cat' and 'animal' can be represented topologically as: concepts(cat,animal).

DoI Processor: This submodule receives the value for the parameter DoI and controls the subtype relations for each concept in the query graph of a user. There are two types of inheritance, i.e., generalization and specialization. If a user wants to consider more general graph matching for each concept of query graph, he/she can specify the value of DoI as +3. The value '+3' means that the CGMA considers the concepts up to 3 levels upward in the type hierarchy of concepts. If a user wants to consider more specific graph matching for each concept of his/her query graph, he/she can specify the value of DoI as -2. The value '-2' means that the CGMA considers the concepts up to 2 levels downward in the type hierarchy of concepts. The value DoI = 0 means there is no semantic consideration.

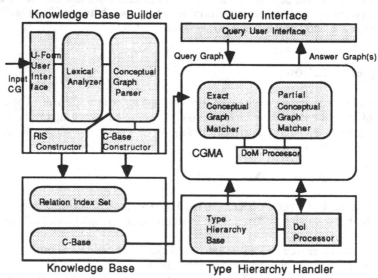

Fig. 1 Architecture of CGMA System

6.2   CGMA Sessions

This subsection shows a sequence of simple user/CGMA interactions to show how the CGMA works.

Exact Matching: After consulting all necessary files, a user types 'query' and supplies necessary information like the following to perform an exact matching of query graph with the answer graph(s) in the C-Base:

```
| ?- query.
Type in the query graph ==>
|: give(agnt(teacher),rcpt(student),obj(X)).
* Type in the Degree of Matching (DoM) : 0 < DoM <= 100% ==> 100.
* Type in the Degree of Inheritance (DoI) : 0 - 7 ==> 0.
* Do you need a specialization or generalization of concepts? ==>
  0. none   1. generalization   2. specialization
  |: 0.
Answer graph = 3
The end of exact graph matching
QG = give(agnt(teacher),rcpt(student),obj(X))
CG = give(agnt(teacher),rcpt(student),obj(cat))
Do you need to see all exact matching graphs? (y/n)
|: n.
```

Partial Matching:

```
1) Generalization
| ?- query.
Type in the query graph ==>
|: give(agnt(teacher),rcpt(student),obj(cat)).
* Type in the Degree of Matching (DoM) : 0 < DoM <= 100% ==> 100.
* Type in the Degree of Inheritance (DoI) : 0 - 7 ==> 1.
* Do you need a specialization or generalization of concepts? ==>
  0. none   1. generalization   2. specialization
  |: 1.
The end of graph matching -  generalization
The final set of conceptual graphs by partial matching = [1]
The DoM of graph 1 --> give(agnt(person),rcpt(person),obj(animal)) = 100%.
The end of partial graph matching
```

```
2) Specialization
| ?- query.
Type in the query graph ==>
|: eat(agnt(professor),loc(place),obj(food),inst(tool)).
* Type in the Degree of Matching (DoM) : 0 < DoM <= 100% ==> 75.
* Type in the Degree of Inheritance (DoI) : 0 - 7 ==> 1.
* Do you need a specialization or generalization of concepts? ==>
  0. none   1. generalization   2. specialization
|: 2.
```

The end of graph matching - specialization
The final set of conceptual graphs by partial matching = [2, 5]
The DoM of graph 2 -> eat(agnt(csprofessor),loc(restaurant),obj(cake)) = 80%.
The DoM of graph 5 -->
eat(agnt(professor),loc(restaurant),obj(cake),inst(fork),manr(quick)) = 100%.
The end of partial graph matching

7   Conclusion

The CGMA can perform exact matching and partial matching.   CGMA allows the
user to choose the DoM and DoI to increase the user's satisfaction in the process of
retrieving the relevant information in the absence of exact match. Depending on the
value of DoI, limited type hierarchy will be considered to reduce the search space. The
DoM indicates the syntactic relevance and DoI indicates the semantic relevance
between query and the answer.

Acknowledgment

We would like to express our special appreciation to Gerard Ellis for many comments
and suggestions which forced us to look into some crucial questions.

References

1.   Rau, L.F. Exploring the Semantics of Conceptual Graphs for Efficient
      Graph Matching. Third Annual Workshop on Conceptual Graphs, Boston,
      1990.

2.   Levinson, R. and G. Ellis. Multi-Level Hierarchical Retrieval.   To be
      appeared in Knowledge Based Systems Journal, 1992.

3.   Sowa, J. Conceptual Structures: Information Processing in Mind and
      Machine. Addison Wesley, Massachusetts, 1984.

4.   Yang, G.C. and J.C. Oh. An Efficient Dictionary Management System for
      Machine Translation. Seoul International Conference on Natural Language
      Processing, Seoul, 1990.

An X-Windows Toolkit for Knowledge Acquisition and Representation Based on Conceptual Structures*

Michel Wermelinger** and José Gabriel Lopes

Centro de Inteligência Artificial/UNINOVA
Quinta da Torre, 2825 Monte da Caparica, PORTUGAL
{mw|gpl}@fct.unl.pt

Abstract. This paper describes GET (Graph Editor and Tools), a tool based on Sowa's conceptual structures, which can be used for generic knowledge acquisition and representation. The system enabled the acquisition of semantic information (restrictions) for a lexicon used by a semantic interpreter for Portuguese sentences featuring some deduction capabilities. GET also enables the graphical representation of conceptual relations by incorporating an X-Windows based editor.

Keywords: conceptual structures, knowledge representation, graphical interfaces, natural language processing.

1 Introduction

Conceptual structures [9] are an ambitious attempt to represent knowledge in a natural and expressive way. An implementation of the necessary machinery would enable us to test their practical suitability for semantic representation of natural language sentences, for conceptual modeling of relational databases, etc. So we decided to program a prototype in X-Prolog [1], a result of the ESPRIT project "Advanced Logic Programming Environments". One of the main reasons for this choice was the possibility to access the X-Windows functionalities in order to display conceptual structures making use of their easy to read graphical notation.

The resulting implementation, called Graph Editor and Tools (GET), is currently a *generic* tool for knowledge acquisition and representation based on conceptual structures and consists of two distinct parts: the Conceptual Graph Tools (CGT), a portable collection of Prolog predicates implementing the most important operations on conceptual graphs, and the Conceptual Graph Editor (CGE) working under X-Windows and using the primitives provided by CGT.

This paper describes CGT, CGE, and a semantic interpreter for Portuguese sentences, focussing on CGE. Finally, possible enhancements as well as some insights gained with this work regarding the utilization of conceptual graphs for Natural Language Processing are given. For a more detailed account see [12] and [13].

* This work was partially supported by JNICT, under contracts PMCT/P/TIT/167/90 and PMCT/MIC/87439, by INIC, under project CALIPSO, by FCT/UNL, and by Gabinete de Filosofia do Conhecimento.

** Owns a scholarship (PMCT/BIC/114/90) from Junta Nacional de Investigação Científica e Tecnológica.

2 The Conceptual Graph Tools

The Conceptual Graph Tools (CGT) are a portable collection of Prolog predicates implementing the most important operations on conceptual graphs, a simple mark-&-sweep memory management system, and a linear notation parser and generator using Definite Clause Grammars [6]. CGT also provides facilities to manipulate graph databases made up of:

- a type hierarchy
- a set of graphs, where each may have some descriptive text associated to it
- for each concept type, a (possibly empty) set of schemata
- for each type, the associated canonical graph and/or definition

A sample database comes with the toolkit; it contains all relations defined in the Conceptual Catalog [9, Appendix B], and several basic concept types. CGT enables the user to easily create new types with their associated definitions, schemata, and canonical graphs in order to build several new databases on top of the given one.

The linear notation of conceptual graphs as parsed and generated by CGT is a bit different from the one used by Sowa. The formal definition in [12] extends the one given in [9, Appendix A.6], especially in what concerns the type and referent fields, including nested contexts. The minor differences are due to efficiency concerns and implementation restrictions (like using '\' for 'λ' and 'V' for '∀' to use just ASCII). Major changes or restrictions were motivated by unclear aspects of the formalism, specially regarding coreference links. For example, should one permit any two concepts to corefer? How can inconsistencies be detected? Therefore, it was decided that coreferenced concepts must have compatible types, i.e. one is a subtype of the other. Furthermore, contexts may not corefer if their referents contain graphs because it would be difficult to check them for incompatibility.

Currently CGT has the following implementational restriction: once a concept or relation type is defined it isn't possible to change neither its definition nor its associated canonical graph. The reason is simple: if the canonical graph associated to a type X changes, all graphs with a concept or relation of type X must probably change, too. And if a type Y is defined in terms of X, the graphs involving concepts or relations of type Y would have to change too, and so on. This weakness can be repaired by adding a specific maintenance tool.

CGT features a SAFE[3] linear notation parser: it doesn't perform any error recovery, stopping with the first error found. It copes quite well with semantic errors (e.g. undefined referent variables, unknown type labels) but it still needs to be made more robust regarding syntax errors (e.g. a missing ']'). Furthermore, it forces the graphs entered by the user to be meaningful by checking them against the canonical graphs of the ocurring relation types.

As one should expect, some parts of Sowa's formalism have not yet been implemented in this first version of the Graph Tools. The most notable omissions are the first-order rules of inference and the φ operator, which translates graphs into first-order logic formulas. The latter could be modified to assert graphs as Prolog clauses in order to use Prolog's inference engine for deductions. Other things still

---

[3] Stop At First Error

need to be improved, specially referents and coreference links. Both will require theoretical work; the former, particularly, will need some reworking while the latter must be carefully analysed with respect to their side-effects on operations such as the canonical formation rules.

3   The Conceptual Graph Editor

CGE enables the user to create and manipulate conceptual structures in a graphical way, using the primitives provided by the Tools. It can be considered to be a kind of "syntax-oriented" editor, as most commands correspond to operations provided by the formalism, thereby enforcing the resulting graph to be canonical. The alternative would be to have a "visual" editor allowing to operate on single nodes and arcs, but its implementation would be more difficult because incomplete and ill-formed graphs would have to be taken into account.

The Conceptual Graph Editor was coded in X-Prolog, a superset of Prolog including the Widget[4] Description Language [1] which enables the programmer to access the X Windows functionalities in a declarative way. Therefore, the editor takes advantage of the underlying graphical interface, providing an easy way to edit graphs using windows, dialog boxes, icons, buttons, selections, and the combination of mouse and keyboard. Furthermore, the choice of the X-Windows standard increases portability and decreases the learning time for users already familiar with other graphical interfaces.

3.1   The Editor Window

A Conceptual Graph Editor is a window comprising five areas (see Fig. 1):

header This area consists of a single line of text displaying a description of the shown graph. The text may be a user defined string (as in the figure), the usual description (e.g. 'canonical graph for BUS(x) is', 'relation AGNT(x, y) is', etc.) or simply the word 'graph'.

graphical display It is under the header and shows the edited graph(s).

linear display It shows the same graph as the graphical display but in linear notation. It is a normal text widget, enabling the user to edit its contents using normal Emacs commands [11] in order to create graphs which cannot be obtained with the menu commands.

menus Under the two display areas, all possible commands to (visually) edit graphs and their nodes are provided. Most operations are directly supported by CGT.

buttons Located to the left of the graphical display, the two top buttons provide access to two commands without keyboard shortcuts ('Restrict Type' and 'Restrict Referent') while the other buttons provide an easy control over the way graphs are drawn in the graphical display.

The relative sizes of the graphical and linear displays may be changed by dragging the small rectangle between them with the mouse. There is also a 'modified' label in

---

[4] Window gadget,—a graphical object in X-Windows terminology.

265

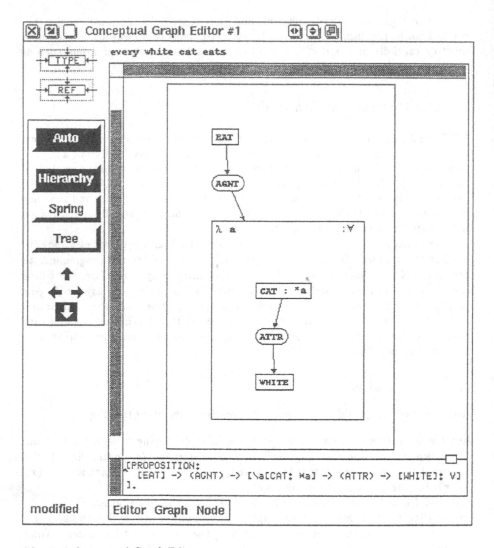

Fig. 1. A Conceptual Graph Editor

the bottom left corner, appearing only when a graph has been added to the database but the latter hasn't been saved on disk.

Several editor windows may be open at the same time. To make better usage of the display area of the monitor, the editor windows may be iconified. In order to distinguish the various editors in an easy way, both the icons and the windows are numbered.

3.2 The Graphical Display

The main area within an editor window is occupied by the graphical display of graphs. In CGE, the visual appearance of graphs may be controlled by the user, either semi-automatically or manually for full control.

To make the display of conceptual graphs easier, a generic widget to handle the visualization of arbitrary graphs was used [8]. *All* graphs in the *same* context, and only them, are displayed with the *same* Graph Widget. This gives a lot of flexibility, as graphs in different contexts may be displayed in different ways. The relevant attributes of the Graph Widget for CGE users are:

layout mode It indicates if the display of the graph is to be done automatically (by the widget) or manually (by the user, dragging the nodes with the mouse to the desired position).

layout function This is the algorithm used in automatic mode to calculate the positions of the graph's nodes.

layout style It may be one of the four available styles (left-right, right-left, top-down, bottom-up) for automatic layout.

There are three pre-defined layout functions:

Hierarchy This function is mainly used for hierarchical graphs and it is the one that provides the best results for conceptual graphs.

Tree This function can only be used for a single graph that is in fact a tree (see Fig. 2. If used to display disconnected graphs, a mess will appear on the screen.

Spring This is the only iterative function, i.e. the visual appearance of the graph will depend on the original position of its nodes, whereas the other functions always display the same graph in the same way.

To the left of the graphical display there's a layout control box consisting of eight buttons. The top one controls the layout mode, the next three control the layout function and the bottom four arrows control the layout style. The buttons have a twofold purpose. By clicking on them with the mouse, the user may set the corresponding attributes in the selected context(s). On the other hand, the state of the buttons reflects the attributes of the selected context(s).

3.3 The Editor Commands

The commands available in CGE may be issued from the keyboard, selecting an entry of a menu, or clicking with the mouse on an icon. Often, there are at least two ways to invoke the same command. Most of the commands use dialog boxes (to interact with the user) and selections (to show the graphs or nodes on which to operate). There are several types of the former (acknowledge dialogs to display error messages, choice dialogs to present a set of options, etc.), taking into account the various needs for user input. Also, two kinds of selections are provided to enable some commands to distinguish their operands (e.g. the insertion operation needs to know the graph to insert and the context in which to insert it). All available commands, except those involving the way graphs are displayed, appear in the following three menus:

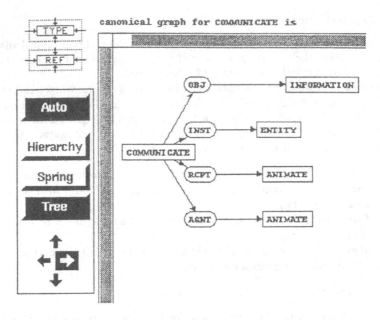

Fig. 2. Tree display

Editor Menu The Editor Menu contains commands that don't belong to the conceptual graph formalism, like loading and saving a graph, change the current graph database, deleting an arbitrary graph, and quitting the editor.

Graph Menu The commands in this menu operate on whole graphs. They include the canonical formation rules and the propositional rules of inference. Furthermore, there are commands to compute the depth of a graph and to check whether one graph is a generalization, a specialization, or a copy of another graph.

Node Menu This menu groups all commands that operate only on relations and concepts. They are divided into three groups: the restrict operation from the canonical formation rules, type expansions, and referent expansions and contractions.

There isn't an "undo" command, but most of the implemented ones have a counterpart, like drawing vs. erasing a double negation, iteration vs. deiteration, etc. To cancel the effect of any operation, the 'Clear Graph' command is provided, but it should be used only in case of a mistake (e.g. the wrong graphs were joined) as it is not a canonical or propositional rule.

3.4 Future Work

The Graph Widget needs some recoding before being of practical use for the display of conceptual graphs: the algorithms must take the size of the nodes into account, make better usage of space, and maybe a new one must be developped for nested graphs. Because of these problems and other implementational details the visualization of coreference links wasn't implemented. Some other possible enhancements are:

- Make a type lattice editor/viewer which would provide an easy way to create new types or to select existing ones.
- Show coreference links and enable the user to edit them in a simple way (e.g. by pointing and dragging).
- Enable the user to choose for each context whether it should be displayed in normal or reduced size, thus improving the effective usage of the available display area.
- Enable the user to do some things (e.g. lambda abstraction in the type field) in a more graphical way, instead of having to write the corresponding linear notation for it.

4   A Semantic Interpreter

A small semantic interpreter for Portuguese sentences was built using conceptual graphs. The approach taken is similar to the one described in [10]: The lexicon associates a canonical graph to each possible meaning of a word and the interpreter proceeds in a bottom-up way when processing the syntactic tree. For each subtree it obtains a graph and its so-called "head". At the next level, the interpreter will try to join the graphs by matching directly the corresponding heads. As one can see, the only operations the interpreter needs from CGT are the canonical formation rules.

The sentences are parsed with a wide coverage Portuguese syntax description [4] using the XG formalism [7]. In its actual state, the semantic interpreter only covers a tiny subset of that description. On the other hand, it performs some deduction on the database constructed from the input sentences. The interpreter accepts three types of sentences:

Declarative sentences. They denote assertions to be added to the database, stating a simple negative or positive fact (e.g. "The cat doesn't eat.") or a *rule* of the form "A if B", where A and B are simple facts (e.g. "The cat eats the mouse if it is hungry."). The former are represented by negative or positive contexts, respectively, and the latter uses the 'IMP' relation [9, section 4.2].

It must be stressed that there is no anaphora resolution. As a consequence, sentences like "The cat eats the mouse if it doesn't run away." must be rewritten into "The cat eats the mouse if the mouse doesn't run away." and even in this case no coreference link between the two 'MOUSE' concepts will be drawn. Therefore, whenever the individual is not specified the interpreter makes the simplifying assumption that the user is always referring to the same one. In the last section of this paper we provide some ideas to work around this problem.

Deductions upon graphs are made according to the following four rules, where G/X is a fact represented by a proposition of polarity X containing graph G, $A \Rightarrow B$ represents the rule "B if A", and $\leq$ denotes specialization:

$$G1/pos \text{ and } G2/pos \Rightarrow G3/X \text{ implies } G3/X \quad \text{if } G1 \leq G2.$$
$$G1/neg \text{ and } G2/neg \Rightarrow G3/X \text{ implies } G3/X \quad \text{if } G2 \leq G1.$$
$$G1/pos \text{ and } G2/X \Rightarrow G3/neg \text{ implies } G2/\neg X \quad \text{if } G1 \leq G3.$$
$$G1/neg \text{ and } G2/X \Rightarrow G3/pos \text{ implies } G2/\neg X \quad \text{if } G3 \leq G1.$$

The first two rules are *modus ponens*, the other two implement *modus tollens*. For example, the third rule states: if the database contains a positive fact A and a rule stating "not C if B", then the fact "not B" will also be in the database if C is a generalization of A.

Whenever a new fact or rule is entered by the user, the interpreter tries to match it with the antecedent and consequent of every rule in order to determine if *modus ponens* or *modus tollens* may be applied. This process is recursively applied to every deduced fact. As soon as a fact (deduced or not) is about to be added, the database is searched for a more general one stating the opposite. If such a fact is found, the sentence must be incoherent with the previously stated premisses, forcing the interpreter to issue a message and to retract all the facts asserted during the deduction process.

Interrogative sentences. They may be simple questions (e.g. "What does the cat eat?") or of the form "A if B" stating hypotheses (e.g. "Does the mouse die if the cat eats?"). The graph representing the question will have the same form as for a declarative sentence, whereby wh-pronouns are simply dropped or substituted by generic concepts.

In the case of a simple question, looking for an answer consists in searching the database for a specialization of the graph representing the question. In the other case ("A if B"), the hypothesis B is temporarily put as a normal assertion in the database (i.e. it is tested for coherence with the known facts and all possible deductions are performed) and then A is treated like a simple question.

Imperative sentences. They are interpreted as instructions to the interpreter. In its actual version, commands consist of a single verb in the imperative form, e.g. "stop!". The graph representing the command must be known to the interpreter, i.e. the program searches its internal command list for an exact copy of the graph.

Only three different commands are known in this version: "mostra!" (show), "apaga!" (erase), "para!" (stop). The first shows one by one on demand the current facts in the database, the second clears the database, and the third is used to quit the interpreter.

Some other aspects of the interpretation process are:

- The backtracking facility of Prolog is used to find alternative syntactic and semantic representations for the sentences and multiple answers for the questions.
- Fillmore's order of preference (agent, instrument, object) is used to join the verb and subject graphs.
- Relative clauses are translated into abstractions of the type corresponding to the noun they modify, like in [10].

– Verb arguments and modifiers are distinguished in that the former restrict the concepts of the verb graph while the latter join new graphs, namely those of the modifiers. This means that the graph for the verb must already make provision for all possible arguments. Furthermore, the interpreter prevents arguments and modifiers from having the same semantic role.

Last, but not least, the interpreter can be used with or without CGE, the difference being how user input and program output is handled. In the former case, the CGE window is used to show the graphs, while dialog windows handle the user's input sentences. In the latter case, the linear notation and Prolog's basic I/O facilities are used.

The current state of the interpreter is not completely satisfactory as far as speed and flexibility are concerned. The coverage can also be much improved, but that wasn't the purpose of this application. The lack of speed is mainly due to the complexity of graph operations and to the constant copying of the intermediate graphs during the process to make backtracking possible. To increase flexibility and semantic coverage the interpreter could use schemata and the supertypes' canonical graphs.

5 Conclusions and Future Perspectives

As far as we know, GET is the first collection of tools to work easily with conceptual graphs (CG) in a logic programming environment with a graphical interface based on X-Windows. It is quite easy to build new types and relations with their associated background knowledge, especially using the CG Editor. The knowledge databases constructed in this manner could then be incorporated into other programs which would call the predicates provided by the CG Tools.

There were several advantages in using Prolog, in particular X-Prolog: an easy access to the X Toolkit C functions was possible, thus enabling the existence of a graphical editor; a linear notation parser and generator could be quickly built with a partially bidirectional DCG; and finally, the existing Portuguese extraposition grammar could be directly used for a toy semantic interpreter.

Unfortunately, X-Prolog is no longer supported. As such, we intend to port the Editor to APPEAL, an X-Windows based programming environment for SICStus Prolog. APPEAL also integrates the Widget Description Language and has the advantage of being supported by a company.

The main disadvantage of using Prolog is poor efficiency. The operations on conceptual graphs are extremely complex, mainly because of contexts (enabling the nesting of graphs) and coreference links (connecting nodes over arbitrary contexts), and the used data structures must be quite dynamic.

As far as it concerns the formalism itself, our overall feeling is that the conceptual structures' main source of expressiveness is also their main source of problems and fuzziness, namely the referents and coreference links. Therefore, some options had to be taken concerning some less clear points, others were deliberately postponed until the CG researching community agrees on them.

Nevertheless, the semantic interpreter showed that the mapping between natural language sentences and conceptual graphs is relatively straightforward. But there are problems such as anaphora resolution that require theoretically backgrounded treatment that we cannot find within the conceptual structures theory. Discourse Representation Theory [2] is the formalism we have chosen at the AI Centre (CRIA) of UNINOVA for handling some of those problems raised by text understanding or intentional participation in conversations by computers [3]. To concile the best of two worlds we envisage a Discourse Representation Structures (DRS) processor. as it already exists at CRIA. with anaphora resolution [5], etc., whereby the graphical visualization can be achieved using conceptual structures which are undoubtebly superior for expressing DRS conditions. The DRS—CG mapping could turn out to be easier than expected, as both formalisms have notions for contexts and referents.

Acknowledgements

We would like to thank Irene Rodrigues for many fruitful discussions, Salvador Abreu for providing X-Prolog and Paulo Quaresma for providing his Graph Widget, both of which made CGE possible, and Claudia Ventura and Sabine Grüninger for reviewing this document.

References

1. Salvador Pinto Abreu. *ALPES X-Prolog Programming Manual.* Centro de Inteligência Artificial, UNINOVA, 1989.
2. Hans Kamp and Uwe Reyle. *From Discourse to Logic: An Introduction to Modeltheoretic Semantics of Natural Language, Formal Logic and Discourse Representation Theory*, volume I. Kluwer, Dordrecht, 1991.
3. José Gabriel Lopes. Architecture for intentional participation of natural language interfaces in conversations. In C. Brown and G. Koch, editors, *Natural Language Understanding and Logic Programming III.* Elsevier Science Publishers, 1991.
4. José Gabriel Lopes and Irene Pimenta Rodrigues. Descrição parcial da sintaxe do Português. Technical report, CRIA/UNINOVA, June 1990.
5. José Gabriel Lopes and Irene Pimenta Rodrigues. Reasoning in resolution of temporal anaphores. Technical Note NT-1/91-CIUNL, Centro de Informática da UNL, January 1991.
6. Fernando Pereira and Stuart Shieber. *Prolog and Natural Language Analysis*, volume 10 of *CSLI Lecture Notes.* Center for the Study of Language and Information, 1987.
7. Fernando C. N. Pereira. Extraposition grammars. *American Journal of Computational Linguistics*, 7(4):243–255, 1981.
8. Paulo Quaresma. Graph widget: A tool for automatic data visualization. Technical Report RT-6/91-CIUNL, Centro de Informática da UNL, April 1991.
9. John F. Sowa. *Conceptual Structures: Information Processing in Mind and Machine.* The System Programming Series. Addison-Wesley Publishing Company, 1984.
10. John F. Sowa and Eileen C. Way. Implementing a semantic interpreter using conceptual graphs. *IBM Journal Res. Develop.*, 30(1):57–69, January 1986.
11. Richard Stallman. *GNU Emacs Manual.* Free Software Foundation, 1985.
12. Michel Wermelinger. GET: Graph Editor and Tools—the incomplete reference. Technical Report RT-3/91-CIUNL, Centro de Informática da UNL, January 1991.
13. Michel Wermelinger. GET—some notes on the implementation. Technical Report RT-4/91-CIUNL, Centro de Informática da UNL, January 1991.

VI.   Natural Language and Applications

Assembly of Conceptual Graphs from Natural Language by Means of Multiple Knowledge Specialists

Graham A. Mann

Artificial Intelligence Laboratory
School of Computer Science & Engineering
University of New South Wales
P.O. Box 1, Kensington NSW 2033, Australia.
mann@spectrum.cs.unsw.oz

Abstract. A method of generating conceptual graphs representing the meanings of real natural language texts is advanced. Three language specialists are employed, each of which may make a contribution to the final conceptual structure. A syntactic specialist tries to assemble a graph by maximally joining graphs accumulated while recursively ascending a phrase structure tree from a syntactic parser. A semantic specialist tries to fill thematic role slots in template graphs by the successive application of selectional constraints to alternatives. A pragmatic specialist tries to classify text according to its intended purpose as a speech act, and handle idioms. Methods of deciding between policies for controlling the specialists and integrating the contributions of the language specialists, to be used by a control specialist, are described. An implementation capable of handling real direction texts for physical navigation is under development.

1 Introduction

Robust mapping of realistic natural language into conceptual graphs is a difficult and perhaps impossible task for a heuristic or algorithm based on any one method. I argue that knowledge from syntactic, semantic and pragmatic sources can be brought to bear on the problem by means of a set of specialists, each of which may make some contribution to the final graph. Collectively, the set of specialists should be capable of handling more cases than any single specialist acting by itself. The texts in question are paragraph-length transcriptions of actual spoken directions offered to help lost persons. The goal is to convert these texts into a graph or set of graphs for later inference and route planning.

Formally, a process of conceptual graph assembly can be specified by 1) a set of the (canonical or schematic) graphs G_W associated with each content-bearing word in the natural language text [1] and 2) an ordered list M of maximal join operations, the parameters for which include the two graphs to join and the specific initial concept-pairs on which to join. Given G_W, each of the syntactic and semantic specialists can be specified as procedures that use their methods on the input text to generate an ordered list of maximal join parameters. The pragmatic specialist uses other methods to make its contribution to the assembly process. The final order must be decided from the contributions according to some kind of policy, which must resolve potential conflicts between the specialists.

2   Syntax Specialist

The syntax specialist consists of a syntactical parser capable of generating a parse tree (or trees) for sentences in the corpus, and a method of using the parse tree to guide access to and joining of partial graphs into an evolving whole. The current implementation uses a chart parser originally written by Bill Wilson, together with a lexicon of over 30,000 English words and a phrase structure grammar especially written to handle simple declarative phrases, simple imperative phrases and other expressive forms typical of direction texts. Advantages can be gained by having the control specialist divide the text into phrase units, which are generally smaller than sentences, by simply cutting on conjunctions and certain punctuation marks. Use of these smaller units simplifies processing without adding problems of conceptual integration that did not already exist for sentences. The parser is at present capable of handling 48% of the current test corpus of 81 such phrases, but better coverage is likely with more work on the grammar.

The specialist uses the Sowa & Way bottom-up recursive algorithm [2] to assemble the graphs from canonical graphs accessed by the terminal symbols and the accumulations of maximally joined graphs at higher nodes. Some difficulty has been encountered in generating representations for complex noun groups such as "entrance of Io Myers studio". The methods for dealing with noun groups offered by Gershman [3] may be able to be applied here. The algorithm tackles word-sense ambiguity by propagating a set of possible candidate graphs for each ambiguous word to the next level of maximal joins. Candidates are either eliminated by improper concept matches at the next highest maximal join, or else accumulate to be ranked for suitability according to the number of concepts they contain. There are usually not a large number of candidate graphs, and since canonical graphs tend not to contain many elements, this method can disambiguate word-senses without excessive computational demands. Sometimes however, such as with ambiguous simple prepositional attachments, all the candidate graphs may happen to contain exactly the same number of concepts. Problems like this may be tackled by depending on disambiguating information from the pragmatic specialist.

The Sowa & Way algorithm does not handle structural ambiguity: this was left to the PNLP parser that they used. Since the chart-parser used in the current implementation does not support the node moving technique it used, other ways to deal with structural ambiguity must be used. One method being investigated is to add grammar rules in such a way as to minimise the overall use of non-terminal symbols in the right hand side of the rules, especially those with many expansions. This reduces the depth of the parse trees and effectively cuts down on alternative parses. Apart from the extra rules, the cost is that the parse trees become less structurally descriptive. Examination of a set of alternative trees offered when using a conventional grammar on a given text unit suggests that many will produce identical results when used for graph assembly; therefore some tradeoff in this direction can be tolerated.

3 Semantic Specialist

When working with real text it is sometimes difficult to believe that any syntactic grammar can completely cover the possibilities of human expression. Some new form which defies analysis always seems to appear. More complex grammatical rules and more complete lexicons lead to more combinatorial possibilities, with subsequent lowering of performance. Although simple syntactical grouping (of noun and verb groups, for example) appears to be an intuition possessed by all competent speakers of a given language [4], it is evident that detailed syntactical parsing of real sentences can be a difficult and controversial matter, even for skilled linguists. Yet humans are able to construct conceptualisations using no syntax at all. Consider the word list (PARK, BOY, PLAY, CAR, HIT, HOSPITAL, DOCTOR OKAY, HOME). Although critical elements of meaning can sometimes be found in syntax, since the Yale group's identification of scripts, plans and other memory patterns as organising principles [5] it comes as no surprise that a complete story can often be made out without much difficulty, even if the word order is randomised.

Noble's construction kit model [6] is particularly useful in this context. This model states that natural language is like the instructions in a construction kit in that it contains information useful for putting together a conceptual structure. Assembly information is of two basic kinds: information for identifying the parts, and information describing how the parts fit together. In natural language, content-bearing words (nouns, adjectives and verbs) and their phrasal structures access conceptual building blocks from the catalog of conceptual types, while other words (such as determiners, adverbs and prepositions) help clarify relationships between the building blocks. In a Sowa-style conceptual catalog, the building blocks (members of set G_W) may be definitions, canonical graphs or schemata. The graphs of principal verb-group and non-groups can be thought of as small templates containing slots into which the graphs for other elements of the text should fit.

The current design of the semantic specialist chooses a set of small template graphs, roughly corresponding to subject, verb and object and attempts to fill them using the thematic role filling heuristics described by Winston [7]. When using conceptual graphs as templates, "filling a slot" actually means restricting the graph by joining general concepts to more specific ones, specifying an instance and/or joining on extra relations and concepts. Filled templates are joined by looking for the relations in non-content-bearing words outside the phrases corresponding to the templates.

Problems faced in the design of the semantic specialist include whether to use canonical or schematic graphs as the elements of G_W, how to choose the initial templates, and how to resolve the ambiguity that arises when not all of the candidates can be eliminated.

4 Pragmatic Specialist

The pragmatic specialist has two responsibilities: 1) classification of text by intended use and 2) handling of idioms. The specialist must be able to be called at any stage of processing, and thus must be able to work on both the raw text and on conceptual graphs produced by the other specialists. Most of the work involves matching input patterns against stored patterns. If no graph is available, the specialist can use a simple flexible matching technique that measures the length of intersection of the set of words in the text unit with especially prepared content-word sets, and checks for caveat words such as negatives like "don't" and "not". This method is effective and fairly simple to implement. If a graph is available, a more reliable conceptual method using the graph similarity measure such as that of Wuwongse & Niyomthai [8] to compare the graphs with a set of meta-schematic graphs. The highest scoring meta-schematic graph represents the role class.

If a piece of text can be classified by its use, i.e. the effect that the speaker intended the communicative act to have on the hearer, then the natural language understander as a whole can be better behaved. This contribution does not always provide a conceptual graph, but may offer information that the other specialists could use to build theirs. The chosen meta-schematic graph mentioned above could, for example, be used by the syntax specialist to help choose between alternative graphs generated during parsing. The classification itself may also be useful for high level control.

Text is regarded as comprehensible if it can be identified as belonging to a pattern of expression common to the chosen domain; in this case, direction texts for physical navigation. Riesbeck's analysis [9] suggests that three kinds of communicative acts occur frequently in such texts: motions, descriptions and comments.

Motions. These are imperative statements about where to go (e.g. "Go up this walkway here."). At the surface level they feature movement verbs such as "walk", "follow" and "turn". Using conceptual types based on primitive act theory [10], graphs generated by motions are dominated by [PTRANS] acts and [DO-ACT]s with [PTRANS] as their instrumental cases.

Descriptions. These tell how to recognise places along the proposed route (e.g. "There's some steps at the end and a sculpture."). They tend to be declarative and are rich in noun phrases of all kinds. If the conceptual type lattice includes the classes of objects [LANDMARK], [BUILDING], [PATHWAY], [PLACE] and [GEO-FEATURE], then these will be common in graphs computed from descriptions. The relations (LOC), (ATTR) and (PATH) also appear frequently.

Comments. These are either metalanguage about the directions themselves ("This is the simplest thing."), warnings ("Look out for cars"), or idioms ("Good luck." or "Can't miss it."). At present metalanguage and warning comments are not distinguished by the pragmatic specialist, and their content is discarded. Reisbeck's heuristic rules reject comments as unimportant to judging the clarity of directions, but sometimes they do play a part. Idioms at least can be included because they are easy to recognise due to their relatively fixed expressive forms. Like metaphors (which do not appear in direction texts) they cannot generally be interpreted directly and so require special handling. There are probably few enough such forms in direction texts to simply store them on a lookup table and deliver a canned conceptual graph for each. However, the current implementation uses the set intersection word-matching method to good effect.

Finally, a piece of text is classified as incomprehensible if all specialists fail their attempt to generate a conceptual graph from it. A null graph is useless, but the detection of incomprehensible texts may signal the need for a request to reform part of the input. This judgement must wait until the verdict from all the specialists is in, and thus is best left to the control specialist.

5   Integration and Control

Each piece of text processed will result in a set of contributions from the specialists. Responsibility for combining the contributions of each specialist into a single conceptual graph, and for deciding on the disposition of that graph, falls to a control specialist.

Assuming some result from each of the specialists was available, then theoretically a policy for integration could be represented by a table like that in Figure 1. A control specialist is the specified as a procedure for carrying out the action specified by each line on the table, according to the nature of the available results. Such tables are partly theoretical statements about the value of the contributions and partly practical suggestions about what to do to get the job done with incomplete results, based on known performance levels. Because the system is not yet fully implemented, all this information is not available. Worse, it may change as the system develops. A solution is to write a special experimental control loop that scores each specialists' performance over an entire corpus. Success/failure rates for the specialists could then be automatically computed for new revisions of the system, to see if a new policy should be implemented for the performing program.

At present the control specialist effectively implements the policy shown in Figure 1 without actually consulting a real action table.

Ideally, each piece of text would map into a single graph, which would then be asserted into a database of graphs for later inference and planning before beginning on another, but things don't always work out so neatly. The integration process must account for the fact that some graphs must be merged with the previous graph, and some must be discarded altogether. To do this the symbol dividing the text fragment from the last, the canonical graphs for conceptual relations in the graphs and information about the role of the communicative act class must be applied. The need for domain specific information means that it might be better to make the integration process the responsibility of the pragmatic specialist. To Reisbeck's rules judging the clarity and cruciality of text units, I add a set of completeness rules, which are used to decide if a the graph representing a motion or description may stand alone or should be

PRAGMATIC	SYNTACTIC	SEMANTIC	ACTION TAKEN
Fail	Fail	Fail	Announce total failure
Fail	Fail	Succeed	Use graph from semantic specialist
Fail	Succeed	Fail	Use graph from syntactic specialist
Fail	Succeed	Succeed	If dissimilar use syntax specialist graph If similar use maximal join of both
Succeed	Fail	Fail	If class is IDIOM output idiomatic graph Else report failure, report classification
Succeed	Fail	Succeed	If class is IDIOM output idiomatic graph Else use graph from semantic specialist
Succeed	Succeed	Fail	If class is IDIOM output idiomatic graph Else use graph from syntactic specialist
Succeed	Succeed	Succeed	If class is IDIOM output idiomatic graph Else If dissimilar use syntax specialist graph Else use maximal join of both

Figure 1. Action table implementing a policy about how contributions from the specialists should be applied.

combined with another. Motions are judged to be complete if the (DEST) slot of its [PTRANS] act is filled with a specific concept. Descriptions are judged to be complete if a (LOC) or (ATTR) relation can be found between [PLACE] and a [T]. Idioms are always complete, and other comments are at present disgarded.

An example will make this process clearer. Consider the processing of the description "There's some steps and a sculpture, I think." The sentence is divided on the "and" since that is a conjunction, and on the comma. Processing the first unit results in the graph

[CURRENT-PLACE] -> (LOC) -> [STAIR:{*}]

which being judged to be a complete descriptive assertion is pushed onto the conceptual stack. The next unit "a sculpture" affords the graph

[SCULPTURE]

which is judged not be a descriptive assertion by itself. Since the word "and" separates the two phrases, the control specialist checks around the concept at the end of the last graph, [STAIR:{*}], for relationships in which [SCULPTURE] might also participate. Since set coercion is blocked, there is only one in this case, (LOC), and its canonical graph admits [SCULPTURE]. The maximal-join of these gives

[T] -> (LOC) -> [SCULPTURE] which, joined to the previous graph, results in

[CURRENT-PLACE] - (LOC) -> [STAIR:{*}]
                (LOC) -> [SCULPTURE].

The next unit "I think" generates the graph [MBUILD] -> (AGNT) -> [SPEAKER]. The comma makes the specialist check the last graph for a connection, but none is found. The graph, classified as a comment, is then discarded.

6 Implementation

The experimental domain in the current work is that of navigation on maps of the physical world. An intelligent agent must read and comprehend transcriptions of real speech intended to direct humans to some destination. It must convert the real language into conceptual graphs and then use those graphs to generate a plan to move the agent from a start point to the destination on in a graphical simulation of the complex of roads and buildings in a university. This serves as a useful experimental platform for experiments in natural language understanding using conceptual graphs. The system is written in Common LISP version 2.0b1 and runs on a Macintosh IIfx with 20M of RAM.

References

1. J.F. Sowa: Using a lexicon of canonical graphs in a sematic interpreter. In M.W. Evens (ed.): Relational models of the lexicon. Cambridge: Cambridge University Press (1988), pp. 113-137.

2. J.F. Sowa & E.C. Way: Implementing a semantic interpreter using conceptual graphs. IBM Journal of Research & Development 30, 1, 57-96 (1986).

3. A.V. Gershman: Conceptual analysis of noun groups in English. Proceedings of the Fifth International Joint Conference on Artificial Intelligence. Cambridge, Mass: (1977), pp. 132-138.

4. V. Fromkin, R. Rodman, P. Collins & D. Blair: An Introduction to Language. Sydney: Holt, Rinehart and Winston 1984, pp. 244-255.

5. R.C. Schank & R.P. Abelson: Scripts Plans Goals and Understanding. Hillsdale, New Jersey: Lawrence Earlbaum Associates (1977).

6. P.H. Winston: Artificial Intelligence. Reading, Mass: Addison-Wesley Publishing (1988), pp. 314- 325.

7. H.M. Noble: Natural Language Processing. Edinburgh: Blackwell Scientific Publications 1988.

8. V. Wuwongse, S. Niyomthai: Conceptual graphs as a framework for case-based reasoning. Proceedings of the 6th Annual Workshop on Conceptual Graphs, pp. 119-133 (1990).

9. C.K. Riesbeck: "You can't miss it!": Judging the clarity of directions. Cognitive Science 4, 285-303 (1980).

10. R.C. Schank Inside Computer Understanding. Hillsdale, New Jersey: Lawrence Earlbaum Associates (1981) pp.10-26.

A SYSTEM THAT TRANSLATES CONCEPTUAL STRUCTURES INTO ENGLISH

Sait Dogru
dogru@cs.umn.edu

James R. Slagle
slagle@cs.umn.edu

Department of Computer Science
University of Minnesota
Minneapolis, MN 55455

Abstract. Conceptual Structures (CS) is a modern, structural knowledge representation language that is gaining popularity across diverse applications. Although simple in notation, large graphs are a potential source of confusion for the human user. In this paper, we describe an implemented system that can translate arbitrarily complex CS graphs into plain English statements. The system can assist the user in interpreting graphs that result from long reasoning chains of the Unified Reasoning Process' (URP) reasoning mechanism, and in validating the correct parsing of the graphs when entering propositions to the URP. The system has been successfully incorporated into the URP system support tools.

1.   Introduction

Sowa's CS is a modern knowledge representation language that has been gaining popularity across diverse applications. Since its conception, it has been researched and modified extensively. The URP reasoning mechanism developed at the University of Minnesota is based on the original CS, extended with theoretically more sound constructs, while taking care to preserve its attractive and simple but powerful notation. Complex graphs, however, can still look ambiguous, even to advanced users of the system. The user thus has difficulty interpreting graphs resulting from several kinds of reasoning strategies used by the URP. Another difficulty arises in entering the initial beliefs, or propositions, into the URP system. The user may not be so sure that the URP system correctly understood what he/she meant by certain input graphs, as it is easy to make mistakes when entering complex graphs, especially when extended over several lines of input.

As a solution to these problems, we have developed a system for the URP. The implemented system can take arbitrarily complex graphs as input and produce a corresponding English translation. By checking the English translation against his/her own interpretation of a particular input graph, the user can easily determine any discrepancies, and take action to correct the problem. The system has been implemented on a Texas Instruments Explorer II Lisp Machine in Common Lisp, and has been incorporated into the URP system support tools.

2. Conceptual Structures: A Brief Introduction

Conceptual Structures (CS) is a modern network representation that brings together ideas from such diverse fields as psychology, linguistics, philosophy, and artificial intelligence. It is designed to be applicable to a wide range of applications, from natural language processing to Databases and Expert Systems. The introduction of CS as given here is based on the extended version, and more information can be found in the references.

A CG (Conceptual Graph) is a finite, connected, bipartite graph. The nodes in the graph can be either concepts or conceptual relations. A conceptual relation node represents a relation between concepts. The two kinds of nodes are connected by arrows, or pointed arcs. A concept can have arcs only to conceptual relations. A conceptual relation can have arcs only to concepts. Each CG represents a single proposition. A typical knowledge base will contain many propositions.

2.1 Concepts

Concepts are enclosed in square brackets, [ ]. Each concept has two fields, separated by a colon (:). To the left of the colon is the type field that specifies the class, or the type, of the concept. To the right of the colon is the referent field. A concept, depending on its referent field, can be either instantiated or constrained. If instantiated, the referent field is simply an individual marker, a string, a number, or a graph set. All the following concepts are instantiated:

[PERSON: #JOHN]
[NAME: "JOHN DOE"]
[MILE: 150]
[PROPOSITION: #{ [PERSON: #JOHN] -> (LOVE) -> [PERSON: #JANE]}]

A constrained concept's referent field has the following sub-fields:

scope-list variable domain @ cardinality control-mark

The scope list is a list of zero or more concept variables enclosed in parantheses. This list shows functional dependencies between concepts. The default scope list is null, i.e., ().

The variable uniquely identifies the concept. If two concepts have the same variable, all other fields in the referent field must also be the same or else, at least one of the concepts must be a pronoun concept. The default variable is unique.

The domain is a set of referents that are candidates for instantiation. One can specify the domain by:
• Explicitly enumerating every member of the set,
• The generic domain, {*}, which contains all possible referents, or,
• A function that returns a set of referents. The default domain is the generic domain.

The cardinality field specifies the number of referents that exist for each value of the scope list variables. It is basically used to constrain the set referents. It can take the form of a single number, a range of two numbers, or fuzzy quantifiers like some, most, or many. The default cardinality is "1 - ∞" (infinity).

The control mark directs the reasoning process in instantiation of the concept. Controls marks are not used during translation, and will not be discussed further in the paper.

The following concepts are constrained:

[COMPUTER: () c {*} @5-∞]
[PERSON: ()p {#JOHN, #JANE, #JIM} @2-3]

2.2   Relations

A relation is enclosed in parantheses, (). A relation node has only a type field, which is usually abbreviated. They indicate a relation involving one or more concepts, connected by arrows. An agent relation, for instance, might be represented as (AGNT). The proposition "John is talking" can similarly be represented as [PERSON: #JOHN] <- (AGNT) <- [TALK].

Typically, a relation is binary. That is, it connects two concepts. However, there are also unary relations that point to a graph set, consisting of one or more CS graphs. Such a unary relation is inherently different in meaning and function in that it affects the interpretation, or meaning of the graphs it points to. Such a unary relation is (PAST), which restricts interpretation of the graph set to English language past tense. There will be more discussion on this topic later in the report.

More complex graphs can be extended over several lines, simply by ending a line with an arrow or a dash. The dash is used to show the connection of a concept to several others, which is terminated by a comma and a period. In the next example, the concept [PERSON: #JOHN] is connected to both agent relations:

[PERSON: #JOHN] -
                <- (AGNT) <- [WEAR] -> (PTNT) -> [BLUE-JEANS]
                <- (AGNT) <- [TALK],.

2.3   Type Hierarchy and Individuals

In CS notation, every concept is a unique individual of a particular type. The type field of a concept node indicates the class, or type, of the individual represented by the node. Especially to be used in reasoning mechanisms, types are organized into a hierarchy. The type PERSON, for instance, is a subtype of type MAMMAL, which in turn is a subtype of type LIVING-THING, and so on.

The type hierarchy defines a partial ordering on the set of types. It is, therefore, possible for some types to have more than one supertype and subtypes. The type DOG, for example, is a subtype of both types PET and CARNIVORE. In this sense, the name hierarchy is a misnomer; it actually defines a lattice. In order to fill the gaps caused by types with no supertypes like FEELING, two special types are introduced. The universal type, represented as T, is supertype of all types. The absurd type, represented as ⊥, is a subtype of all types.

3.   Primary Data Structures

In this section, we shall go into the details of the major data structures and parameters of the system.

3.1   Dictionary

The project is one of natural language processing. It is, therefore, natural that some sort of dictionary be used. This concept of dictionary, in terms of organization and function, is much like an English language dictionary used by people. It provides information as to whether a particular word is a verb, a noun, an adjective, a pronoun, etc. In case of verbs, it also provides past and past participle forms of the verb. It contains prepositions, if any, that usually go with the verb. Our dictionary, however, contains some extra information that is considered common knowledge and hence not found in a regular dictionary. For example, it is awkward to say "John is loving Jane" when what is really meant is the statement "John loves Jane." No dictionary warns against such blunders. Our dictionary, therefore, must possess some common sense knowledge in order to avoid such traps.

For the purposes of the project, we have divided dictionary entries into two categories. The first is verbs, and the second contains others, including nouns, adjectives, and adverbs. Verb entries, for instance, include information such as the full unabbreviated verb, its past and past participle forms, whether it could be used in forming passive sentences and so on. The dictionary has been implemented as a hash table, and indexed by the abbreviation of the entries. When not in use, the dictionary is kept on a disk file for long term storage.

3.2   The Type Hierarchy

This is a list of 2-element sublists, known in Artificial Intelligence literature as an association list (a-list). Both elements of each sublist are type indicators, with the first being a subtype of the second, or equivalently, the second a supertype of the first. This structure is mainly used during noun and verb phrase generation stages of the system to select an appropriate connecting word. Consider the following graphs:

[PERSON: #JOHN] -> (LOVE) -> [CITY: #MINNEAPOLIS] <- (LIVE)
                                              <- [PERSON: #JANE]
[PERSON: #JOHN] -> (LOVE) -> [PERSON: #JANE] -> (LIKE) -> [ICE-CREAM]

By using the type hierarchy, we determine that city is a place, which takes "WHERE" in noun phrases. Thus, the system translates the first graph as "John loves Minneapolis where Jane lives." Following the same idea in processing the second graph, and after determining that "WHO" is used for people, the system translates the graph as "John loves Jane who likes ice cream."

3.3   Parameters of the System

We shall now cover the concept of unary relation nodes avoided in Section 2.2. Such a unary relation node differs from ordinary relation nodes in two ways. First, it has only one arc, as opposed to a regular relation node, which usually, has two arcs, one coming in, and the other going out. The second and more interesting difference, is that its single arc points to a graph set, whereas a regular node connects two concept nodes. It defines the context in which each graph in the graph set is to be interpreted. There are two such relation nodes in extended CS notation. They are time, which can be any English language tense, and under which the graph set is to be translated, and truth and negation, which simply indicate the truth value of the graphs in the graph set. Consider the following examples:

(PAST) -> [PROPOSITION: #{[PERSON: #JOHN] -> (LIVE) -> [CITY: #MINNEAPOLIS],
                            [PERSON: #JANE] -> (LOVE) -> [PERSON: # JIM] }]
(TRUE) -> [PROPOSITION: #{[PERSON: #JOHN] -> (LOVE) -> [PERSON: #JANE] }]

Note that a graph set is introduced by a proposition-type concept node. The first example says that John used to live in Minneapolis and that Mary was in love with Jim while the second asserts as true the statement "John loves Jane."

During translation, we realized, there is no need to remember the truth indicator of given graphs. It suffices to precede the whole result of the translation with "It is not true that ...." Thus we can do away with such unary relation nodes by taking care of them, as noted above, whenever we come across them.

Time relation nodes, on the other hand, must be remembered and passed on to subordinate functions during the translation process. There are many functions that put together verbs and phrases in accordance with the specified tense of the graph, and they constantly check its current value and proceed accordingly. Thus, we have chosen time as a parameter of the system.

Another parameter is voice, specifying whether the remaining graph is to be translated as an active, or passive statement. This actually comes from the flexibility of the translator. For any CS graph, there are as many ways to start the translation as there are nodes in the graph. Depending on the initial start position of the graph, the translation can be either active or passive. The system does have a preference toward the construction of active sentences, however.

The third and final parameter is more application dependent. As the process of translation proceeds, phrases will be constructed which in turn are used to form larger phrases and finally complete sentences. During this process, a relation node needs specific information about the concepts that immediately follows it (in the active case), or the concepts that immediately precedes it (in the passive case). Specifically, the relation node needs to know the type of the relevant concept, whether it is singular or plural, and the result of its translation, which is done recursively. We have created a node structure that contains just these pieces of information, and made it the third parameter.

4. How The System Works

When the system is loaded, it tries to construct the dictionary hash table from the default dictionary file, namely, "main-dictionary.dict". It also attempts to construct the type hierarchy again from the default type hierarchy file, "type-hierarchy.dat". It gives no error messages if one or both of the default files are missing.

The system then presents the user with a 5-option menu. The user can,
- give a CS graph, and have it translated,
- load the dictionary from a given file,
- enter a new entry to the dictionary at run time,
- load the type hierarchy from a given file, or
- enter a new pair(s) of (type super-type) into the type hierarchy.

5. The System at Work

We shall illustrate several aspects of the system. Each feature is accompanied with several examples. Unfortunately, space limitations do not permit more complex graphs that show how various aspects combined can result in high quality translations.

5.1 Subject-Verb Agreement: The translator obeys all basic English grammar rules, including the subject and verb agreement rule. Contrast the following examples:

A) [PERSON: #JOHN] -> (LIV) -> [CITY: #MINNEAPOLIS]
"THE PERSON JOHN LIVES IN THE CITY MINNEAPOLIS "

B) [PERSON:()P {#JOHN, #JANE}@2] -> (LIV) -> [CITY: #MINNEAPOLIS]
" PERSONS JOHN AND JANE LIVE IN THE CITY MINNEAPOLIS "

5.2 Tenses: The default tense of the translator is Present (Progressive) Tense. However, the system is capable of handling other English language tenses as well. Here, we shall limit the examples to the past, present, and present progressive tenses:

A) (PAST) -> [PROPOSITION: #{[PERSON: #JOHN] -> (SPEAK).}]
"THE PERSON JOHN WAS SPEAKING "

B) (PAST) -> [PROPOSITION: #{[PERSON: #JOHN] -> (LIV)
-> [CITY: #MINNEAPOLIS]}]

"THE PERSON JOHN LIVED IN THE CITY MINNEAPOLIS "

C) (PRES-PROGRES) -> [PROPOSITION: #{[PERSON: #JOHN] -> (SPEAK).}]
   "THE PERSON JOHN IS SPEAKING "

D) (PRES-PROGRES) -> [PROPOSITION: #{[PERSON: #JOHN] -> (TALK)
                                            -> [PERSON: #JANE]}]
   "THE PERSON JOHN IS TALKING TO THE PERSON JANE "

5.3   Active vs. Passive Sentences: By rearranging the input graph, or depending on the start node of the graph, the translator will choose to form a sentence in either active, or passive voice, whichever is more convenient. Contrast the examples:

A) [PERSON: #JOHN] -> (LOV) -> [PERSON: #JANE]
   "THE PERSON JOHN LOVES THE PERSON JANE "

B) [PERSON: #JANE] <- (LOV) <- [PERSON: #JOHN]
   "THE PERSON JANE IS LOVED BY THE PERSON JOHN "

C) [BODY: @∞-∞] -> (KNOW) -> [PRESIDENT: #BUSH]

 " EVERY BODY KNOWS THE PRESIDENT BUSH ", if we start from the first node, and

 "THE PRESIDENT BUSH IS KNOWN BY EVERY BODY ", if we start from the last.

5.4   When There are no Verbs: The system will precede the translation with the phrase "There exist(s) ..." as shown in the next examples:

A) [PERSON: #JOHN]
   " THERE EXISTS THE PERSON JOHN"

B) [BOOK:]
   " THERE EXIST SOME BOOKS "

5.5   When the Passive Voice is not Meaningful: There are sentences that can not be put into passive form. "John lives in Minneapolis" is an example. In this case, the translator can either choose a substitute verb to form a passive form of the sentence, or it can create a phrase instead. The examples illustrate:

A) [CITY: #MINNEAPOLIS] <- (LIV) <- [PERSON: #JOHN]
   "THE CITY MINNEAPOLIS IS WHERE THE PERSON JOHN LIVES"

B) [STUDENT: ()S {*}@100-∞] <- (CONSIST) <- [GROUP: #CULTURAL-DIVERSITY]
   " AT LEAST 100 STUDENTS FORM THE GROUP CULTURAL-DIVERSITY "

C) [GROUP: #CULTURAL-DIVERSITY] -> (CONSIST) -> [STUDENT: ()S {*}@100-∞]
   "THE GROUP CULTURAL-DIVERSITY CONSISTS OF AT LEAST 100 STUDENTS "

5.6   One Graph - Many Translations: As can be seen from the examples so far, the translator is sensitive to the start node of the input graph.

This may result in several interpretations, or rather, translations where different concepts are emphasized.

5.7   About Cardinality Information: The examples below show the importance of this field in natural language translation:

A) [PERSON: ()P {*}@1]
   " THERE EXISTS A PERSON "

B) PERSON: ()P {*}@1-5]
   " THERE EXIST 1 TO 5 PERSONS "

C) [PERSON: ()P {*}@3-∞]
   " THERE EXIST AT LEAST 3 PERSONS "

D) [PERSON: ()P {#JOHN, #JANE, #JILL}@2-3]
   " THERE EXIST 2 TO 3 PERSONS AMONG JOHN, JANE AND JILL"

E) [YEAR: #1991] -> (CONSIST) -> [DAY:()D {*}@365]
   "THE YEAR 1991 CONSISTS OF EXACTLY 365 DAYS "

F) [PERSON: ()P {#JOHN, #JANE, #JILL}@2-∞]
   " THERE EXIST AT LEAST 2 PERSONS AMONG JOHN, JANE AND JILL"

5.8   Cardinality is not Always Needed: There are some cases, however, where cardinality information can be omitted, resulting in more natural translations. Following is such a case:

[PERSON:()P {#JOHN, #JANE}@2]
" THERE EXIST PERSONS JOHN AND JANE", as opposed to "2 persons among John and Jane."

5.9   Simple Disjunctions: The CS notation is capable of handling disjunctions as well, again through the cardinality information. The translator again correctly handles these situations:

[PERSON: ()P {#JOHN, #JANE}@1]
" THERE EXISTS THE PERSON JOHN OR JANE"

5.10 Complex Disjunctions: A complex disjunction is a concept whose domain contains more than two elements, and whose maximum cardinality is less than the number of the domain elements. Such graphs are translated as in the following example:

[PERSON: ()P {#JOHN, #JANE, #JILL}@1]
" THERE EXISTS EXACTLY ONE PERSON AMONG JOHN, JANE AND JILL"

5 .11 How Cyclic References are Handled: The linear form, being more easily readable than the traditional graph form, is the preferred representation of CS graphs. Hence, all graphs are put into a linear form before processing by the URP system. This transformation means that some nodes will be duplicated in the linear form. Duplicated nodes are implemented in the URP system by only mentioning full information about them once, and subsequently referring to them by their variable. The translator system is able to tell the difference between the two occurrences of the same concept. The system precedes the succeeding appearances with the definite article "the" to separate the two occurrences of a concept. (To play it safe, the system also presents the original translation enclosed in parantheses.) On the initial encounter, the following structure is translated as " A PERSON ". Successive encounters, however, are translated as " THE PERSON (A PERSON) "

[PERSON: ()P {*} @1]

6. Future Work

6.1 Propositions and other unary relations

As we have seen earlier, a proposition concept node represents a graph set that may contain more than one graph. The system, as of the date of the completion of this paper, is not equipped to handle proposition nodes. This is again the case for new, unexpected unary relations. It should also be pointed out that it is easy to add the few lines of code to do this, since all that needs to be done is iteratively calling the translator on each graph in the graph set. This can be accomplished elegantly in a mapcar statement.

6.2 Scope Lists

Scope lists are a fundamental extension to Sowa's original CG notation. They provide interdependency information between concept nodes of the graph. Although they play an essential role in reasoning mechanisms of the URP system, we have avoided their use and have used null scope lists throughout the report. As stated in [4] and [7], they are found to be too difficult to express in a natural language statement.

7. Summary

We have developed a natural language interface for Conceptual Structures, one of the latest structural representation languages that has been used across numerous applications. The implemented translator obeys English language grammar rules. It is capable of handling various features, including the subject and verb agreement rule. The user of the URP system now has a tool that will help eliminate ambiguity over CS graphs. The translator can assist the user in both interpreting graphs that result from long reasoning chains of the URP, and in validating the correct parsing of the graphs when entering propositions to the system. The translator system has been successfully incorporated into the URP system support tools.

8. Acknowledgements

The URP System implementation is derived from CONSTRUCT. CONSTRUCT is a tool for creating and manipulating conceptual structures developed by John

Esch, Tim Nagle, Morgan Yim, and others from Unisys Corporation and Control Data Corporation. The source code is licensed to the University of Minnesota for research purposes. We would like thank all those involved in the development of CONSTRUCT, without which this work would not have been possible.

David Gardiner implemented the extended notation for CS, and we are also grateful for his comments on syntax. Bosco Tjan was of invaluable help in maintaining the TI Explorer II Lisp machines.

8. References

[1] John F. Sowa. Conceptual Structures: Information Processing in Mind and Machine. Addison-Wesley, Reading, MA, 1984.

[2] David A. Gardiner, Bosco S. Tjan, James R. Slagle. Extended Conceptual Notation. University of Minnesota Computer Science Department technical Report TR 89-88.

[3] Bosco S. Tjan and James R. Slagle. A conceptual structure semantic theory based on a semantic game with partial information. Proceedings of the Third Annual Workshop on Conceptual Graphs. August, 1988.

[4] James R. Slagle, David A. Gardiner, Bosco S. Tjan. Extended Conceptual Structures and The Unified Reasoning Process. University of Minnesota Computer Science Department technical Report TR 89-69.

[5] Bosco S. Tjan, David A. Gardiner, James R. Slagle. Direct Inference Rules for Conceptual Graphs with Extended Notation. University of Minnesota Computer Science Department technical Report TR 90-28.

[6] James R. Slagle, David A. Gardiner, Bosco S. Tjan. Reasoning with Conceptual Structures: Year Two Final Report. University of Minnesota Computer Science Department technical Report TR 90-22.

[7] David A. Gardiner and James R. Slagle. Extended Conceptual Structures for Intelligent Document Retrieval. Proceedings of the Sixth Annual Workshop on Conceptual Graphs, July 1991, pp. 169-185.

[8] Guy L. Steele, Jr. Common Lisp: The Language, 2nd ed. Digital Press, Reading, MA, 1990.

Knowledge Based Analysis of Radiology Reports Using Conceptual Graphs

Martin Schröder

University of Hamburg, Computer Science Department
Natural Language Systems (NatS), Bodenstedtstr. 16
W-2000 Hamburg 50, Germany
email: martin@nats4.informatik.uni-hamburg.de

Abstract. The telegraphic language found in radiological reports can be well understood by a natural language system using the underlying domain knowledge. We present the METEXA system, which emphasizes the use of radiological domain knowledge to determine the semantics of utterances. Syntactic and semantic analysis, lexical semantics and the structure of the domain model are described in some detail. A resolution-based inference engine answers relevant questions concerning the contents of the reports. As knowledge representation formalism the Conceptual Graph Theory by John Sowa has been chosen.

1 Introduction

Medicine is a domain where large amounts of routine texts play a significant role. One example is radiology reports that are an important source of patient data. The reports are typically spoken into a dictating machine and afterwards typed into a machine readable and printable format. The automatic processing of qualitative findings can be a substantial aid for clinical documentation and decision making. Over the last few years, there have been several approaches to the processing of clinical narratives, each with a different emphasis. To name a few, the Linguistic String Project (Sager et al. 1987), a system for the acquisition of radiological facts from reports (Ranum 1988), or a dialogue system for querying and updating medical databases (Mery et al. 1987).

In this paper the METEXA system ("MEdical TEXt Analysis") for the knowledge based analysis of radiological reports is described. The importance of the domain model, which is used as the central source of knowledge in all processing stages, is reflected in the design of the system. This approach is suggested by the restricted style and content of radiological reports. The *Conceptual Graph Theory* by John Sowa (1984) has been chosen as the semantic and knowledge representation language, because it provides a theoretically sound formalism that is suited for practical applications. Related approaches to semantic interpretation with conceptual graphs can be found in Sowa and Way (1986), Fargues et al. (1986), Sowa (1988), and Velardi et al. (1988).

Another goal of the project is the generation of expectations to support the recognition of continuous speech. While isolated speech recognizers are already applied in the domain of radiological report generation (e. g. Billi et al. 1991), analyzing continuous speech is still a very hard problem. Missing word boundaries cause a combinatorical explosion of the number of word hypotheses. It is the goal of the METEXA project to guide the selection of appropriate word hypotheses by exploiting domain specific knowledge.

2 Sublanguage Analysis

The language style found in radiology reports shows characteristics of a sublanguage (Lehrberger 1986). Sublanguages have a restricted and simplified structure compared to language as a whole. The radiological reports, which are the basis for the following studies, can be characterized as follows:

- The texts are often written in *telegraphic style*. Verbs are frequently omitted. Utterances typically consist of coordinated noun phrases, prepositional phrases and adverbs. Many utterances consist of standard phrases, but with non-standard findings the language can become surprisingly rich.

- For the understanding of an utterance the *semantic structure* is more important than the syntactic structure. An utterance often consists only of constituents without "syntactic glue", but the semantics of the words and their semantic relationship to each other makes it easily understandable. Even morpho-syntactic features, which usually play a crucial role in highly inflected languages like German, are "overridden" by semantic structure. For example, it is common to say "Fraktur rechter Oberschenkel" (engl. "fracture of the right thigh") instead of the correct form "Fraktur des rechten Oberschenkels".

- The understanding, as well as the production of a report is highly determined by specialized domain knowledge. When reading a report, the physician usually expects certain anatomical locations and their radiological characteristics to occur, so that potential ambiguities or misunderstandings are avoided.

We work with a corpus of about 1.500 radiological reports, about half of them describing X-ray images of the chest (thorax). The corpus contains about 8.000 different word forms with about 120.000 occurences.

3 The METEXA System

The fact that the understanding of the reports is guided to a large degree by semantics and domain knowledge has lead to an architecture that puts the domain model into the center of the system. The following components form the backbone of the system:

- **Lexicon:** At present a fullform lexicon is used. For each word the uninflected form, the syntactic category and the lexical semantics are specified. The semantics is defined by means of a reference to the domain model.

- **Parser and Grammar:** The parser is a bottom-up parser in Prolog ("BUP"), developed by Matsumoto et al. (1983), augmented by a unification mechanism to produce a feature matrix of attribute-value-pairs. The grammar rules are coded in Prolog in a quite readable fashion. The parser is described in (Schröder 1991). Up to now about 70 grammar rules have been written.

- **Semantic Analysis:** A conceptual graph for a sentence is constructed compositionally by joining relations incrementally. The construction of the semantic representation, including knowledge base access, works interleaved with the syntactic analysis. The analyser conforms to the framework proposed by Sowa (1988) on an abstract level.

- **Domain Model:** A type lattice and canonical graphs are used during analysis. The type lattice includes anatomical objects, pathological alterations, radiological terms, and other more general concept types. Schemata will be used for the application-dependent interpretation of graphs.

- **Inference:** The resolution-based inference method by Fargues (1986) has been implemented. A rule base serves to answer interesting questions about a radiological report.

- **Planning:** Checklists, plans and scripts are used to generate expectations about the next incoming utterance. These structures contain knowledge about the typical contents of radiological reports.

3.1 Syntactic and Semantic Analysis

An important issue, concerning the architecture of the system, is the different weights of the components: While many traditional systems first analyse the complete syntactic structure of a sentence and afterwards construct a semantic representation or map the results to a backend system (e.g. LSP, Sager 1987), the METEXA system emphasizes semantics, so that the task of the parser is only to tentatively propose subconstituents that might be connected by semantic relations. Whether a phrase is admissable or not is decided by semantic means. This is illustrated by an example phrase that has two syn-

```
    NP
        |NP
        |        |NP
        |        |        |ADJ      regelrecht
        |        |        |N        Zwerchfellstand
        |KOORD
        |        |KONJ     und
        |        |NP
        |        |        |NP
        |        |        |        |ADJ      normal
        |        |        |        |N        Transparenz
        |        |        |MOD
        |        |        |        |NP
        |        |        |        |        DET    der
        |        |        |        |        N      Lunge
```

Fig. 1. Feature matrix "regelrechter Zwerchfellstand und normale Transparenz der Lunge" (engl. roughly "correct position of the diaphragm and normal translucency of the lung")

tactic interpretations. Fig. 1 shows the correct interpretation: The conjunction "und" connects two independent noun phrases on the top level. The distributive reading (not shown here) is blocked, because a semantic relationship between the concepts DIA_-POSITION ("Zwerchfellstand", the position of the diaphragm) and LUNG does not exist. Between TRANSLUCENCY ("Transparenz") and LUNG, however, a semantic relationship can be found (TRANSLUCENCY is defined as a "radiological characteristics" of LUNG). Fig. 2 shows the canonical graphs that are verified during the analysis of the example sentence. (Please note that the METEXA system up to now treats only the German language. Example sentences and type labels have been translated for illustration purposes only.) The example sentence results in the conceptual graphs shown in

```
[DIA_POSITION]->(SATTR)->[CORRECT]
[LUNG]->(XCHAR)->[TRANSLUCENCY]
[TRANSLUCENCY]->(XATTR)->[NORMAL]
```
Fig. 2. Canonical graphs verified during the analysis

Fig. 3. All possible interpretations are produced via backtracking, in this case only one.

```
[DIA_POSITION: #2]-
      (STATE)<-[DIAPHRAGM: #1]
      (SATTR)->[CORRECT: #3].

[TRANSLUCENCY: #5]-
      (XATTR)->[NORMAL: #4]
      (XCHAR)<-[LUNG: #6].
```
Fig. 3. Conceptual Graphs for the sentence of Fig. 1

Syntactic and semantic processing work in an interleaved fashion: the semantic relationship is checked as soon as two syntactic constituents have been detected. This way no syntactic interpretation without a valid semantic interpretation is generated.

The semantic representation is based on the idea that the conceptual relation between each pair of conceptual entities has to be determined, if there is any. The process of semantic interpretation can be described as detecting all the valid semantic relations between syntactic constituents. Explicit and implicit relations are distinguished according to Pazienza and Velardi (1987). Explicit relations are those that have a lexical realisation in the utterance; implicit relations correspond to syntactic roles (see Fig. 4). In general, explicit relations are easier and faster to verify, because the relation type is already suggested by the lexical level.

Explicit Relation:	Implicit relation:
"Luft unter dem Zwerchfell" (engl. *"air below the diaphragm"*)	*"Elongation der thorakalen Aorta"* (engl. *"elongation of the thoracic aorta"*)
[AIR]->(LOC_BELOW)->[DIAPHRAGM]	[THORACIC_AORTA]->(PATHO)->[ELONGATION]

Fig. 4. Examples of explicit and implicit relations

The parser does not use morpho-syntactic features to determine possible interpretations. This is due to the telegraphic sublanguage that often lacks verbs and correct inflectional endings, as discussed in section 2. It would be easy to augment the feature matrix by morpho-syntactic features, because the parser already provides unification for the construction of feature matrices, but experience has shown that generally semantic verification is sufficient to find the correct interpretation.

Let us briefly discuss one specific representation problem: In the domain of medical reporting you frequently find noun phrases with conjunctions that are used as subject or object. The problem can be illustrated by the simple sentence "A cat and a dog are running". What should the conceptual graph for this utterance look like? John Sowa (1991) has proposed three different alternatives, not being satisfied with any of them:

```
1.[RUN]-
        (AGNT)->[CAT]
        (AGNT)->[DOG].
2. [RUN]->(AGNT)->{[CAT], [DOG]}.
3. [RUN]->(AGNT)->[COLLECTION: [CAT] [DOG]].
```

We have decided to provisionally use the first alternative, because it does not need a new syntax and avoids the use of contexts. This makes processing easier and faster. However, it is questionable whether a relation like (AGNT) should be duplicated. It is more intuitive to keep only a single instance of the relation. In this case the syntax for describing a set of instances, like in the sentence "The cats Miau and Mao are running"

```
[RUN]->(AGNT)->[CAT: {Miau,Mao}].
```

could be extended for concepts with different types, resulting in alternative 2.

3.2 Lexical Semantics

The guideline for the definition of lexical semantics is to define semantic information only as a "pointer" to the domain model. This pointer alone does not fully describe the semantics of the lexical entity, but after accessing the domain model the canonical graph for that concept can be determined. The advantage of this approach is the fact that the knowledge is not "scattered around" in the lexicon, but is stored in a central place. Fig. 5 shows the various kinds of pattern types, which are coded as macros. These are

Pattern type	Example (macro)	Semantic pattern (exp. macro)
Concept	[C,DIAPHRAGM]	[DIAPHRAGM]->(_)->[_]
Relation/Concept	[RC,AATTR,FLATTENED]	[_]->(AATTR)->[FLATTENED]
Concept/Relation	[CR,INDICATION,THEME]	[INDICATION]->(THEME)->[_]
Concept2	[C2,NORMAL]	[_]->(_)->[NORMAL]
Relation	[R,LOC_ON]	[_]->(LOC_ON)->[_]
Relation, inverse	[RINV,THEME]	[_]<-(THEME)<-[_]

Fig. 5. Semantic patterns in the lexicon

expanded when accessing the lexical entries. The underscores are Prolog variables. They can be read as holes that have yet to be filled by appropriate labels. The mappings are based on the assumption that the head concepts of each two constituents can be connected via a semantic relation (see Pazienza and Velardi 1987):

```
[CONCEPT1]<->(RELATION)<->[CONCEPT2]
```

The concept in the first position is always considered to be the head concept of the resulting phrase. The expanded semantic expressions serve as building blocks for the construction of a canonical graph. For example, for the phrase "abgeflachtes Zwerchfell" (engl. roughly "flattened diaphragm") the two uncomplete graphs would be put one over the other to yield

```
[DIAPHRAGM]->(AATTR)->[FLATTENED]
```

This graph has to be verified in the knowledge base. The knowledge base contains the canonical graph

```
[DIAPHRAGM]->(AATTR)->[DIAPHRAGM_ATTR]
```

so that via DIAPHRAGM_ATTR > FLATTENED the graph can be proved as being correct. The result is the same as if we had joined two canonical graphs for DIA-PHRAGM and FLATTENED:

```
DIAPHRAGM:   [DIAPHRAGM]->(AATTR)->[DIAPHRAGM_ATTR]
FLATTENED:   [DIAPHRAGM]->(AATTR)->[FLATTENED]
```

Implicit relations are resolved in the same way, with the only difference that the relation type is still unknown. For example, "Fraktur des Oberschenkels" (or simply "Fraktur Oberschenkel", engl. "fracture of the thigh") results in

```
[FRACTURE]->(_)->[THIGH]
```

and the relation PATHO is found. The interpretation does not rely on morphological or syntactic features, but only on the assumption that two adjacent noun phrases modify each other.

The advantages of our approach are:

- The definition of lexical semantics is concise. The names of concept types and relations serve as pointers to the domain model.

- The knowledge that is expressed by a canonical graph is stored only in a single place: in the knowledge base, and not in the lexicon.

- The number of canonical graphs is small, because not for each concept type a canonical graph has to be defined. It is defined for a more general concept type and is inherited to all subtypes.

The main disadvantage of our approach is that the semantic interpretation has to take into account which types of semantic patterns have to be combined in what way. For example, connecting the patterns of two noun phrases by a semantic relation is structurally different from connecting the patterns of an adjective and a noun. The difference does not directly depend on the syntactic constituents, but on the macro types involved. However, different variations of the semantic interpretation procedure have to be defined that take care of the different combinations of semantic patterns.

German is a language with many compound words. In contrast to English, they are written in one word (e.g. "compound noun" vs. "Nominalkompositum"). To describe their semantics, additional macro types have been defined that expand to more complex graphs. On the other hand it is sometimes necessary to map lexemes that consist of multiple words to a single concept type. In the medical domain this is often the case for anatomical locations and deseases that have Greco-Latin names, e.g. "Sinus phrenicocostales" (engl. "phrenicocostal sinus"). Idiomatic and other domain specific phrases fall into this category as well. Multiple word lexemes are detected in a lexical preprocessing phase.

3.3 The Structure of the Domain Model

During syntactic-semantic analysis the type lattice and canonical graphs are accessed. A selection of the relations that are used are shown in Fig. 6. These are the definitions of conceptual relations according to Appendix B.3 in (Sowa 1984). These definitions are not accessed during interpretation. Their purpose is to ensure consistency when defining canonical graphs.

Concept type	is connected by relation type	to concept type
[ANAT]	PATHO	[PATHO_ALT]
[ANAT]	SIDE	[SIDE]
[ANAT]	LOC	[ANAT_LOC]
[ANAT]	AATTR	[AATTRIBUTE]
[ANAT_STATE]	SATTR	[SATTRIBUTE]
[PATHO_ALT]	PATTR	[PATTRIBUTE]
[XFEATURE]	XATTR	[XATTRIBUTE]
[GEOMETRY]	GATTR	[GATTRIBUTE]
[ANAT]	XCHAR	[XFEATURE]
[ANAT]	STATE	[ANAT_STATE]
[PATHO_ALT]	LOC	[ORGAN_LOC]
[PATHO_ALT]	LOC_UNDER	[ANAT]

Fig. 6. Conceptual relations

Some difficulties have been observed when creating a type lattice and conceptual relations during domain modelling. One difficulty concerns the condition that a relation type should be defined only once. For example, the concept type NORMAL may be used as an attribute for an anatomical object as well as for a radiological characteristics. This would lead to two different definitions for the relation type ATTR. Therefore, two different relation types, AATTR (anatomical attribute) and XATTR (radiological attribute), have to be introduced:

```
[ANAT]->(AATTR)->[AATTRIBUTE]
[XFEATURE]->(XATTR)->[XATTRIBUTE]
```

Another solution would be to introduce a new superclass of ANAT and XFEATURE, named "ALL_THINGS_THAT_HAVE_ATTRIBUTES". But this is not a natural ontological category, it is only born out of technical reasons. We think that such a concept type should be avoided.

We have observed that the structure of the type lattice has changed during the development of the system in a systematic way: For the analysis of utterances it is sufficient to have a rather flat structure. For example, defining a canonical graph

```
[ANAT]->(PATHO)->[PATHO_ALT]
```

is sufficient to verify that the PATHO relation holds between LUNG and CONGESTION. However, for the purpose of predicting the semantics of incoming utterances it

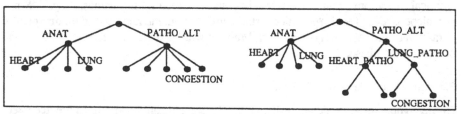

Fig. 7. Flat vs. deep type lattice

is necessary to have a much more specific type lattice, e.g. pathological alterations must be defined only for those anatomical objects where they really occur:

[LUNG] -> (PATHO) -> [LUNG_PATHO]

Since this principle is applied to pathological alterations, radiological characteristics, attributes etc. as well, the type lattice becomes deeper and will have more concept types (see Fig. 7). More importantly, many more canonical graphs have to be defined now, e.g. one graph for each group of pathological alterations.

We experienced that the construction of a hierarchy of concept types which is not a lattice is very likely. For example, NORMAL and CORRECT are subtypes of AATTRIB-UTE (anatomical attribute), but of XATTRIBUTE (radiological attribute) as well (depicted in Fig. 8 (a)). This structure is constructed during the incremental adding of new subtype relations to the hierarchy.

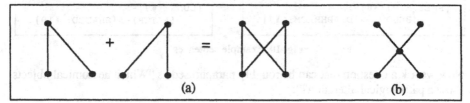

(a)   (b)

Fig. 8. Lattice structures

In a lattice any pair of type labels must have a unique maximal common subtype (see Sowa (1984), Ass. 3.2.5). This condition ensures that the result of unification-like operations is well-defined. It can be hurt easily. Yet, it is possible to transform any partial order into a lattice automatically (see e.g. Aït-Kaci et al. 1989) by introducing new nodes into the hierarchy (depicted in (b)). This solution, however, may be problematic:

- The automatic creation of additional nodes can lead to an exponential growth of the number of nodes. This depends on the structure of the partial order.

- The semantics of the newly created nodes may be unclear, at least they have no user-defined name. A user might be confused about their meaning and purpose.

Therefore, we favor an interactive solution that is based on an "intelligent" workbench. A demon has to watch out for possible violations of the lattice structure, display all necessary information and make proposals to the user about how to solve the problem.

3.4 Inference

The system should be able to answer interesting questions concerning the contents of the analysed utterances. This is one of the most important advantages of producing a semantic representation at all. Therefore, we have implemented the Prolog-like resolution method by Jean Fargues et al. (1986). A small example shows how the rule-based mechanism can be used. In Fig. 9 three example rules are shown.

```
*** RULE P1:            *** RULE P2:            *** RULE X1:
>>> IF                  >>> IF                  >>> IF
  [FREE_AIR: _x]->        [ANAT: _x]->            [ANAT: _x]->
    (BELOW)->              (XCHAR)->               (AATTR)->
    [DIAPHRAGM: _z].       [SHADE: _y].            [SHADED : _y].
>>> THEN                >>> THEN                >>> THEN
  [ABDOMEN: _y]->         [ANAT: _x]->            [ANAT: _x]->
    (PATHO)->              (PATHO)->               (XCHAR)->
    [FREE_AIR: _x].        [PATHO_ALT: _z].        [SHADE: _z].
```

Fig. 9. Example rules

Let us now assume that the following two utterances have been analysed:

"freie Luft unter dem Zwerchfell" ("free air below the diaphragm")	"Lunge verschattet" ("shaded lung")
`[FREE_AIR: #8]-> (BELOW)->[DIAPHRAGM: #7].`	`[LUNG: #9]-> (AATTR)->[SHADED: #10].`

Fig. 10. Example sentences

Now we ask a question that can be roughly paraphrased as "Which anatomical objects show a pathological alteration?":

```
QUESTION: [ANAT]->(PATHO)->[PATHO_ALT].
*** Selected Rule: P1
RESULT: [ABDOMEN: _79]->(PATHO)->[FREE_AIR: #8].
*** Selected Rule1: P2
*** Selected Rule1: X1
RESULT: [LUNG: #9]->(PATHO)->[PATHO_ALT: _58].
```

Two solutions are found: the abdomen has a pathological alteration "free air", and the lung has some alteration that cannot be specified at the moment. The system finds the solutions by applying the rules backwards in a goal-driven fashion. The graphs are matched by the symmetric matching operation as defined by Fargues (1986, p. 77).

3.5 Generating Expectations

For the generation of expectations 3 knowledge structures have been defined, ranging from general to specific:

- **Checklist:** unordered set of topics that may occur in a specific radiological examination.
- **Plan** (or "speaker-specific checklist"): ordered sequence of topics or subtopics.
- **Script:** ordered sequence of propositions (without individual referents) that describe a standard finding (speaker-specific).

A checklists and a plan have the syntactic form of a schema, but are interpreted in a specific way. A script is a sequence of conceptual graphs. Such a graph is matched with an incoming graph by the *matching operation* of Fargues (1986).

The knowledge structures defined above form the basis for a multi-layered prediction.

After a prediction is generated on the conceptual level, the concepts are mapped to the lexical level. The resulting set of words is used to constrain the set of word hypotheses that are produced by the lower levels of speech recognition components. For a more detailed description see Schröder (1992).

4 Implementation

The METEXA system is implemented in Prolog (ProLog by BIM) on a SUN SparcStation.. Current activities concentrate on the development of more sophisticated planning strategies. The basic operations for conceptual graphs have been implemented by Shahram Gharaei-Nejad. I am greatful to Lutz Euler for many helpful contributions, especially concerning lattices and related structures.

References

Aït-Kaci, H. et al. (1989) Efficient Implementation of Lattice Operations. *ACM Transactions on Programming Languages and Systems 11 (1)*, pp. 115 - 146.

Billi, R., Buttafava, P., De Stefani, P., Gamba, M., and Voltolini, D. (1991) Computer-Aided, Voice-based, Medical Report Preparation: An Application To Radiology. In *EURO-SPEECH-91*, pp. 961.

Fargues, J., Landau, M.C., Dugourd, A., and Catach, L. (1986) Conceptual graphs for semantics and knowledge processing. *IBM Journal of Research and Development 30*, (1), pp. 70-79.

Lehrberger, J. (1986) Sublanguage Analysis. In *Analyzing Language in Restricted Domains: Sublanguage Description and Processing*. Lawrence Erlbaum Associates, Grishman, R. and Kittredge, R., pp. 19-38, Hillsdale, NJ.

Matsumoto, Y., Tanaka, H., Hirakawa, H., Miyoshi, H., and Yasukawa, H. (1983) BUP: A Bottom-Up Parser Embedded in Prolog. *New Generation Computing 1*, 145-158.

Mery, C., Normier, B., and Ogonowski, A. (1987) 'INTERMED': A Medical Language Interface. In *AIME 87*, Fox, J., Fieschi, M., and Engelbrecht, R., Springer-Verlag, Berlin, pp. 3-8.

Pazienza, M.T. and Velardi, P. (1987) A structured representation of word-senses for semantic analysis. In *European ACL 87*, pp. 249-257.

Ranum, D.L. (1988) Knowledge Based Understanding of Radiology Text. In *The 12th Annual Symposium on Computer Applications in Medical Care*, Greenes, R.A., IEEE Computer Society Press, pp. 141-145.

Sager, N., Friedman, C., and Lyman, M.S. (1987) *Medical Language Processing: Computer Management of Narrative Data*, Addison-Wesley, Reading, MA.

Schröder, M. (1991) Ein semantisch-gesteuerter Bottom-Up-Parser in Prolog. Tech. Rept. FBI-HH-M-195/91, Mitteilung, Universität Hamburg, FB Informatik, Mai 91.

Schröder, M. (1992) Supporting Speech Processing by Expectations: A Conceptual Model of Radiological Reports to Guide the Selection of Word Hypotheses. *KONVENS-92, 1. Konferenz "Verarbeitung natürlicher Sprache"*, 7. - 9. Oktober 1992.

Sowa, J. F. (1984) *Conceptual Structures: Information Processing in Mind and Machine*. Addison-Wesley, Reading (Mass.).

Sowa, J. F. and Way, E. C. (1986) Implementing a semantic interpreter using conceptual graphs. *IBM Journal of Research and Development 30*, (1), pp. 57-69.

Sowa, J. F. (1988) Using a Lexicon of Canonical Graphs in a Semantic Interpreter. In *Relational Models of the Lexicon*. Cambridge Univ. Press, Evens, M., pp. 113-137.

Sowa, J. F. (1991) in a contribution to the Conceptual Graph mailing list, 8. May 1991.

Velardi, P., Pazienza, M.T., and Giovanetti, M.D. (1988) Conceptual graphs for the analysis and generation of sentences. *IBM Journal of Research and Development 32*, (2), 251-267.

Open Systems Interconnection Abstract Syntax Notation: ASN.CG

Timothy R. Hines

Computer Science Telecommunications Program
University of Missouri-Kansas City
4747 Troost
Kansas City, Mo. 64110-2499
THINES@CSTP.UMKC.EDU

Abstract. Open Systems Interconnection (OSI) enables communication between computers using a common set of protocols (rules) and structured data [5,9]. ASN.1 (Abstract Syntax Notation One) is a formal notation allowing communication between two different (or same) computer systems; each system agrees on what the data will mean when transmitted even though each system can choose its own local representation of the exchanged information [3]. ASN.1 has several drawbacks including usability of the language and its macro facility.

A new notation ASN.CG (Abstract Syntax Notation using Conceptual Graphs) using conceptual graphs for representing abstract syntaxes is proposed. ASN.CG enhances usability and eliminates the macro facility.

1   ASN.1 Concepts and Problems

ASN.1 is a compilable formal notation for specifying data passed between the Application and Presentation layers in OSI [1, 2, 5, 9]. ASN.1 provides the application writer the ability to define types and values (abstract syntaxes) and develop applications to manipulate those abstract syntaxes. *Abstract syntax* describes a collection of related values independent of any hardware. For example, when we represent an integer we are not concerned whether a short or long integer is being represented. Abstract syntaxes are input to the ASN.1 compiler, Figure 1. If macros are included in the abstract syntaxes they are expanded before being compiled. The compiler outputs a set of declarations or *concrete syntaxes* (data structures expressed in the local machine, or target language) and *encode* and *decode routines*, Figure 1. The encode routines receive an instance of a local declaration (local value) and apply a basic encoding rule (BER) converting it from the target language to a transfer syntax [4]. *Transfer syntaxes* are bit streams transmitted over the communication line. When the end system receives the transfer syntax, the decode routine maps it to a local value of its target language. BER encodes values into a triple -- type, length, and value (TLV). Figure 2 shows two end systems connected by a communication line with transfer syntaxes transmitted between these two systems.

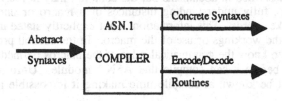

Fig. 1. ASN.1 compiler inputs and outputs.

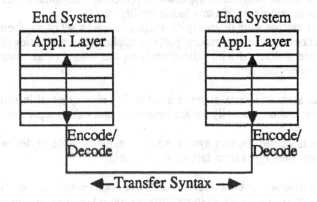

Fig. 2. Transfer syntaxes passed between two end systems.

Modules are used to encapsulate new data type definitions, possibly value assignments, and macro definitions. Modules may export these types, values, or macros, and may also import types, values, or macros exported by other modules. New types or values defined by the modules are constructed from predefined ASN.1 types or new user-defined types. The predefined ASN.1 types are *simple types, string types, time types, any types, object types, structured types,* and *tagged types* [3]. Additional keywords *optional, default,* and *components of* may be used in conjunction with these ASN.1 data types.

Macros are used to define templates (i.e. information about objects), and parameterized types, and to indicate constraints on values. Macros do textual substitution and can generate types and values. Macros include a type and value notation defining the grammar rules to constrain the types and values. The type and value notation can be defined using predefined or user-defined ASN.1 types and values.

One of the major drawbacks of ASN.1 has been the macro facility. Even though the macro facility is very powerful and useful, it is difficult to fully specify a macro compiler. There are two issues involved in the macro problem: semantics and grammar rules. In the first problem, the underlying semantic information about the macro may be described in documents but the compiler may not have access to this information, e.g. fully understanding instances of a macro or understanding the meaning of the keywords. This information is not explicitly stated in the macro but is important to the workings or use of the macro. In the second problem, a macro compiler needs to know the grammatical rules of the macros defined and used in an ASN.1 module before it can compile that ASN.1 module. Unfortunately, all the macros may not be known at compile time making it impossible to know all the grammar rules.

The standards use macros extensively for RO-Notation and ASDC. RO-Notation (Remote Operation Notation) and ASDC (Abstract Service Definition Convention) provides a set of tools, specified in ASN.1 macros, for specifying the abstract syntax of an interactive application protocol. Elimination of the macro facility requires development of a translation facility.

ASN.1 macros were originally developed to meet a set of requirements which have arisen in conjunction with general-purpose application capabilities [6, 8]. These requirements are concerned with fully specifying the values of an abstract syntax. There are four requirements:

1. The abstract syntax notation must define new types of information objects, and must specify the attribute types and values representing it.

2. Users of the abstract syntax notation must be able to define their own attribute types from information objects.

3. The abstract syntax notation must leave appropriately shaped "holes" in its protocol to carry arbitrary information relating to an information object.

4. The abstract syntax notation must be able to link between fields in the protocol and information relating to an object, i.e. an attribute type would have a constraint of the type of format that the a value can have.

ASN.1 is only adequate in meeting requirement 3, i.e. leaving holes in the protocol by using the ASN.1 type ANY. ASN.1 partially meets requirement 4, linkage between the fields of a protocol, by use of ANY DEFINED BY. The ASN.1 type EXTERNAL also contributes to meeting requirement 3. Although, current researchers who are modifying ASN.1 are eliminating the use of the EXTERNAL type because it provides no notational indication of what should be stored in it, creating an infinite list of possible values.

ASN.1 macros were developed to partially meet requirements 1 and 2. Remote operations (RO-Notation and ASDC) were originally invented to meet these requirements. However, the main problem is there is no connection between these macro definitions and the role they have in defining the linkage in requirement 4.

The main thrust in current research with abstract syntax notation is the elimination of macros because of the difficulty in understanding macros by designers and the difficulty in processing them by machine [6, 8]

2 ASN.CG

In theory more than one abstract syntax notation can be used in OSI [5]. ASN.CG (Abstract Syntax Notation using Conceptual Graphs), a new notation for representing abstract syntaxes, enhances usability and eliminates the macro facility. ASN.1 does not capture as much semantic information as is possible with conceptual graphs. Conceptual graphs can express all constructs available in ASN.1.

The requirements listed in section one are fulfilled by the notation of ASN.CG in an independent manner from ASN.1. An informal discussion of the constructs and techniques of ASN.CG are presented by providing ASN.1 examples and ASN.CG equivalents. ASN.CG provides a notation for defining abstract syntaxes for types and values. Simply defining a new notation for defining abstract syntaxes is not adequate, there must be a translation facility translating previously defined macros into this new notation. The translation facility is necessary because macros have been used extensively in OSI. Section one briefly discussed RO-Notation and ASDC which defines a set of macros for use in designing remote operations of an OSI application context and for describing a distributed application both in structure and operationally. RO-Notation and ASDC has been used to define ISO OSI application standards, e.g., the directory service standard.

ASN.CG's notation meets all four of the requirements listed in section one. Conceptual graphs are used to fullfill requirements 1 and 2, i.e. defining new types and variations from these types. Concepts in conceptual graphs meets requirement 3, i.e. specifying arbitrary concept types. The notation used by conceptual graphs meets requirement 4, i.e. constraining the type of format for a concept.

Figure 3 is an example of an ASN.1 module defining new abstract types and values. The module "Company" contains three new types Name, Employees and Country. The words INTEGER and STRING are predefined ASN.1 types. A new value called valNameA is defined from the type Name and having the value "IBM".

```
Company DEFINITIONS ::=
BEGIN
        Name ::= STRING
        valNameA Name ::= "IBM"
        Employees ::= INTEGER
        Country ::= STRING
END
```

Fig. 3. ASN.1 Company module defining types and values.

Figure 4 shows the conceptual graph representing the types and values in the module Company in Figure 3.

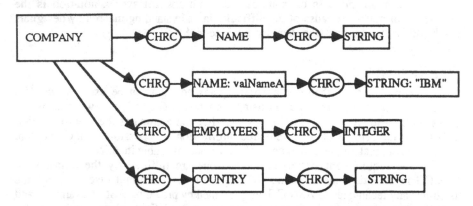

Fig. 4 Conceptual graph for the ASN.1 module COMPANY.

Figure 5 and 7 are examples of ASN.1 macros. The macro in Figure 5 is called OPERATION which is used to specify an operation for a particular application. Operations are invoked in application associations. The macro contains arguments passed to an object, results returned, possible errors, and linked operations which specify possible child operations.

```
OPERATION MACRO ::=
BEGIN
    TYPE NOTATION       ::= Argument Result Errors LinkedOps
    VALUE NOTATION   ::= value(VALUE CHOICE {
                                    localValue INTEGER,
                                    globalValue OBJECT IDENTIFIER})
    Argument            ::= "ARGUMENT" NamedType I empty
    Result              ::= "RESULT" ResultType I empty
    ResultType          ::= NamedType I empty
    Errors              ::= "ERRORS" "{" ErrorNames "}" I empty
    LinkedOps           ::= "LINKED" "{" LinkedOpNames "}" I empty
    ErrorNames          ::= ErrorList I empty
    ErrorList           ::= Error I ErrorList "," Error
    Error               ::= value(ERROR) I type
    LinkedOpNames       ::= OperationList I empty
    OperationList           ::= Operation I OperationList "," Operation
    Operation           ::= value(OPERATION) I type
    NamedType           ::= type I identifier type
END
```

Fig. 5. ASN.1 OPERATION macro.

The OPERATION macro in Figure 5 can be represented by a conceptual graph. Figure 6 is the conceptual graph representing the OPERATION MACRO. The conceptual graph depicts the type in a cleaner and more concise form. For example, there are no grammar rules for defining a list, the notation { } specifies a set.

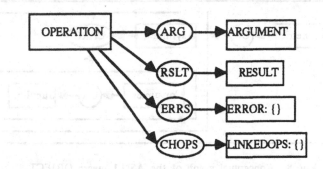

Fig. 6. Conceptual graph for the ASN.1 macro OPERATION.

Figure 7 provides another example of an ASN.1 macro. The macro is called OBJECT which is used to define new abstract objects. Object types and values can be defined from the OBJECT macro. The object macro is described by a list of ports. Each port can be symmetric or asymmetric. If it is asymmetric then it can be a consumer or supplier port.

```
OBJECT MACRO ::=
BEGIN
        TYPE NOTATION      ::= "PORTS" "{" PortList "} l empty
        VALUE NOTATION     ::= value(VALUE OBJECT IDENTIFIER)
        PortList           ::= Port "," PortList l Port
        Port               ::= value (Port) PortType
        PortType           ::= Symmetric l Asymmetric
        Symmetric          ::= empty
        Asymmetric         ::= Consumer l Supplier
        Consumer           ::= "[C]"
        Supplier           ::= "[S]"
END
```

Fig. 7. ASN.1 OBJECT macro.

The OBJECT macro can be represented with a conceptual graph. Figure 8 is the conceptual graph representing the OBJECT macro.

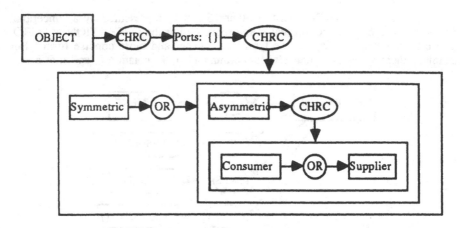

Fig. 8. Conceptual graph of the ASN.1 macro OBJECT.

The OPERATION macro in Figure 6 is an abstraction defining a template for an operation. Operation types and values can be defined using the OPERATION macro. For example, the directory service standard defines an operation called SEARCH. This SEARCH operation, in ASN.1 syntax, specifies values for argument, result, and errors, Figure 9.

```
Search ::= OPERATION
             ARGUMENT    SearchArgument
             RESULT      SearchResult
             ERRORS      {attributeError, nameError, serviceError,
                          referral, abandoned, Security, Error}
```

Fig. 9. Search operation defined using the ASN.1 OPERATION macro.

The SEARCH operation is represented by the conceptual graph shown in Figure 10. The referents of the concepts in the graph are filled by the specific values from ARGUMENT, RESULT, and ERRORS. The concepts in the conceptual graph take on the values listed in the Search macro definintion. The concepts ARGUMENT, RESULT, and ERRORS take on the values SearchArgument, SearchResult, and the list of errors attributeError, nameError, serviceError, referral, abandoned, Security, and Error, respectively.

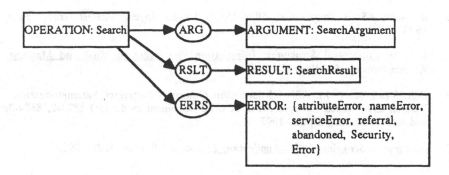

Fig. 10. Search operation defined by a conceptual graph.

3 Summary

This research proposes a new abstract syntax notation for OSI based on conceptual graphs. Several drawbacks of ASN.1 were discussed including the semantics and grammar rules of macros, usability of the language, and the EXTERNAL type. In section one, ASN.1 was shown that it does not meet the four requirements for fully specifying an abstract syntax notation. ASN.1 macros were introduced to partially meet some of these requirements, but failed to meet all of the requirements and introduced additional problems. Section two provided an informal discussion of ASN.CG showing that the notation meets all four of the requirements listed in section one. The notation of ASN.CG, elimination of the macro facility and the EXTERNAL type, and greater semantics capture were addressed by this work.

Work in progress includes the formalization of the notation used in ASN.CG, a translation facility to translate macros to ASN.CG, and a compiler which takes as input ASN.CG conceptual graphs and returns as output concrete syntaxes and encode/decode routines.

References

1. P. Gaudette: A Tutorial on ASN.1, May 1989.

2. J. Henshall, S. Shaw: OSI Explained, Ellis Horwood, 1990.

3. ISO 8824, Information Technology - Open Systems Interconnection - Specification of Abstract Syntax Notation One (ASN.1), 1990.

4. ISO 8825, Information Technology - Open Systems Interconnection - Specification of Basic Encoding Rules for ASN.1, 1990.

5. M.T. Rose: The Open Book: A Practical Perspective on OSI, Prentice-Hall, 1990.

6. B. Scott: ASN.1 Presentation, OIW Upper Layers Special Interest Group, June, 1991.

7. J. Sowa: <u>Conceptual Structures: Information Processing in Mind and Machine</u>, Addison-Wesley, Reading, MA., 1984.

8. D.A. Steadmam (Ed.): ASN.1 Information Objects, Constraints, Parameterization - Tutorial and Worked Examples, ISO Tutorial Document to the ISO 8824-2, 8824-3, and 8824-4 working Drafts, 1992.

9. A. Tang, S. Scoggins: <u>OSI - Application Protocols</u>, Prentice-Hall, 1992.

Bridging Accounting and Business Strategic Planning Using Conceptual Graphs

Simon Polovina

Department of Computer Studies, Loughborough University of Technology (LUT),
United Kingdom LE11 3TU

Abstract. This paper reveals the existence of a gateway between the two seemingly disparate subjects of *accounting* and *strategic planning*. This is achieved by examining two important existing methodologies, *events accounting* and *cognitive mapping*, and restating them both in *conceptual graphs*. Also, for the bridge to be viable the framework must be usable by accountants and strategic planners. To achieve this usability a conceptual graph-based *conceptual analysis and review environment* (*CARE*) is proposed. From all this, a way forward exists whereby a business can develop and maintain a corporate knowledge-base that permeates throughout more of its diverse activities.

1. Introduction

To enable an organisation to have more control over its affairs, it would be very desirable to bring together into one positive framework two seemingly disparate yet important parts of business activity. The first area, *accounting*, essentially provides an established basis for controlling numerically-based problems[1] whereas the second, *strategic planning*, is characterised by fundamental but highly subjective problems that do not lend themselves to accountancy's quantitative techniques. The paper reveals the existence of a gateway between the two by examining an influential methodology in each area, and then restating them both in *conceptual graphs* [15]. The first methodology, *events accounting*, is a significant attempt to capture the conceptual basis underlying accountancy [5,9]. The second, based on *cognitive mapping*, is a leading knowledge-based strategic planning analytical tool [3].

2. Events Accounting

The major benefit of adopting a structured model of a problem is so that such a model, by its inherent nature, draws out all the problem's relevant parameters from which a solution can be investigated fully. Contrast this with a written or spoken text discussion of the problem where it is well known that ambiguity and obfuscations can occur easily. This 'natural language' interpretation of problems may be the most flexible and easily followed, but without at least a basis in some structured form it can be dangerously erroneous. Hence the emergence of disciplines such as accounting that attempt to model the dynamics of economic activity in a structured way. The model,

---

[1] Although the author has previously discussed the modelling of non-numeric elements within accountancy [10,11].

of course, must also be structured on a suitably principled basis. Otherwise it will omit or misinterpret the salient issues of the problem situation.

2.1 The Double Entry Bookkeeping Model

It was with the above in mind that the traditional model of accountancy, the *bookkeeping model*, was developed in the Middle Ages [7]. The principle behind this model is *economic scarcity*. In other words for every benefit a sacrifice has to be made. For example, the benefit of a business owning its office is sacrificing $1,000,000 that could be employed elsewhere; a book prepared by its author researching a new exciting area in semantic understanding may have involved that author deciding against many complex yet important alternatives, such as the costs of, say, not participating in his or her growing family. These 'transactions' occur because the decision-maker makes a *value judgement* that the benefits outweigh the costs.

The bookkeeping model appears simple but rigourous. Fundamentally, instead of recording one amount per transaction it records two: A 'debit' and a 'credit'. Moreover these amounts are complementary to one another, hence they 'balance' against each other. An accounting 'balance sheet' is merely the aggregate of all these debits and credits. The rigourousness derives from this principled 'double entry' structure so that each benefit is accounted for by a cost and vice versa. Hence every gain is matched to a sacrifice.

2.2 Problems with the Bookkeeping Model

However on deeper investigation the double entry bookkeeping model is unlikely to capture all these economic value trade offs. Say the business in the first example above decides to sell its office. This transaction can be recorded easily by the elementary bookkeeping entries "DEBIT Cash $1,000,000, CREDIT Fixed Assets $1,000,000". The second, preparing the book, is simply too qualitative to be recorded by the bookkeeping model yet the author may want to know clearly about *all* the actual costs and benefits of such a transaction. This neglect on the part of the bookkeeping model is elaborated on below.

Errors of Omission. The threshold where the bookkeeping model may break down is perhaps lower than may be thought. Reconsidering the first example about the office, the value of selling the current office may be the purchase of cheaper offices for $500,000. In this case the bookkeeping is basically "DEBIT Fixed Assets $500,000, CREDIT Cash $500,000". Now say, by spending the remaining $500,000 elsewhere, the business generates a revenue of $600,000. On aggregate in the balance sheet the business's money worth then increases by $100,000 (Represented primarily as "DEBIT Profit and Loss Account $100,000, CREDIT Reserves $100,000"). However if the value of the current office is retaining key employees through a comfortable work environment then, as in the author example above, the bookkeeping model is inappropriate. Therefore the double entry bookkeeping model is easily liable to make significant *errors of omission*.

Errors of Commission. Furthermore the bookkeeping model could mislead. Reconsidering the 'preparing the book' example the value may be viewed as the more easily quantified cost of the author ceasing to conduct consultancy work at $2,500 a time instead. This revenue would have been recorded by the bookkeeping model on an ongoing basis. However the book might bring its author satisfaction of a deep desire for an enhanced reputation amongst peers. Unless this can be translated into a cash benefit the bookkeeping model would not record these judgements and thereby leave a 'loss' of $2,500. By choosing to author the book the decision-maker *qualitatively* has to justify, *against the grain of the bookkeeping model's assessment of value*, why that $2,500 has been forsaken even though this may the lesser value item. Therefore the double entry bookkeeping model, taken too literately, can also readily lead to significant *errors of commission*.

2.3 The Nature of the Events Accounting Model

The above problems are familiar to most accountants. The *Economist*, for instance, summarises the difficulties accountants have in attaching a monetary value to 'intangible' assets such as product brand names [4]. The *events accounting* model [5,9] is intended to overcome the above obstacles. Unlike the bookkeeping paradigm, events accounting attempts to capture the qualitative dimensions of economic scarcity. The model is shown by the diagrams that make up *Fig. 1. Fig. 1(a)* shows the events accounting model as an *entity-relationship diagram* [1]. In line with the earlier discussion about the desirability of structured models, Chen argues that the pictorial nature of entity-relationship diagrams are particularly useful in structuring problems qualitatively stated in natural language [2]. Given this actuality, events accounting represents a powerful means of recording scarcity as more than a monetary measure. Setting aside its 'dotted' part for the moment, *Fig. 1(a)* reveals the fundamental links between an 'economic resource', which means some exchangeable item of value, and the parties which create the 'economic event' that causes the economic resource to be exchanged.

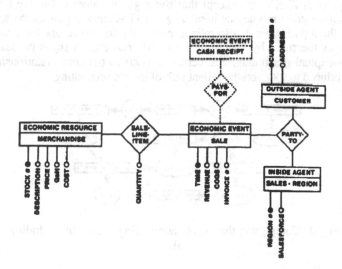

Fig. 1(a). The *events accounting* model in entity-relationship form [7].

2.4 The Events Accounting Model as Conceptual Graphs

Why. Sowa, in response to Chen's argument, adds conceptual graphs can further structure the dimensions of natural language-based problems [16]. Conceptual graphs extend entity-relationship diagrams by adding the capacity of first-order logic in a pictorial way [17]. It is therefore sensible to transform the events accounting model into conceptual graphs form.

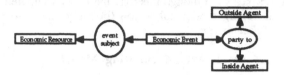

Fig. 1(b). Transformation of the events accounting model into supertype conceptual graphs.

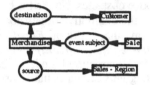

Fig. 1(c). Transformation of the events accounting model into subtype conceptual graphs.

Restating the Events Model. The conceptual graph in *Fig. 1(b)* represents this transformation. This graph is basically a conceptual graphs reproduction of the general supertype parts of *Fig. 1(a)*, except that the arguably more definitive term 'event subject' is substituted for 'sale line item'. *Fig. 1(c)* basically expresses the specialised subtypes of the top diagram. It also refines 'party to' into 'source' and 'destination', and thereby shows the route by which the economic resource changes possession. For both the conceptual graph diagrams, certain aspects in the entity-relationship model are not reproduced to focus on the salient nub of events accounting.

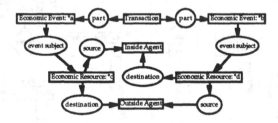

Fig. 2. Completing the events accounting model by including duality.

Duality. However the conceptual graphs within *Fig. 1* do not yet fully capture events accounting. To achieve this the *duality* principle of economic scarcity needs to be considered. Duality is defined by the statement "for every benefit there has to be a

316

sacrifice made". The events accounting model indicates duality in the 'dotted part' of *Fig. 1(a)*. This 'cash receipt pays for the sale' is really a shorthand to make that top diagram concise. For instance 'party to' should also connect to 'cash receipt' because it is also part of the exchange. *Fig. 2* reveals this duality in full at the supertype level. Both sides of duality are shown in *Fig. 2* by linking the economic events to the same transaction. As already indicated, the 'Economic Resource' concepts may be specialised to any quantitative or qualitative concept describing an item of value. As one value describes the benefit in the transaction, the other value depicts what had to be sacrificed for that benefit. An example scenario is portrayed in *Fig. 3*.

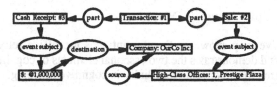

Fig. 3. Apparent duality of selling offices.

Intangible Values. *Fig. 3* develops the earlier example of a business wishing to realise cash by selling its expensive offices, with a view perhaps to purchasing cheaper ones. The diagram shows a '$1,000,000' cash benefit to 'OurCo Inc.' for the loss of their high-class offices at '1, Prestige Plaza'. As such, it may easily be recorded by the bookkeeping model. However this may not be the full picture. In addition to the loss of the present offices, there may be an intangible loss due to disgruntled key employees leaving the corporation as a consequence. As discussed earlier, such intangible values remain outside the bookkeeping model. Nevertheless the business would require its inclusion into the sphere of consideration should this factor, say, determine the survival of the business itself. The conceptual graph of *Fig. 4*, manages to record this qualitative cost.

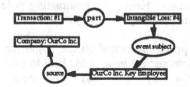

Fig. 4. Further cost-side duality of selling offices.

2.5 Summary: Present Accounting in a Knowledge-Based Form

The events accounting model, in summary, does not restrict the recording of transactions to only those that can be captured by numeric monetary measures. Thereby its wider scope enables qualitative value judgements to be modelled and hence be gauged along with the quantitative elements. In databases, such parameters are defined in data fields which can be then subjected to some selection criteria. Within any knowledge-base, in addition to this 'lookup' facility, there should exist the ability to make adequate inferences for new knowledge. Hence if accounting transactions can be presented in a knowledge-based form they can be subjected to this inferencing.

Peirce logic, which is the inference mechanism associated with conceptual graphs, will be seen to be a powerful part of the pivotal area that performs this role. This area, strategic planning, is addressed in the following discussion on *cognitive mapping*.

3. Cognitive Mapping

Cognitive mapping is a practical technique used by strategic planners to structure the highly subjective problems that characterise strategic planning [3]. This technique stems from Kelly's theory of *personal constructs* [6].

3.1 The Nature of Cognitive Mapping

The small cognitive map shown in *Fig. 5* continues the office relocation example discussed so far, and demonstrates the two essential elements of cognitive maps. These elements are referred to as 'concepts' and 'links' in cognitive mapping.

buy down-market offices at 13, Sidestep Row
... retain high-class offices at 1, Prestige Plaza ────────▶ disgruntled key employees

Fig. 5. An example showing the essential elements of a cognitive map.

Concepts in Cognitive Mapping. Each cognitive mapping concept is represented as one 'emergent pole', which describes one side of the problem, and a 'contrasting pole' which is meant to focus the concept by a meaningful contrast to the emergent pole. In *Fig. 5* the left-hand concept contains the emergent pole 'buy down-market offices at 13, Sidestep Row'. The contrasting pole, separated from the emergent pole by '...', is 'retain high-class offices at 1, Prestige Plaza'. Where a contrasting pole is unspecified, as in the case of 'disgruntled key employees', then the contrasting pole is determined to be 'not <emergent pole>'. Hence the contrasting pole for the latter concept becomes 'not disgruntled key employees'.

Links in Cognitive Mapping. Attached to the cognitive mapping concepts are links which are used to demonstrate how poles lead to other poles. The link is the arrow which leads from the 'buy cheaper offices at 13, Sidestep Row ... retain current offices at 1, Prestige Plaza' concept to the concept with the emergent pole 'disgruntled key employees' and default contrasting pole 'not disgruntled key employees'.

3.2 Cognitive Mapping as Conceptual Graphs

From the above the down-market offices lead to disgruntled key employees whereas the current offices lead to these employees not being disgruntled. After refining this statement from the elementary form illustrated by *Fig. 5* into the more logically advanced conceptual graphs, the choice of offices and its consequence may be remodelled as given by the conceptual graphs of *Fig. 6*.

Restating the Cognitive Map Concepts. In *Fig. 6(a)* the pair of specified poles denoting the choice of offices can be restated as a conceptual graph by placing

318

each pole into a separate conceptual graph concept and together surrounding them within a Peirce negative context. The single specified disgruntled employee pole and its unspecified 'not' contrast is also restated in *Fig. 6(a)* although, as can be seen, this graph turns out to be a tautology.

Fig. 6(a). Conceptual graphs for the concepts of the example cognitive map.

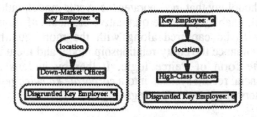

Fig. 6(b). Conceptual graphs for the links of the example
cognitive map.

Restating the Cognitive Map Links. *Fig. 6(b)* demonstrates, in conceptual graphs, the cognitive map links as two implications based on the Pierce logic operations of deiteration and double negation. The conceptual graphs show both implications to be generalised, assuming this step is valid, to show that for any down-market offices the key employees located there are disgruntled whilst in any high-class offices they are not (*modus ponens*). Similarly if such employees are not disgruntled then they are not located in down-market offices (*modus tollens*).

Why. Conceptual graphs provide the contrast in cognitive map concepts by stating that one pole must be false if the other is true. However when both poles are specified the converse, that if one is false the other is true, cannot be determined. This distinction, which is not clear from the elementary cognitive mapping model, is intended. The business (referred to earlier as 'OurCo Inc.') may, in the event, dispose of 1, Prestige Plaza and obtain offices other than at 13, Sidestep Row or even not any other down-market offices for that matter. Hence it may be quite wrong to assert that they *will* be at either 13, Sidestep Row or 1, Prestige Plaza. Furthermore conceptual graphs allow each pole to be generalised or specialised to differing degrees so enabling a potential continuum of contrast. This dimension happens to accord more precisely with Kelly's personal construct theory mentioned earlier [6]. There may be, for example, a more general contrasting pole to a more specific emergent one. Ultimately the most general contrasting pole is the 'not <emergent pole>'. As for the cognitive mapping links, the modus tollens inference is arguably not evident in the elementary form given by *Fig. 5*.

3.2 Summary: Conceptual Graphs Enrich Cognitive Maps

As well as the ability to a) generalise and specialise problem scenarios to the appropriate degree and b) vividly express modus tollens, conceptual graphs further reveal that c) there is no real distinction between the concepts and links of cognitive mapping anyway. As should be evident from a careful examination of *Fig. 6*, these links turn out to be like emergent and contrasting poles. Hence conceptual graphs remove the arbitrary choice as to what a link or a pole. Moreover strategic planning problem situations can now be captured alongside the accountancy domain through the single medium of conceptual graphs. This opens the way towards suggesting a bridge between the two techniques.

4. Suggesting a Bridge

An accounting methodology that incorporates the events model, rather than merely the bookkeeping model, enables the more qualitative elements of transactions based on economic scarcity to be captured along with the more quantitative elements. Conceptual graphs enhance the entity-relationship model and include a visual method of inference in the form of Peirce logic. It thereby widens an organisation's numeric-based financial record systems into a *corporate knowledge-base* that can now include the parameters of strategic planning.

4.1 The Technical Bridge

However, given the discussion so far in this paper, it is not expected that inference will play a major role for the qualitative elements in accounting transactions. Such inference becomes significant instead within strategic decision-making where, as also discussed, it in fact plays the major part. Yet presenting accounting events as conceptual graphs because of the inference capability remains worthwhile. This is because their qualitative aspects can then be interrogated and reviewed by the conceptual graphs derived from strategic analysis. For example, should the graph ¬[[Disgruntled Key Employee]] be asserted as a parameter of a strategic decision then, given the graphs so far for the office location example (and OurCo Inc. Disgruntled Key Employee < Disgruntled Key Employee), the transaction disposing the offices at 1, Prestige Plaza would be blocked. Such operations describe the technical half of the bridge.

4.2 The Usability Bridge

As the nature of much business activity is not only qualitative but also constantly changing, the corporate knowledge-base is unlikely to fully keep up with the tacit and implicit parts of business knowledge [8]. Therefore, to facilitate sufficiently rapid analysis and review, the corporate knowledge-base must be usable by accountants and strategic planners themselves to complete the remaining half of the bridge.

CARE. To achieve the usability bridge the author proposes a human-computer interface which may most aptly be referred to as a *conceptual analysis and review environment*, or *CARE*, through which the accountant or strategic planner can interact with the corporate knowledge-base. The author suggests CARE could be conceptual graph-based. This is because a) this paper identified that accountants and strategic planners evidently appreciate the need for structuring problems, b) business information professionals such as accountants use structured diagrams in the general course of their work [14,18] and c) negative contexts happen to be similar to the way the accountant's bookkeeping model negates numeric values by enclosing them within rounded brackets [7].

Safeguards. There must be, nevertheless, certain safeguards. Terms like ∀, ∃ and ¬, for instance, are unlikely to be understood by these non-logicians and should therefore be avoided. A study conducted by the author discovered that even the linear form of conceptual graphs is too abstruse for users [13]. Hence CARE would employ conceptual graphs in display form. In addition, points of the theory itself must be as clear as possible. The author has jointly attempted such a clarification [12]. Ideally, intricate parts of theory should be handled by the machine and be transparent to the domain expert without making the power of CARE too trivial. Allied to this, CARE should prevent the domain expert from drawing incorrectly structured graphs.

Fundamental Problems. Finally there will be, of course, fundamentally unavoidable problems through allowing the user to build more advanced models. For instance, 'A' affects 'B', 'B' affects 'C', 'C' affects 'D', and 'D' affects 'A'. In some cases this may be a valid recursion but in others it may not. This dilemma is especially significant in conceptual graphs given that knowledge can be generalised and specialised at many levels. Such obstacles should be mitigated as far as is possible given that it going to be very difficult to determine valid from invalid cases. For this example some kind of user-warning facility may help. Accountants, through their experience in the use of spreadsheets for instance, tend to be acquainted with the nature of circular arguments albeit at a much more straightforward level.

5. Concluding Remarks

From all the above, conceptual graphs have revealed the fundamental difference between accounting and strategic planning is the former is essentially *descriptive* in nature and can thereby be captured by knowledge representation alone whereas the latter is *prescriptive* and requires the dynamics of inference. However the benefit is not one way: Accounting can offer strategic planning a firmer conceptual basis from which its dynamic models can be built. With CARE, conceptual graphs are potentially usable by domain experts themselves. These domain experts would be accountants and strategic planners. Hence a way forward exists whereby a business can develop and maintain a corporate knowledge-base that permeates throughout more of its diverse activities.

Acknowledgements

The author would like to thank Paul Finlay, Loughborough Business School, LUT; John Heaton, Computer Studies, LUT; Chris Hinde, Computer Studies, LUT; and Malcolm King, Loughborough Business School, LUT for their invaluable help towards the content of this paper.

References

1.   Chen, Peter P. (1976) "The Entity-Relationship Model–Toward a Unified View of Data", *ACM Transactions on Database Systems*, 1(1), 9–36.

2.   Chen, Peter P. (1985) Position Statement. *From* Sowa, John F.; Chen, Peter P.; Freeman, Peter; Salveter, Sharon C.; Schank, Roger C. (1985) "Mapping Specifications to Formalisms: Panel Session", *Proceedings of the 4th International Conference on Entity-Relationship Approach*, IEEE Computer Society Press, 100–101.

3.   Eden, C. (1991) "Working on Problems Using Cognitive Mapping", *Operations Research in Management*, Littlechild, S.; Shutler, M. (eds.), Prentice-Hall.

4.   Economist, The (1992) February 8, p.104.

5.   Geerts, Guido; McCarthy, William E. (1991) "Database Accounting Systems", *Information Technology Perspectives in Accounting: An Integrated Approach*, Williams, B.; Sproul, B.J. (eds.), Chapman and Hall Publishers, June.

6.   Kelly, George (1955) *The Psychology of Personal Constructs*, Norton, New York.

7.   Lee, G. A. (1986) *Modern Financial Accounting*, Van Nostrand Reinhold, UK.

8.   Leith, Philip (1990) *Formalism in AI and Computer Science*, Ellis Horwood.

9.   McCarthy, William E. (1987) "On the Future of Knowledge-Based Accounting Systems", *Artificial Intelligence in Accounting and Auditing: The Use of Expert Systems*, Vasarhelyi, M. (ed.), Markus Wiener Publishing, October 1987.

10. Polovina, Simon (1991) "Towards the Thought Machine", *Accounting Technician: The Journal of the Association of Accounting Technicians, UK.*, October, 25–26.

11. Polovina, Simon (1991) "Tomorrows Spreadsheets: Conceptual Graphs as the Knowledge Based Decision Support Tool for the Management Accountant", *Proceedings of the Sixth Annual Conceptual Graphs Workshop*, July 11-13, State University of New York at Binghamton, 375–385.

12. Polovina, Simon; Heaton, John (1992) "An Introduction to Conceptual Graphs", *AI Expert*, 7(5), May, 36–43.

13. Polovina, Simon (1992) "An Initial User Experiment with Linear Form Conceptual Graphs for Strategic Planning", *Loughborough University of Technology Department of Computer Studies*, Technical Report No. 720.

14. Sizer, John (1989) *An Insight into Management Accounting*, Penguin Business, UK, pp. 411–420.

15. Sowa, John F. (1984) *Conceptual Structures*, Addison-Wesley.

16. Sowa, John F. (1985) Position Statement. *From* the same as [2] above.

17. Sowa, John F. (1990) "Knowledge Representation in Databases, Expert Systems, and Natural Language", *Artificial Intelligence in Databases and Information Systems (DS-3)*, Meersman, R. A.; Shi, Zh.; Kung, C-H: North-Holland Publishing Co., Amsterdam, 1990, 17–50.

18. Woolf, Emile (1990) *Auditing Today*, Prentice Hall International (UK),pp.100–121.

Skeletal Plans Reuse:
A Restricted Conceptual Graph Approach

Wu Zhaohui, Bernardi Ansgar and Klauck Christoph

German Research Center of Artificial Intelligence,
Postfach 2080, Kaiserslautern 6750,
Germany
email: {wu, bernardi, klauck}@dfki.uni-kl.de
tel: 0049-631-205-3477

Abstract. In order to reuse the existing skeletal plans in the manufacturing process planning system *PIM*, in this paper, we propose a plan reuse framework, in which Restricted Conceptual Graphs are used as the internal representations of these skeletal plans and reusing these skeletal plans is approached by retrieving the most specific general candidate and effectively modifying.

1 Introduction

Generating plans from scratch is a computationally expensive process and computation complexity will be greatly reduced through reusing existing plans. The most percent of planning problems such as 80 percent of all mechanical engineering tasks, are solved by adapting old plans to new situations[1]. A Major obstacle to successful deployment of plan reuse mechanism schemes is how to map/fix/retrieve the appropriate plans that can be efficiently reused in the new situation through almost minimum cost modification. Recent researches[2][3] [4][5] show that efficiency can be greatly improved if the reuse mechanism for plans has the ability to retrieve and locate the applicable part of existing plans, and also to effectively modify the inapplicable part plan to refit the new situations.

In our proposed *PIM* system[6][7], a manufacturing process planning system, we utilize skeletal plans to describe expert knowledge about process planning on the part or whole of a workpiece. Reusing these existing skeletal plans for new workpieces or in new working environments becomes more important, especially when the complexity of workpieces and the amount of existing skeletal plans increase, or when the working environment changes. This requires us to introduce an appropriate internal representation of skeletal plans which can address certain relations between them such as the generalization relation.

Conceptual graph[8] which provides powerful knowledge representation mechanisms, is one adoptable candidate for our applications. However, a general conceptual graph is expensive and inefficient in our application domain, e.g. the classification of the general conceptual graph is intractable and expensive[10]. On the other hand, a given domain only needs a domain-specific representation framework instead of a general and intractable one. Restricting the general representation leads to better efficiency of processing and better fitness of domain. Therefore, for our domain applications, Restricted Conceptual Graph (*RCG*), which inherits and restricts some mechanisms of conceptual graph, is proposed; and a *RCG* based reuse framework is presented, in which a restricted conceptual graph is used to represent the internal structure of a skeletal plan and is classified into a hierarchy. Reusing these existing skeletal plans is performed through selecting the most specific general conceptual graphs and modifying their skeletal plans.

After giving the overview of the *PIM* system in the second section, we propose the Restricted Conceptual Graph in the third section. The plan reuse framework including the conservative controlling strategy about retrieval and its application in PIM are described in section 4 and conclusion is drawn in section 5.

2 Overview of *PIM*

The PIM system (Planning in Manufacturing) is a knowledge-based Computer Aided Process Planning (CAPP), which provides a group of representation languages and formalisms to bridge the gap between CAD and CAM. In these formalisms, SKEP_REP (SKEPeletal plans REPresentation) is a skeletal plan representation language to describe expert knowledge about manufacturing process planning. A skeletal plan about *"shaft-crest"* is illustrated in the left side of figure 1, in which the feature about the *"shaft-crest"* workpiece is specified in the operational feature structure of this skeletal plan. Skeletal plans are features associated and dependent. The features associated with these skeletal plans are to describe the manufacturable parts of the workpiece. A complete production plan is generated by performing a sequence of operations of skeletal planning such as selection, refinement and merging.

The plan reuse in *PIM* is to reuse previous skeletal plans for a different workpiece or in new working environment. In many cases in *PIM*, the reused skeletal plan can be used to adapt to new situations or to improve the performance, e.g. acquiring new skeletal plans through reusing existing skeletal plans, and speeding the procedure of skeletal planning by reusing some previously planned skeletal plans. Therefore, introducing an effective skeletal plans reuse framework becomes more important.

3 Restricted Conceptual Graph

The proposed Restricted Conceptual Graph (*RCG*) is a special kind of conceptual graph, which is to restrict the representation mechanism of Sowa's conceptual graph [8] such as the number of relations, referent mapping, type function and formation rules. The definition of *RCG* is given in section 3.1. The formation rules about *RCG* are introduced in section 3.2 and the taxonomy of *RCGs* is described in section 3.3.

3.1 Definition

The definition about *RCG* is given as follows:

– Definition: Restricted Conceptual Graph (*RCG*) is a finite, connected, graph (N,R) with labeled nodes and labeled relations. Every labeled relation has one or more arcs, each of which must be linked to one labeled node. A single labeled node can form a *RCG*. The labels of node are any arbitrary strings included in named N set. The labels of relation are a limited set R which depends on the applicable domain. Each of these relations has its own formations rules.

In our application domain, the domain-specific relation set R only includes three relations, R = { *IS_PART_OF, IS_CONTEXT_OF , IS_INSTANCE_OF* }. The *IS_PART_OF* relation represents the "part of" relation among these linked labeled nodes. The *IS_CONTEXT_OF* relation represents that the labeled node has meaning only under the context of the linked labeled node. The *IS_INSTANCE_OF* relation means that the labeled node is a instance of the linked labeled node.

A Restricted Conceptual Graph of a skeletal plan of *PIM* is illustrated in figure 1.

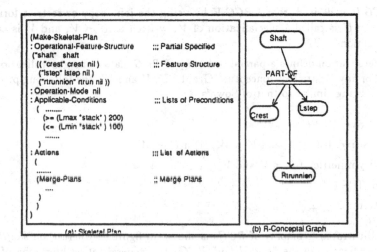

Fig. 1. A *RCG* and its Corresponding *PIM*'s Skeletal Plan

We define the labeled node, which *always* is linked from others, as the IN labeled node such as the *"Lstep"* labeled node in figure 1; and call the node as OUT labeled node, which *always* links to other nodes and is not linked from any other labeled nodes, e.g. the *"Shaft"* labeled node in the figure 1.

3.2 Relation-Based Formation Rules

The formation rules in Restricted Conceptual Graph (*RCG*) are special forms of those of Sowa's, which are relation based.

For the relation *IS_PART_OF* , there are three formation rules for deriving a restricted conceptual graph X from others U and V:

- Copy: X is an exact copy of U.
- Top-Down Join: If a OUT labeled node n in U is identical to a IN labeled node m in V, then let X be the graph obtained by joining U and V and deleting m node.
- Bottom-up split: If we want to split the subgraph X from the graph U, copys the interface labeled nodes and split them. It is reverse operation of join.

The formation rules make guarantee that the formed *RCG* is right. There exists similar formation rules such as Copy, Restrict and Extend rules for IN relation in the *RCGs*.

3.3   Taxonomy of *RCGs*

If a *RCG U* is derived from a *RCG V* by using the join, copy or restrict formation rules, then U is called a specialization of V, written as $U \preceq V$, and V is called a generalization of U.

Generalization defines a partial ordering of *RCG* called the generalization hierarchy. For any Restricted Conceptual Graphs U, V and W, the following properties are true, being similar to that of Sowa's:

- Reflexive: $U \preceq U$
- Transitive: If $U \preceq V$ and $V \preceq W$, then $U \preceq W$.
- Antisymmetric: If $U \preceq V$ and $V \preceq U$, then $V = U$.
- Top: $\top \succeq U$
- Bottom: $U \succeq \bot$

An Illustrated taxonomy of *RCGs* is shown in figure 2, where the *"shaft"* RCG is a generalization of the *"shaft-lstep"* RCG. The generalization hierarchy of *RCGs* is built by effective classification. The classification of RCGs is more efficient than that of general one.

4   Reuse Framework in *PIM*

4.1   Problem Statement

The Skeletal Plans Reuse Problem in *PIM* is addressed as follows: Given a new workpiece in a situation (recognized as Restricted Conceptual Graph S(n)) and a group of existing pairs of skeletal plan and its *RCG* described as {[S0,P0],....[S(n-1),P(n-1)]}. Produce: A skeletal plan called P(n) will be generated by effectively modifying the retrieved *RCG* set {P(i),....,P(j)}.

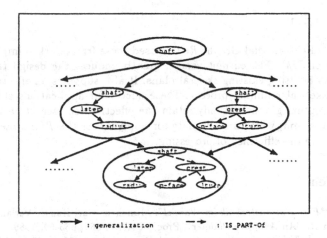

→ : generalization ⇢ : IS_PART-Of

Fig. 2. A Taxonomy of *RCGs*

4.2 Basic Cycle of Reuse

The reuse procedure in *PIM* proceeds in the following steps:

- Recognition: to recognize the *RCG* out of a given workpiece.
- Retrieval: A group of *RCGs* named {S(i),...,S(j)}, is retrieved from the partial *RCG* space {S(0),....,S(n-1)}.
- Modification: The modification of the retrieved skeletal plans P(i),..,P(j) taken from these pairs is to be completed by merging or splitting these skeletal plans.

4.3 Controlling the reuse

Controlling the reuse procedure is to get the minimal cost modification of these existing plans. In our framework, the retrieval process is an important step in the basic reuse cycle. The retrieval process proceeds the following steps:

- the classification process - to classify the target *RCG* x in the hierarchy of *RCGs* and to find the partial relations $u \preceq x \preceq v$. This process is similar to the taxonomic reasoning in KL-ONE[9]. A breadth-first search of the taxonomic generalization space is implemented [10].

- the selecting process – to choose the best candidate from the partial *RCG* space. The selective strategy is listed as follows: If $x = u$ or v, this process selects u or v as the best candidate and ends this retrieval process. If $v = \top$ and $u \neq \bot$, then this process has to select the u and to exit from the retrieval process. If $v = \top$ and $u = \bot$, then this process selects nothing and goes to the interviewing process. Otherwise, this is most cases, the conservative strategy for retrieval is to take v as the candidate and goes to the projecting process.

5 Conclusion

The Restricted Conceptual Graph (*RCG*) based reuse framework is implemented as one module of *PIM*. The current version greatly reduces the design task for new skeletal plans by using existing skeletal plans. It also supports an effective retrieval during the skeletal planning process. These successful applications show that restricted conceptual graphs not only retain the effective representation mechanisms of conceptual graph, which are adequate for applying for the *PIM* system; but also greatly improve the efficiency of processing.

References

1. Spur G.:"*Producktionstechnik im Wandel*", Munchen, Carl Hanser Verlag
2. Alterman R.:"*An Adaptive Planner*", Proc. of AAAI'86, pp65-69, 1986
3. Tenenberg J.:"*Planning with Abstraction*", Proc. of AAAI'86, pp76-80, 1986.
4. Kambhampati S.:"*A Theory of Plan Modification*", Proc. AAAI'90, pp176-182, 1989.
5. Kambhampati S.:"*Mapping and Retrieval During Plan Reuse: A Validation Structure Based Approach*", Proc. AAAI'90, pp170-175,
6. Bernardi A, Boley H., et al:"*ARC-TEC:Acquisition,Representation and Compilation of Technical Knowledge*", In Expert Systems and Their Applications:Tools, techniques and Methods,1991,pp133-145.
7. Legleitner, R., Bernardi, A, et al:"*PIM: Skeletal Plan based CAPP*", Proceeding of Int' Conference on Manufacturing Automation, 1992.
8. Sowa.J.F:"*Conceptual Structures: information Processing in man and machine*", Addison-Wesley Publishing Company. 1984
9. Brachman,R.J., Schmolze, J.G. :"*An Overview of the KL-ONE Knowledge Representation System*", Cognitive Sci. 16, 1985.
10. Ellis G:"*Compiled Hierarchical Retrieval*", Proceeding of the Sixth Annual Workshop on Conceptual Graphs. Binghamton, July 11-13,1991.
11. Wu Zhaohui and Bernardi Ansgar:"*VSKEP-EDITor: A Visual Tool for Reusable Editing Skeletal Plans*", DFKI Document 1992.

Lecture Notes in Computer Science

Lecture Notes in Artificial Intelligence (LNAI)

Vol. 590: B. Fronhöfer, G. Wrightson (Eds.), Parallelization in Inference Systems. Proceedings, 1990. VIII, 372 pages. 1992.

Vol. 592: A. Voronkov (Ed.), Logic Programming. Proceedings, 1991. IX, 514 pages. 1992.

Vol. 596: L.-H. Eriksson, L. Hallnäs, P. Schroeder-Heister (Eds.), Extensions of Logic Programming. Proceedings, 1991. VII, 369 pages. 1992.

Vol. 597: H. W. Guesgen, J. Hertzberg, A Perspective of Constraint-Based Reasoning. VIII, 123 pages. 1992.

Vol. 599: Th. Wetter, K.-D. Althoff, J. Boose, B. R. Gaines, M. Linster, F. Schmalhofer (Eds.), Current Developments in Knowledge Acquisition - EKAW '92. Proceedings. XIII, 444 pages. 1992.

Vol. 604: F. Belli, F. J. Radermacher (Eds.), Industrial and Engineering Applications of Artificial Intelligence and Expert Systems. Proceedings, 1992. XV, 702 pages. 1992.

Vol. 607: D. Kapur (Ed.), Automated Deduction – CADE-11. Proceedings, 1992. XV, 793 pages. 1992.

Vol. 610: F. von Martial, Coordinating Plans of Autonomous Agents. XII, 246 pages. 1992.

Vol. 611: M. P. Papazoglou, J. Zeleznikow (Eds.), The Next Generation of Information Systems: From Data to Knowledge. VIII, 310 pages. 1992.

Vol. 617: V. Mařík, O. Štěpánková, R. Trappl (Eds.), Advanced Topics in Artificial Intelligence. Proceedings, 1992. IX, 484 pages. 1992.

Vol. 619: D. Pearce, H. Wansing (Eds.), Nonclassical Logics and Information Processing. Proceedings, 1990. VII, 171 pages. 1992.

Vol. 622: F. Schmalhofer, G. Strube, Th. Wetter (Eds.), Contemporary Knowledge Engineering and Cognition. Proceedings, 1991. XII, 258 pages. 1992.

Vol. 624: A. Voronkov (Ed.), Logic Programming and Automated Reasoning. Proceedings, 1992. XIV, 509 pages. 1992.

Vol. 627: J. Pustejovsky, S. Bergler (Eds.), Lexical Semantics and Knowledge Representation. Proceedings, 1991. XII, 381 pages. 1992.

Vol. 633: D. Pearce, G. Wagner (Eds.), Logics in AI. Proceedings. VIII, 410 pages. 1992.

Vol. 636: G. Comyn, N. E. Fuchs, M. J. Ratcliffe (Eds.), Logic Programming in Action. Proceedings, 1992. X, 324 pages. 1992.

Vol. 638: A. F. Rocha, Neural Nets. A Theory for Brains and Machines. XV, 393 pages. 1992.

Vol. 642: K. P. Jantke (Ed.), Analogical and Inductive Inference. Proceedings, 1992. VIII, 319 pages. 1992.

Vol. 659: G. Brewka, K. P. Jantke, P. H. Schmitt (Eds.), Nonmonotonic and Inductive Logic. Proceedings, 1991. VIII, 332 pages. 1993.

Vol. 660: E. Lamma, P. Mello (Eds.), Extensions of Logic Programming. Proceedings, 1992. VIII, 417 pages. 1993.

Vol. 667: P. B. Brazdil (Ed.), Machine Learning: ECML – 93. Proceedings, 1993. XII, 471 pages. 1993.

Vol. 671: H. J. Ohlbach (Ed.), GWAI-92: Advances in Artificial Intelligence. Proceedings, 1992. XI, 397 pages. 1993.

Vol. 679: C. Fermüller, A. Leitsch, T. Tammet, N. Zamov, Resolution Methods for the Decision Problem. VIII, 205 pages. 1993.

Vol. 681: H. Wansing, The Logic of Information Structures. IX, 163 pages. 1993.

Vol. 689: J. Komorowski, Z. W. Raś (Eds.), Methodologies for Intelligent Systems. Proceedings, 1993. XI, 653 pages. 1993.

Vol. 695: E. P. Klement, W. Slany (Eds.), Fuzzy Logic in Artificial Intelligence. Proceedings, 1993. VIII, 192 pages. 1993.

Vol. 698: A. Voronkov (Ed.), Logic Programming and Automated Reasoning. Proceedings, 1993. XIII, 386 pages. 1993.

Vol. 699: G.W. Mineau, B. Moulin, J.F. Sowa (Eds.), Conceptual Graphs for Knowledge Representation. Proceedings, 1993. IX, 451 pages. 1993.

Vol. 723: N. Aussenac, G. Boy, B. Gaines, M. Linster, J.-G. Ganascia, Y. Kodratoff (Eds.), Knowledge Acquisition for Knowledge-Based Systems. Proceedings, 1993. XIII, 446 pages. 1993.

Vol. 727: M. Filgueiras, L. Damas (Eds.), Progress in Artificial Intelligence. Proceedings, 1993. X, 362 pages. 1993.

Vol. 728: P. Torasso (Ed.), Advances in Artificial Intelligence. Proceedings, 1993. XI, 336 pages. 1993.

Vol. 743: S. Doshita, K. Furukawa, K. P. Jantke, T. Nishida (Eds.), Algorithmic Learning Theory. Proceedings, 1992. X, 260 pages. 1993.

Vol. 744: K. P. Jantke, T. Yokomori, S. Kobayashi, E. Tomita (Eds.), Algorithmic Learning Theory. Proceedings, 1993. XI, 423 pages. 1993.

Vol. 745: V. Roberto (Ed.), Intelligent Perceptual Systems. VIII, 378 pages. 1993.

Vol. 746: A. S. Tanguiane, Artificial Perception and Music Recognition. XV, 210 pages. 1993.

Vol. 754: H. D. Pfeiffer, T. E. Nagle (Eds.), Conceptual Structures: Theory and Implementation. Proceedings, 1992. IX, 327 pages. 1993.

Lecture Notes in Artificial Intelligence

This subseries of Lecture Notes in Computer Science reports new developments in artificial intelligence research and teaching, quickly, informally, and at a high level. The timeliness of a manuscript is more important than its form, which may be unfinished or tentative. The type of material considered for publication includes

- drafts of original papers or monographs,
- technical reports of high quality and broad interest,
- advanced-level lectures,
- reports of meetings, provided they are of exceptional interest and focused on a single topic.

Publication of Lecture Notes is intended as a service to the computer science community in that the publisher Springer-Verlag offers global distribution of documents which would otherwise have a restricted readership. Once published and copyrighted, they can be cited in the scientific literature.

Manuscripts

Lecture Notes are printed by photo-offset from the master copy delivered in camera-ready form. Manuscripts should be no less than 100 and preferably no more than 500 pages of text. Authors of monographs and editors of proceedings volumes receive 50 free copies of their book. Manuscripts should be printed with a laser or other high-resolution printer onto white paper of reasonable quality. To ensure that the final photo-reduced pages are easily readable, please use one of the following formats:

Font size (points)	Printing area (cm)	(inches)	Final size (%)
10	12.2 x 19.3	4.8 x 7.6	100
12	15.3 x 24.2	6.0 x 9.5	80

On request the publisher will supply a leaflet with more detailed technical instructions or a T_EX macro package for the preparation of manuscripts.

Manuscripts should be sent to one of the series editors or directly to:

Springer-Verlag, Computer Science Editorial I, Tiergartenstr. 17,
D-69121 Heidelberg, FRG

ISBN 3-540-57454-9
ISBN 0-387-57454-9